AF869941

Tourism, Change and the Global South

This significant volume is the first to focus on both the changing nature of tourism and the capacity of tourism to effect change, especially in the Global South.

Geographically, this changing nature of tourism is based on the transforming relationships between demand, supply and location. While this is nothing new in tourism, recent decades have intensified the changing characteristics of global tourism. From another perspective, tourism represents a change, and nowadays many localities and regions aim to use tourism as a tool for positive change, i.e. development. However, this has turned out to be a challenging task in practice, especially in the Global South context where the relationship between tourism growth and local development has often been controversial. This book looks at a host of critical concepts in one volume, such as growth and development, adaptation and resilience, sustainability and responsibility, governance and planning and heritage and destination management strategies. By understanding the drivers of change, this book sheds new insight into the promise and role of sustainability and responsibility in tourism development.

This book will be of great interest to all upper-level students, academics and researchers in the fields of Tourism, Geography and Cultural and Heritage studies.

Jarkko Saarinen is Professor of Geography at the University of Oulu, Finland, and Distinguished Visiting Professor (Sustainability Management) at the University of Johannesburg, South Africa. His research interests include tourism and development, sustainability in tourism, climate change adaptation, tourism–community relations and nature conservation studies. He is Editor for *Tourism Geographies*. His recent publications include edited/co-edited books: *Tourism Enclaves* and *Resilient Destinations*.

Jayne M. Rogerson is Associate Professor of Tourism at the School of Tourism and Hospitality, University of Johannesburg, South Africa. Her research interests include tourism accommodation, urban tourism, tourism history and sustainable tourism. Her most recent publications are two co-edited books: *New Directions in South African Tourism Geographies* and *Urban Tourism in the Global South.*

Contemporary Geographies of Leisure, Tourism and Mobility

Series Editor: C. Michael Hall, *Professor at the Department of Management, College of Business and Economics, University of Canterbury, Christchurch, New Zealand*

The aim of this series is to explore and communicate the intersections and relationships between leisure, tourism and human mobility within the social sciences.

It will incorporate both traditional and new perspectives on leisure and tourism from contemporary geography, e.g. notions of identity, representation and culture, while also providing for perspectives from cognate areas such as anthropology, cultural studies, gastronomy and food studies, marketing, policy studies and political economy, regional and urban planning, and sociology, within the development of an integrated field of leisure and tourism studies.

Also, increasingly, tourism and leisure are regarded as steps in a continuum of human mobility. Inclusion of mobility in the series offers the prospect to examine the relationship between tourism and migration, the sojourner, educational travel, and second home and retirement travel phenomena.

The series comprises two strands:

Contemporary Geographies of Leisure, Tourism and Mobility aims to address the needs of students and academics, and the titles will be published simultaneously in hardback and paperback.

Routledge Studies in Contemporary Geographies of Leisure, Tourism and Mobility is a forum for innovative new research intended for research students and academics, and the titles will initially be available in hardback only. Titles include:

Sense of Place and Place Attachment in Tourism
Ning Chris Chen, C. Michael Hall and Girish Prayag

Tourism, Change and the Global South
Edited by Jarkko Saarinen and Jayne M. Rogerson

The Power of New Urban Tourism
Policies, Representations and Contestations
Edited by Claudia Ba, Sybille Frank, Claus Müller, Anna Laura Raschke, Kristin Wellner and Annika Zecher

For more information about this series, please visit: www.routledge.com/Contemporary-Geographies-of-Leisure-Tourism-and-Mobility/book-series/SE0522

Tourism, Change and the Global South

Edited by Jarkko Saarinen and Jayne M. Rogerson

Routledge
Taylor & Francis Group
LONDON AND NEW YORK

First published 2021
by Routledge
2 Park Square, Milton Park, Abingdon, Oxon OX14 4RN

and by Routledge
605 Third Avenue, New York, NY 10158

Routledge is an imprint of the Taylor & Francis Group, an informa business

© 2021 selection and editorial matter, Jarkko Saarinen and Jayne M. Rogerson; individual chapters, the contributors

The right of Jarkko Saarinen and Jayne M. Rogerson to be identified as the authors of the editorial material, and of the authors for their individual chapters, has been asserted in accordance with sections 77 and 78 of the Copyright, Designs and Patents Act 1988.

All rights reserved. No part of this book may be reprinted or reproduced or utilised in any form or by any electronic, mechanical, or other means, now known or hereafter invented, including photocopying and recording, or in any information storage or retrieval system, without permission in writing from the publishers.

Trademark notice: Product or corporate names may be trademarks or registered trademarks, and are used only for identification and explanation without intent to infringe.

British Library Cataloguing-in-Publication Data
A catalogue record for this book is available from the British Library

Library of Congress Cataloging-in-Publication Data
Names: Saarinen, Jarkko, 1968- editor. | Rogerson, Jayne M., editor.
Title: Tourism, change and the global South / edited by Jarkko Saarinen and Jayne M. Rogerson.
Description: Abingdon, Oxon; New York, NY: Routledge, 2021. |
Series: Contemporary geographies of leisure, tourism and mobility |
Includes bibliographical references and index.
Identifiers: LCCN 2020056929 (print) | LCCN 2020056930 (ebook)
Subjects: LCSH: Tourism–Developing countries.
Classification: LCC G155.D44 T6827 2021 (print) |
LCC G155.D44 (ebook) | DDC 306.4/81–dc23
LC record available at https://lccn.loc.gov/2020056929
LC ebook record available at https://lccn.loc.gov/2020056930

ISBN: 978-0-367-54953-4 (hbk)
ISBN: 978-0-367-54955-8 (pbk)
ISBN: 978-1-003-09132-5 (ebk)

Typeset in Times New Roman
by Deanta Global Publishing Services, Chennai, India

Contents

List of figures viii
List of tables x
List of contributors xi
Preface xiii

PART I
Introduction and frameworks 1

1 **Tourism and change: Issues and challenges in the Global South** 3
JARKKO SAARINEN AND JAYNE M. ROGERSON

2 **Tourism for change: Change management towards sustainable tourism development** 15
JARKKO SAARINEN

3 **Democratising the cultural past: Western values, the Global South and cross-cultural perspectives in heritage tourism** 33
DALLEN TIMOTHY

4 **Antecedents of tourism vicarious nostalgia: Conceptual model, systematic review and research agenda** 48
OSCAR HENGXUAN CHI AND CHRISTINA GENG-QING CHI

PART II
Change in tourism 63

5 **Images, instruments and the governance capacity of local governments in tourism development planning: Evidence from Ghana** 65
EMMANUEL AKWASI ADU-AMPONG

6 **From small island developing states to large ocean states: Tourism in the changing periphery of island states in the Global South** 86
C. MICHAEL HALL

7 **The generative power of nurturing new connections: Indigenous tour operators learning across, learning deeply** 101
MARNIE GRAHAM, SANDIE SUCHET-PEARSON, UNCLE LEXODIOUS DADD AND DIRK PIENAAR

8 **Roots tourism and the Year of Return campaign in Ghana: Moving belonging beyond the history of slavery** 123
MATTHIAS GEBAUER AND MARIE UMSCHEID

9 **"The culture of the souq is lost": Valuing social capital and a legacy within tourism development** 135
MANUELA GUTBERLET

10 **The evolution of township tourism in South Africa** 151
IRMA BOOYENS

PART III
Tourism for change 167

11 **Impacts of tourism on the livelihoods of women in Victoria Falls, Zimbabwe** 169
TARIRO SIBANDA AND JOSEPH M. CHEER

12 **Economic development, tourism and conservation in developing countries** 187
JOSEPH E. MBAIWA

13 **Municipal assets for tourism and small-town local economic development: Caravan parks in South Africa** 205
CHRISTIAN M. ROGERSON AND JAYNE M. ROGERSON

14 The gig economy and employment in tourism in Southern Africa: A global finger in the informal sector pie? 222
TOM BAUM AND JULIA K. GIDDY

15 "African" representations: Culturalisation of Poverty in Township tourism 237
MALTE STEINBRINK AND INA VOSHAGE

Index 259

Figures

4.1	An integrative conceptual model of tourism vicarious nostalgia. Source: Authors	51
5.1	The deserted and empty Elmina arts and craft market. Source: Author	74
5.2	The deserted and empty Elmina arts and craft market. Source: Author	74
7.1	Dirk Pienaar (second from left) sharing some of his knowledge with Uncle Lex (left), Marnie (middle), Sandie (taking the picture) and their families on Erin Game Ranch. Source: Authors	103
7.2	Uncle Lex checking out the trees at Erin Game Ranch for which he felt such familiarity. Source: Authors	108
7.3	#Khomani San tour guide Xopan and Uncle Lex discussing the Kalahari trees that Uncle Lex felt so connected to. Source: Authors	109
7.4	Uncle Lex made a particular new friend in a yellow mongoose. This connection was a highlight of his trip. Source: Authors	109
7.5	Uncle Lex, #Khomani San tour guide Xopan and Dirk connect in the shade of a shepherd tree, Erin Game Ranch. Source: Authors	111
7.6	Erin Game Ranch, safari tents. Source: Authors	112
7.7	Looking out from the Hide at Erin Game Ranch towards the waterhole	113
7.8	In 2017, #Khomani San efforts for recognition paid off with World Heritage listing of the #Khomani Cultural Landscape. Source: Authors	114
7.9	Sandie's, Marnie's and Dirk's kids learning together in the red sands of Erin Game Ranch. Source: Authors	118
9.1	A shop of the well-established trader community in Souq Muttrah. Source: Gutberlet (2013)	142
9.2	Flashy tourist shops at the entrance of Souq Muttrah. Source: Gutberlet (2019)	146

10.1	Freedom Square, Kliptown Precinct. Source: Author	159
10.2	Township tourism development paths. Source: Author	161
11.1	Livelihood activity by proportion of men and women. Source: Authors	176
11.2	Length of men's and women's involvement in tourism. Source: Authors	177
11.3	Impacts of tourism on men and women. Source: Authors	178
11.4	Challenges of tourism by proportion of men and women. Source: Authors	178
11.5	Tourism planning involvement by proportion of men and women. Source: Authors	180
13.1	The geography of municipal caravan parks in the Western Cape on a local municipality basis. Source: Authors	212
15.1	Expectations of the tour. Source: Adapted from Steinbrink et al., 2016, p. 46	244
15.2	Tourists' township associations before the tour. Source: Adapted from Steinbrink et al., 2016, p. 49	245
15.3	Semantic differential: Tourists' notions before the tour, n=70. Source: Adapted from Steinbrink et al., 2016, p. 51	247
15.4	Tourists' township impressions after the tour. Source: Adapted from Steinbrink et al., 2016, p. 89	252
15.5	Semantic differential: Tourists' preconceptions before versus their impressions after the tour. Source: Adapted from Steinbrink et al., 2016, p. 90	253

Tables

2.1 The UN Sustainable Development Goals (SDGs) 23
5.1 Stakeholder governance capacity 67
5.2 Financing of associated projects implemented under the Elmina 2015 Strategy 77
6.1 Characteristics of SIDS 88
9.1 Framework: Dimensions of bonding and bridging social capital in Souq Muttrah 148
10.1 Phases in the development of township tourism 152
10.2 Township visits by foreign dignitaries, politicians and celebrities (1990s to present) 156
10.3 Key cultural and struggle heritage products in townships 158
11.1 Description of key informants interviewed 174
11.2 Monthly income of respondents 176
12.1 Case studies of community-based tourism projects and impacts on rural livelihoods 194
12.2 Approaches to community tourism management 197
12.3 Wildlife tourism partnership models 198

Contributors

Emmanuel Akwasi Adu-Ampong, Cultural Geography Chair Group, Wageningen University and Research, the Netherlands; School of Tourism and Hospitality, University of Johannesburg, South Africa

Tom Baum, Department of Work, Employment and Organisation, Strathclyde Business School, University of Strathclyde, Glasgow, Scotland; School of Tourism and Hospitality, University of Johannesburg, South Africa

Irma Booyens, Department of Work, Employment and Organisation, Strathclyde Business School, University of Strathclyde, Glasgow, Scotland; School of Tourism and Hospitality, University of Johannesburg, South Africa

Joseph M. Cheer, Center for Tourism Research, Wakayama University, Japan

Christina Geng-Qing Chi, School of Hospitality Business Management, Washington State University, Pullman, USA; School of Tourism and Hospitality, University of Johannesburg, South Africa

Oscar Hengxuan Chi, School of Hospitality Business Management, Washington State University, Pullman, USA

Uncle Lexodious Dadd, Senior Darug Custodian, Burangilli Cultural Education Services, Sydney, Australia

Julia K. Giddy, School of Biology and Environmental Sciences, University of Mpumalanga, Mbombela, South Africa; School of Tourism and Hospitality, University of Johannesburg, South Africa

Matthias Gebauer, Institute of Geography, University of Bayreuth, Germany

Marnie Graham, Stockholm Resilience Centre, Stockholm University, Sweden; School of Tourism and Hospitality, University of Johannesburg, South Africa; Department of Geography and Planning, Macquarie University, Sydney, Australia

Manuela Gutberlet, Research Associate, School of Tourism and Hospitality, University of Johannesburg, South Africa

C. Michael Hall, Department of Management, Marketing and Entrepreneurship, University of Canterbury, Christchurch, New Zealand

Joseph Mbaiwa, Okavango Research Centre, University of Botswana, Maun, Botswana; School of Tourism and Hospitality, University of Johannesburg, South Africa

Dirk Pienaar, #Khomani San Community Property Association (Tourism and Conservation), Andriesvale, South Africa

Christian M. Rogerson, School of Tourism and Hospitality, University of Johannesburg, South Africa

Jayne M. Rogerson, School of Tourism and Hospitality, University of Johannesburg, South Africa

Jarkko Saarinen, Department of Geography, University of Oulu, Finland; School of Tourism and Hospitality, University of Johannesburg, South Africa

Tariro N. Sibanda, Project Officer, ADESIM Developments, South Africa

Malte Steinbrink, Department of Geography, University of Passau, Germany; School of Tourism and Hospitality, University of Johannesburg, South Africa

Sandie Suchet-Pearson, Department of Geography and Planning, Macquarie University, Sydney, Australia

Dallen Timothy, School of Community Resources and Development, Arizona State University, Phoenix, USA; School of Hospitality and Tourism, University of Johannesburg, South Africa

Marie Umscheid, Department of Geography, University of Passau, Germany

Ina Voshage, Department of Geography, University of Passau, Germany

Preface

The tourism industry has always been dynamic, and for decades, many localities and regions have strategically used tourism as a tool for positive change. However, this has been a challenging task in practice, especially in the context of the Global South where the relationship between tourism growth and local development has often been controversial. This collection aims to offer a range of both theoretical and applied perspectives that address tourism in change/tourism for change in the Global South. Key conceptual and theoretical themes include growth and development, adaptation and resilience, sustainability and responsibility, governance and planning and heritage and destination management strategies.

The genesis of this book was a conference organised in late 2019 by the University of Johannesburg which hosted a number of key international research associates. Over five days, presentations, field trips, sundowners and leisure time provided the spaces for new research and ideas to be shared. The attendees were eager to formalise the discussions and debates, and the result is this edited collection which focuses on the evolving nature of tourism and its capacity to create change, especially in the Global South. More than half of the chapters resulted from papers presented at the conference, with the remainder being contributed by special invitation academics covering a wide range of topics.

It cannot be ignored that in the year 2020, the tourism industry, like many others, was brought to a standstill. While tourism is often considered the world's largest and fastest-growing industry, in 2020 this was certainly not the case. The COVID-19 pandemic and related travel restrictions led to a devastating decline in global tourism with many airlines announcing up to a 90% drop in sales. Destinations highly dependent on international tourists have been the most severely impacted and many destinations, including those in the Global South, experienced serious economic hardships and social crises.

The COVID-19 crisis has forced us to change the way that we live and also the way that we travel, and these changes will certainly affect the future of the tourism industry in ways still to be determined. That said, the pandemic is temporary, whereas the issues highlighted in the following chapters are not. Rather, they are fundamental to the tourism industry and are therefore relevant to a pre- and post-COVID world.

For supporting the publication of this volume, we would like to extend our thanks to Routledge and, in particular, to Emma Travis and Lydia Kessell. Our thanks must also go to the book series editor C. Michael Hall. As the editors of the book, we would like to acknowledge several individuals in the development of this book. Jayne would like to thank the University of Johannesburg for facilitating the 2019 Research Associates Conference without which, this book would never have been conceived. Thanks go to Arabella Rogerson for her editorial expertise and the long nights that she put in to ensure a quality end product. Finally, to her husband, Chris, and to Teddy, Skye and Dawn Norfolk for their inspiration, love and support, she is most grateful. Jarkko would like to thank all the above and many colleagues, postgraduate students, friends and institutions and other actors who have contributed to the ideas related to change and sustainability in tourism. In particular, he would like to thank C. Michael Hall at the University of Canterbury, Joseph Cheer at Wakayama University, Robin Nunkoo at the University of Mauritius, Alison Gill at Simon Fraser University, Sanjay Nepal at the University of Waterloo and Julius Atlhopheng at the University of Botswana. The Universities of Johannesburg and Botswana and the colleagues there, Lagotto Romagnolos, IGU and AAG, Villa Pablo and Ribeiro Santo have also contributed (positively) to the process.

Collectively, we would like to express our gratitude to all who participated in writing the various chapters and the academics who supported the peer-review process for this book. Each chapter was externally peer-reviewed by two independent experts in their field. Reviewers at the following institutions are acknowledged: ALU School of Wildlife Conservation, Rwanda; Arizona State University, USA and China; Breda University of Applied Sciences, the Netherlands; Cape Peninsula University of Technology, South Africa; Flinders University, Australia; German University of Technology in Oman, Oman; Kwame Nkrumah University of Science and Technology (KNUST), Ghana; Trinity College Dublin, Ireland; University of Botswana, Botswana; University of Dar es Salaam, Tanzania; University of Edinburgh, Scotland; University of Leicester, UK; University of Johannesburg, South Africa; University of Mpumalanga, South Africa; University of Otago, New Zealand; University of Pretoria, South Africa; University of South Africa, South Africa; Wageningen University, the Netherlands. Thank you.

Part I

Introduction and frameworks

1 Tourism and change

Issues and challenges in the Global South

Jarkko Saarinen and Jayne M. Rogerson

Introduction

Change is a constant feature of tourism. For Wall and Mathieson (2006, p. 6), tourism represents "a forceful agent of change" that creates impacts in destination environments. Similarly, for Butler (1980, p. 5), "there can be little doubt that tourist areas are dynamic, that they evolve and change over time". In Butlerian tourism evolution thinking, this change in tourist areas largely results from the nature of the tourism industry and how it operates in destinations. Furthermore, Page (2003, p. 29) considers tourism as an "ever-changing phenomenon" based on "interaction between the demand for and supply of tourism opportunities through time", a process that needs to be *managed for change* by focusing on planning, organising, leading and controlling.

Tourism creates various changes in destinations and the overall tourism system (Butler, 1999; Leiper, 1979; Rogerson & Rogerson, 2021). These changes can be ecological, socio-cultural and/or economic. They can be positive or negative, depending on the perspective, and direct or indirect in time and space (Saarinen, 2019). All this makes tourism-related changes complex and constantly transforming entities, which take place on different scales. Perhaps the most typical scale and a concrete unit of analysis in tourism studies has been the destination level in which the change has been intensively studied and modelled for a relatively long time (see Butler, 1980; Keller, 1987; Miossec, 1977; Stansfield, 1978). This is understandable as destinations are the locations in which the tourism phenomenon takes place in the most visible ways (Saarinen, 2014).

While much of the focus of change in tourism has been based on a destination scale, there is an increasing understanding that tourism is a multi-scalar and relational phenomenon taking place simultaneously in different socio-spatial scales (see Hannam, 2002; Wieckowski & Saarinen, 2019). Although host and guest encounters, for example, may happen in very specific places in tourism destinations, they can bring many other scales of economy, politics, norms and values together. In addition, the role and dimension of change in tourism represent different matters for different actors and from different perspectives. For us, as individuals and members of a family or other social group, tourism is an activity offering relaxation and a potential break from everyday life, routines and norms, and a possibility to connect with new people and places. For entrepreneurs and companies working directly or indirectly with tourists, the tourism phenomenon and related consumption needs are a way of making a living, developing business operations and creating an economic surplus. Local communities can perceive tourism as an opportunity for employment or creating a business in the future (Rogerson, 2013). Similarly, regional authorities and developers, including many governments and supra-national organisations, use the tourism industry as a tool for creating employment and economic activity on a larger geographical scale (Saarinen, 2020; Sharpley, 2002).

Especially in the Global South, the tourism industry is used for economic growth and diversification (see Rogerson & Visser, 2004, 2007; Saarinen et al., 2020) and a potential instrument for sustainable use of natural and cultural resources (Stone, Lenao & Moswete, 2020). The term Global South is usually connected with emerging and developing economies or low- and middle-income countries, which were previously known as developing countries or Third World countries (see Carmody, 2019; Puri, 2010). In this respect, countries with a high UNDP Human Development Index (HDI) are considered as the Global North with the rest belonging to the Global South (see Greve, 2020). However, the division between the Global South and Global North is not geographical in the traditional sense. The edges can be blurred as culturally or economically peripheral, and marginalised parts of a Global North country can be interpreted as representing the context for issues and challenges typical for a Global South country. Similarly, countries in the Global South can have subregions and economies reflecting the wealth usually associated with Global North situations. As noted by Sparke (2007, p. 117):

> the Global South is everywhere, but always also somewhere: accountable, embodied, and highly heterogeneous. It interrupts the flat earth, borderless world, and smooth space conceits of globalists, highlighting asymmetry and inequality amidst intensifying global interdependency.

Thus, the categorisation of the Global South and North is much more complex than it first implies. It involves various issues such as power relations, socio-spatial inequalities, environmental justice and unequal division of benefits and costs of global capitalism and tourism (see Olatunji & Bature, 2019). As such, it calls for highly nuanced analyses both in theory and practice, and this is very much the case in the change, management and development contexts of tourism in the Global South, as this volume fruitfully demonstrates.

Overall, change is truly a multidimensional element in tourism. If simplified, the issue of change has been approached by focusing on the impacts of tourism, the developmental aspects of tourism and changing characteristics of the tourism phenomenon. Obviously, these three perspectives are partly interrelated and they can be highly overlapping in certain cases. Although the impacts of tourism can be negative or positive, previous impact studies have had a main focus on the detrimental changes that tourism causes, especially in the tourism destination environments (see Hall & Lew, 2009; Wall & Mathieson, 2006). Recently, however, the scale of analysis has expanded to a wider system level in the studies on global climate change and tourism, for example (Gössling et al., 2021; Hall et al., 2015; Hall et al., 2012).

The developmental aspect of tourism highlights the positive side of the impacts of development. Thus, instead of mapping, describing or modelling the impacts of tourism, the emphasis is on the potential and active role of the industry as a tool for positive change. Typically, this has involved local or regional development needs, and, since the 1990s, an increasing interest has been given to the potential connections between tourism and sustainable development on a local or regional scale, namely destination level (Hall & Lew, 2009; Saarinen, 2014; Sharpley, 2000, 2009). Recently, this developmental focus has been integrated into the United Nations Sustainable Development Goals (SDGs) (see Hughes & Scheyvens, 2016; Rasoolimanesh et al., 2020), which could and should include a wider scale than destination sustainability alone in tourism (Saarinen, 2020; Scheyvens, 2018). Based on this, the chapter will outline the general nature and perspectives of change with an emphasis given to the impacts of tourism. However, the developmental aspects and connections of tourism with sustainable development and the SDGs will be further discussed in Chapter 2. Finally, the contributions of the book are introduced.

Tourism: Impacts, development and changes

A traditional way to categorise the impacts of tourism is based on the division between *environmental, socio-cultural* and *economic impacts* (Mathieson & Wall, 1982; Wall & Mathieson, 2006). This general division is widely used, although the tourism impacts can involve a vast range of more specific issues such as local administration and politics (Getz, 1983; Hall, 1994; Hollenhorst et al., 2014; Mbaiwa, 2005). There are many excellent books and reviews on the impacts of tourism (see Hall & Lew, 2009; Holden, 2000; Mathieson & Wall, 1982; Wall & Mathieson, 2006). Briefly, physical impacts refer to changes in the ecological environment caused by tourism activities. These activities and changes are based on traffic, construction of tourism facilities (hotels, resorts, amusement parks, golf courses etc.) and related general infrastructure such as roads, airports and housing, for example. Many of these environmental impacts are considered negative, an outcome which is rather paradoxical as it often indicates that the growth of tourism destroys the attraction elements on which tourism development depends (Butler, 1980, 1999; Holden, 2003).

However, tourism also has the potential to benefit the environment, for example, by supporting financially and politically better environmental management and nature conservation in destination communities (Fennell, 1999). Somewhat surprisingly, these positive changes may be interpreted as creating outcomes that can be construed as negative from another perspective by another stakeholder: tourism activities supporting nature conservation are good for the environment and people working in nature-based or ecotourism sectors, but they may cause negative social and economic outcomes for local resource-extractive industries and people employed in those sectors and traditional livelihoods (Saarinen, 2019).

Although the environmental impacts of tourism are well identified and discussed in the literature, there are relatively few empirical studies and especially theoretical models on the overall tourism-environment connections (Dowling, 1993). This is not the case with socio-cultural impacts of tourism, which have been conceptualised and theorised in various ways (see Murphy, 1985), including Doxey's (1975) Irridex Index or Social Exchange Theory and the Altruistic Surplus Phenomenon approach in tourism (see Nunkoo & Ramkissoon, 2011, 2012; Paraskevaidis & Andriotis, 2017; Rasoolimanesh et al., 2015). The social impact of tourism refers to how the industry, tourists and seasonal workers create changes in communities, and to individual and collective behaviour, values, norms, lifestyles and an overall quality of life (Hall, 1994). Many context-specific elements influence how local residents react to and perceive these tourism-based changes (Saarinen, 2019). These include issues such as the nature and scale of tourism development and tourist numbers, activities and behaviour, community characteristics and local economic dependency on tourism (Murphy, 1985; Nunkoo & Ramkissoon, 2011, 2012). Nunkoo and Gursoy (2012), for example, utilised the Social Exchange Theory in Mauritius and found that residents' resource-based occupational, environmental and gender identities influenced their support for the tourism industry on the island.

The economic impacts of tourism have been studied for decades. According to Mitchell (1979, 1984), the first journal article representing geographical tourism research was published in the journal *Economic Geography* as far back as 1933 (Jones, 1933; see also Gilbert, 1939, 1949), and this economic emphasis on tourism dominated the impact studies up to the 1970s. Still today, there is growing and considerable scholarship on the economic impact and benefit creation in tourism (see Bianchi, 2018; Dwyer et al., 2004; Hunt et al., 2015; Mosedale, 2014, 2015; Rogerson, 2013). Economic impacts are seen as "the driving force behind the development of tourism […] particularly so in less developed countries' (Telfer & Sharpley, 2008, p. 180; see Britton, 1982, 1991). Indeed, the economic effects of tourism are usually regarded as positive per se (Saarinen, 2019). In this respect, Telfer and Sharpley (2008) have further stated that tourism may represent an economic lifeline measured by contribution to gross domestic product (GDP), export earnings and employment generation. Thus, the tourism industry is often seen as a beneficial tool for local, regional and national development, serving as a response to economic challenges and changes taking place in the Global South.

Many elements explain the elementary connection of tourism with development. However, this connection is complicated and depends on how we understand the idea of development (see Carmody, 2019; Hannam, 2002; Holden, 2013; Saarinen & Lenao, 2014). Peet and Hardwick (2015, p. 1) have defined development as an evolutionary process in which human capacity increases towards "a better life". In tourism studies, Sharpley (2020) has associated development with the notion of well-being. In general, well-being can refer to different social and spatial scales, including societal, community or individual levels (see Sen, 1999).

At a local community level, the United States Local Government Commission has emphasised the difference between development and growth. For the Commission, "growth means to get bigger, development means to get better – an increase in quality and diversity" (cited by Pike et al., 2007, p. 1253). Thus, development thinking is focused on human well-being and qualitative dimensions in social and economic processes taking place in localities, while growth represents a quantitative measure that may not have an instrumental connection to positive changes in local or community development. Due to its positive connotations, many scholars such as Reyes (2001) have directly linked development to sustainability. From this perspective, development is seen as a social condition in which the needs of people and communities are satisfied by the sustainable use of natural resources. This connection with sustainability also differentiates development from growth: in tourism, growth may lead to wider development in destination systems and communities, but it can also create local inequalities, unsustainable practices and economic leakages (see Mathieson & Wall, 1982; Mbaiwa, 2005, Rogerson, 2012, 2019; Rogerson & Rogerson, 2019).

Change in tourism does not only refer to impacts and developmental aspects caused by the tourism industry. The notion that tourism is a dynamic phenomenon also involves those changes taking place in tourism itself. These changes are contextually based on the interplay between the (global) tourism industry and specific destinations and their socio-cultural, economic and ecological environments. Although externally driven tourism growth may result in placelessness, i.e. homogenised tourism environments with standardised services and infrastructures separated from their surrounding socio-cultural environments and local economies (Saarinen & Wall-Reinius, 2019; Shaw & Shaw, 1999), the same global processes and trends do not necessarily create universally similar outcomes in different destinations (Gordon & Goodall, 2000; Manuel-Navarrete, 2016). In this respect, place matters, which is well demonstrated in this book focusing on the connections between tourism and change in the Global South. Next, we introduce the main section of the book with reference to the individual chapter contributions.

Insights to tourism and change in the Global South

The book includes three sections: Introduction and frameworks; Change in tourism; and Tourism for change. The first section consists of introductory and theoretical chapters. After this opening chapter, Saarinen discusses the changing nature of tourism and the capacity of tourism to create positive change. Furthermore, he focuses on the future drivers of change in tourism, with an emphasis on the links between tourism and the SDGs (United Nations, 2015). Timothy continues the theoretical discussion by examining the changing meanings and values of heritage in society from a changing tourism perspective. According to him, changes in demand for heritage places and cultural experiences is currently extending beyond the normative and conventional mass tourist attractions, towards a greater acceptance of vernacular spaces and ordinary heritages. Based on this, Timothy discusses how different cultures value heritage differently, with the suggestion that many traditional societies in the Global South continue to view conservable and marketable heritage as ancient and extraordinary, while in the Global North, there is a movement towards a more democratised view of human patrimony. Related to the past as an attraction and resource for tourism, Chi and Chi explore the antecedents of vicarious tourism nostalgia by proposing an integrative conceptual model that categorises the antecedents of nostalgia into four dimensions: negative moods, traveller characteristics, travelling motivations and tourism contexts. They conclude by identifying research gaps that could guide future tourism nostalgia research with applications to African tourism development contexts.

Adu-Ampong begins the second section focusing on change in tourism. He focuses on local instruments and governance capacity in tourism development planning with a case from Ghana. His analysis demonstrates how the local governance capacity is often constrained by the dispersal of needed policy instruments, messy institutional arrangements and poor stakeholder interactions in tourism planning. Next, Hall examines the changing discourses of small island developing states (SIDS) over time while simultaneously charting changes with respect to tourism flows and development indicators. He concludes that despite assertions of new regionalism and a stated desire for more Indigenous and green growth as a way of fulfilling the SDGs, the ground reality is one in which tourism is still an economic focus and growth in visitor numbers remains a goal. This may result in undesired environmental impacts in the region.

Graham, Sucher-Pearson, Dadd and Pienaar take the discussion to the local scale, detailing changes in tourism with an emphasis on aspects of learning in tourism operations. Specifically, they focus on the generative power of nurturing new connections in Indigenous tour operators and places. They conclude by emphasising the need to decolonise tourism operations in Indigenous contexts. Gerbauer and Umscheid continue the discussion on colonial legacies of tourism in the African context by analysing the history of slavery used in the development of roots tourism in Ghana. They conclude that roots tourism is facilitated within a mode that can be described as a post-modern eventification of the history of slavery that is built around a concept of making the imagined and performed motherland Africa. Gutberlet discusses the issues of social capital and legacy in local tourism development with the focus on Souq Muttrah in the Sultanate of Oman. Due to international tourism growth, the marketplace has turned into a "tourist bubble", and Gutberlet aims to understand the related social changes within the local trader community. Booyens concludes the section by examining the evolutionary process of township tourism in South Africa. She identifies different phases in township tourism development and demonstrates that township tourism is changing through co-evolution, resulting in the maturation of township destinations.

The final section of the book emphasises the idea of tourism for change. Sibanda and Cheer analyse the impacts of tourism on the livelihoods of women in Victoria Falls, Zimbabwe. Their exploratory study particularly investigates the participation of women. Based on empirical results, they indicate that although tourism has had a positive impact on the income of women, they continue to have limited access to formal employment opportunities within the local tourism industry. Mbaiwa analyses the economic development potential or promise of tourism with links to nature conservation in the Global South. His approach is based on the sustainable development framework. He suggests that the wildlife-based tourism industry in the Global South should try to follow the ideals of sustainable tourism, which would ensure wildlife conservation and improved rural livelihoods and economies. Rogerson and Rogerson investigate the main challenges of leveraging a particular form of tourism: their focus is on caravan parks as a municipal asset with specific planning challenges around their leveraging for local development. In relation to the Province of the Western Cape, South Africa, their analysis demonstrates that administrative responsibility for caravan parks is not usually linked to either tourism or local development planning but instead to parks and recreation or community development, thereby undermining their potential to contribute more significantly to the local tourism economy.

Baum and Giddy focus on the so-called gig economy in Southern Africa. The gig economy refers to an employment environment that is characterised by temporal and short-term tasks as opposed to a stable permanent job market. They overview the nature of employment within both the gig and informal economies in the Southern African tourism sector with a critical assessment of tourism employment within the regional gig economy. Finally, Steinbrink and Voshage conclude the section by examining the nature of township tourism from the perspective of tourists' expectations. Based on their analysis from Katutura, Windhoek (Namibia), it is argued that township tourism should be understood as a form of cultural tourism that in particular uses post-colonial ideas of "real Africa" and "African culture".

Conclusion

The insights of the individual chapters manifest the different perspectives of tourism and change. The chapters demonstrate the dynamic character of tourism that is in a constant process of change that in turn creates changes in its destinations. Indeed, change in tourism involves the various impacts of tourism and tourism development in destinations, while tourism for change emphasises the associated potential of tourism for local and regional socio-economic and environmental development. Obviously, tourism does not take place in a vacuum. Thus, it is important to understand and analyse how the industry operates in destinations with its surrounding communities, economies and societies (see Lew et al., 2016, Novelli & Gebhardt, 2007). More research is needed on local people and communities and their possibilities to access the labour market and business opportunities provided by tourism development in the Global South. Related to this, it is important to examine the hosting societies' tourism and regional development and planning policies and how those strategies are practised and experienced locally (see Hall, 2011; Rogerson & Visser, 2020).

Although there is a rich research tradition on the impacts and changes in tourism, the connections between tourism and sustainable development are still weakly understood and applied in the Global South. Recently, this research need has been highlighted in the SDGs context (Prideaux et al., 2020; Rogerson & Baum, 2020), which forms a key research agenda and framework for tourism and change studies in the Global South for the foreseeable future. This agenda should include post-COVID-19 challenges, but also emerging impacts, issues and resilience needs in relation to global climate change and deepening socio-economic and political inequalities between the Global North and South.

References

Bianchi, R. (2018). The political economy of tourism development: A critical review. *Annals of Tourism Research, 70*, 88–102. doi:10.1016/j.annals.2017.08.005

Britton, S. G. (1982). The political economy of tourism in the third world. *Annals of Tourism Research, 9*, 331–338. doi:10.1016/0160-7383(82)90018-4

Britton, S. G. (1991). Tourism, capital, and place: Towards a critical geography of tourism. *Environment and Planning D: Society and Space, 9*(4), 451–478. doi:10.1068/d090451

Butler, R. W. (1980). The concepts of a tourist area cycle of evolution: Implications for management of resources. *Canadian Geographer, 24*(1), 5–12. doi:10.1111/j.1541-0064.1980.tb00970.x

Butler, R. W. (1999). Sustainable tourism: A state-of-the-art review. *Tourism Geographies, 1*(1), 7–25. doi:10.1080/14616689908721291

Carmody, P. (2019). *Development Theory and Practice in a Changing World*. Abingdon: Routledge.

Dowling, R. (1993). An environmentally-based planning model for regional tourism development. *Journal of Sustainable Tourism, 1*(1), 17–37. doi:10.1080/09669589309450698

Doxey, G. V. (1975). A causation theory of visitor-resident irritants: Methodology and research inferences. *The Impact of Tourism, Sixth Annual Conference Proceedings of the Travel Research Association*, 195–198.

Dwyer, L., Forsyth, P., & Spurr, R. (2004). Evaluating tourism's economic effects: New and old approaches. *Tourism Management, 25*(3), 307–317. doi:10.1016/S0261-5177(03)00131-6

Fennell, D. (1999). *Ecotourism: An Introduction*. London: Routledge.

Getz, D. (1983). Capacity to absorb tourism: Concepts and implications for strategic planning. *Annals of Tourism Research, 10*(2), 239–263. doi:10.1016/0160-7383(83)90028-2

Gilbert, E. W. (1939). The growth of island and seaside health resorts in England. *Scottish Geographical Magazine, 55*(1), 16–35. doi:10.1080/00369223908735100

Gilbert, E. W. (1949). The growth of Brighton. *The Geographical Journal, 114*(1–3), 30–52. doi:10.2307/1789987

Gordon, I., & Goodall, B. (2000). Localities and tourism. *Tourism Geographies, 2*(3), 290–311. doi:10.1080/14616680050082535

Gössling, S, Scott, D., & Hall, C. M. (2021). Pandemics, tourism and global change: A rapid assessment of COVID-19. *Journal of Sustainable Tourism, 29*(1), 1–20. doi:10.1080/09669582.2020.1758708

Greve, B. (2020). *Poverty*. London: Routledge.

Hall, C. M. (1994). *Tourism and Politics: Policy. Power and Place*. Chichester: John Wiley & Sons.

Hall, C. M. (2011). Policy learning and policy failure in sustainable tourism governance: From first- and second-order to third-order change? *Journal of Sustainable Tourism, 19*(4–5), 649–671. doi:10.1080/09669582.2011.555555

Hall, C. M., & Lew, A. (2009). *Understanding and Managing Tourism Impacts: An Integrated Approach.* London: Routledge.

Hall, C. M., Gössling, S., & Scott, D. S. (2012). *Climate Change and Tourism: Impacts, Adaptation and Mitigation.* London: Routledge.

Hall, C. M., Gössling, S., & Scott, D. (Eds.) (2015). *The Routledge Handbook of Tourism and Sustainability.* London: Routledge.

Hannam, K. (2002). Tourism and development: Globalization and power. *Progress in Development Studies, 2*(3), 227–234. doi:10.1191/1464993402ps039pr

Holden, A. (2000). *Environment and Tourism.* London: Routledge.

Holden, A. (2003). In need of new environmental ethics for tourism. *Annals of Tourism Research, 30*, 94–108. doi:10.1016/S0160-7383(02)00030-0

Holden, A. (2013). *Tourism, Poverty and Development.* London: Routledge.

Hollenhorst, S. J., Houge-MacKenzie, S., & Ostergren, D. M. (2014). The trouble with tourism. *Tourism Recreation Research, 39*(3), 305–319. doi:10.1080/02508281.2014.11087003

Hughes, E., & Scheyvens, R. (2016). Corporate social responsibility in tourism post-2015: A development first approach. *Tourism Geographies, 18*(5), 469–482. doi:10.1080/14616688.2016.1208678

Hunt, C., Durham, W., Driscoll, L., & Honey, M. (2015). Can ecotourism deliver real economic, social, and environmental benefits?: A study of the Osa Peninsula, Costa Rica. *Journal of Sustainable Tourism, 23*(3), 339–357. doi:10.1080/09669582.2014.965176

Jones, S. B. (1933). Mining and tourist towns in the Canadian Rockies. *Economic Geography, 9*(4), 368–378. doi:10.2307/140491

Keller, C. P. (1987). Stages on peripheral tourism development – Canada's Northwest territories. *Tourism Management, 8*(1), 20–32. doi:10.1016/0261-5177(87)90036-7

Leiper, N. (1979). The framework of tourism. *Annals of Tourism Research, 6*(4), 390–407. doi:10.1016/0160-7383(79)90003-3

Lew, A. A., Ng, P. T., Ni, C., & Wu, T. (2016). Community sustainability and resilience: Similarities, differences and indicators. *Tourism Geographies, 18*(1), 18–27. doi:10.1080/14616688.2015.1122664

Manuel-Navarrete, D. (2016). Boundary-work and sustainability in tourism enclaves. *Journal of Sustainable Tourism, 24*(4), 507–526. doi:10.1080/09669582.2015.1081599

Mathieson, A., & Wall, G. (1982). *Tourism, Economic, Physical and Social Impacts.* London: Longman.

Mbaiwa, J. (2005). Enclave tourism and its socio-economic impacts in the Okavango Delta, Botswana. *Tourism Management, 26*(2), 157–172. doi:10.1016/j.tourman.2003.11.005

Miossec, J. (1977). Un modele de l'espace touristique. *L'Espace Geographique 6*(1), 41–48.

Mitchell, L. (1979). The geography of tourism: An introduction. *Annals of Tourism Research, 6*(3), 235–244. doi:10.1016/0160-7383(79)90100-2

Mitchell, L. (1984). Tourism research in the United States: A geographic perspective. *GeoJournal, 9*(1), 5–15. doi:10.1007/BF00518313

Mosedale, J. (2014). Political economy of tourism. In A. A. Lew, C. M. Hall, & A. M. Williams (Eds.), *The Wiley Blackwell Companion to Tourism* (pp. 55–65). Chichester: John Wiley & Sons.

Mosedale, J. (Ed.) (2015). *Neoliberalism and the Political Economy of Tourism.* London: Routledge.

Murphy, P. (1985). *Tourism: A Community Approach*. London: Methuen.
Novelli, M., & Gebhardt, K. (2007). Community based tourism in Namibia: "Reality Show" or "Window Dressing"? *Current Issues in Tourism, 10*(5), 443–479. https://doi.org/10.2167/cit332.0
Nunkoo, R., & Gursoy, D. (2012). Residents' support for tourism: An identity perspective. *Annals of Tourism Research, 39*(1), 243–268. doi:10.1016/j.annals.2011.05.006
Nunkoo, R., & Ramkissoo, H. (2011). Developing a community support model for tourism. *Annals of Tourism Research, 38*(3), 964–988. doi:10.1016/j.annals.2011.01.017
Nunkoo, R., & Ramkissoo, H. (2012). Power, trust, social exchange and community support. *Annals of Tourism Research, 29*(2), 997–1023. doi:10.1016/j.annals.2011.11.017
Olatunji, F. O., & Bature, A. I. (2019). The inadequacy of post-development theory to the discourse of development and social order in the Global South. *Social Evolution & History*, 1(2), 229–243.
Page, S. (2003). *Tourism Management: Managing for Change*. Oxford: Butterworth-Heinemann.
Paraskevaidis, P., & Andriotis, K. (2017). Altruism in tourism: Social exchange theory vs altruistic surplus phenomenon in host volunteering. *Annals of Tourism Research, 62*, 26–37. doi:10.1016/j.annals.2016.11.002
Peet, R., & Hartwick, E. (2015). *Theories of Development: Contentions, Arguments, Alternatives* (3rd ed.). New York: Guilford Press.
Pike, A., Rodríguez-Pose, A., & Tomaney, J. (2007). What kind of local and regional development and for whom? *Regional Studies, 41*(9), 1253–1269. doi:10.1080/00343400701543355
Prideaux, B., Thompson, M., & Pabel, A. (2020). Lessons from COVID-19 can prepare global tourism for the economic transformation needed to combat climate change. *Tourism Geographies, 22*(3), 667–678. doi:10.1080/14616688.2020.1762117
Puri, S. H. (2010). *Rise of the Global South and its Impact on South-South Cooperation*. World Bank. Retrieved from https://openknowledge.worldbank.org/handle/10986/6076
Rasoolimanesh, S. M., Jaafar, M., Kock, N., & Ramayah, T. (2015). A revised framework of social exchange theory to investigate the factors influencing residents' perceptions. *Tourism Management Perspectives, 16*, 335–345. doi:10.1016/j.tmp.2015.10.001
Rasoolimanesh, S. M., Ramakrishna, S., Hall, C. M., Esfandiar, K., & Seyfi, S. (2020). A systematic scoping review of sustainable tourism indicators in relation to the sustainable development goals, *Journal of Sustainable Tourism*. doi:10.1080/09669582.2020.1775621
Reyes, G. E. (2001). Four main theories of development: Modernization, dependency, word-system, and globalization. *Nómadas. Revista Crítica de Ciencias Sociales y Jurídicas, 4*(2), 109–124.
Rogerson, C. M. (2012). Tourism – agriculture linkages in rural South Africa: Evidence from the accommodation sector. *Journal of Sustainable Tourism, 20*, 477–495. doi:10.1080/09669582.2011.617825
Rogerson, C. M. (2013). Urban tourism, economic regeneration and inclusion: Evidence from South Africa. *Local Economy, 28*(2), 188–202. doi:10.1177/0269094212463789
Rogerson, C. M. (2019). Peripheral tourism trajectories: Evidence from King Sabata Dalindyebo Municipality, South Africa. *GeoJournal of Tourism and Geosites, 26*(3), 974–992. doi:10.30892/gtg.26323-411
Rogerson, C. M., & Baum, T. (2020). COVID-19 and African tourism research agendas. *Development Southern Africa, 37*(5), 727–741. doi:10.1080/0376835X.2020.1818551

Rogerson, C.M., & Rogerson, J. M. (2019). Tourism, local economic development and inclusion: Evidence from Overstrand local municipality, South Africa. *GeoJournal of Tourism and Geosites*, *25*(2), 293–308. doi:10.30892/gtg.25202-360
Rogerson, C. M., & Rogerson, J. M. (Eds.) (2021). *Urban Tourism in the Global South: South African Perspectives*. Cham: Springer.
Rogerson, C. M., & Visser, G (Eds.) (2004). *Tourism and Development Issues in Contemporary South Africa*. Pretoria: Africa Institute of South Africa.
Rogerson, C. M., & Visser, G (Eds.) (2007). *Urban Tourism in the Developing World: The South African Experience*. London: Transaction Publishing.
Rogerson, J. M., & Visser, G. (Eds.) (2020). *New Directions in South African Tourism Geographies*. Cham: Springer.
Saarinen, J. (2014). Transforming destinations: A discursive approach to tourist destinations and development. In A. Viken & B. Granås (Eds.), *Destination Dynamics: Themes, Turns and Tactics* (pp. 47–62). Aldershot: Ashgate.
Saarinen, J. (2019). Communities and sustainable tourism development: Community impacts and local benefit creation tourism. In S. F. McCool & K. Bosak (Eds.), *A Research Agenda for Sustainable Tourism* (pp. 206–222). Cheltenham: Edward Elgar Publishing.
Saarinen, J. (2020). Tourism and sustainable development goals: Research on sustainable tourism geographies. In J. Saarinen (Ed.), *Tourism and Sustainable Development Goals: Research on Sustainable Tourism Geographies* (pp. 1–10). London: Routledge.
Saarinen, J., & Lenao, M. (2014). Integrating tourism to rural development and planning in the developing countries. *Development Southern Africa*, *31*(3), 363–372. doi:10.1080/0376835X.2014.888334
Saarinen, J., & Wall-Reinius, S. (2019). Enclaves in tourism: Producing and governing exclusive spaces for tourism. *Tourism Geographies*, *21*(5), 739–748. doi:10.1080/14616688.2019.1668051
Saarinen, J., Moswete, N., Athlopheng, J., & Hambira, W. (2020). Changing socio-ecologies of Kalahari: Local perceptions towards environmental change and tourism in Kgalagadi, Botswana. *Development Southern Africa*, *37*(5), 855–870. doi:10.1080/0376835X.2020.1809997
Scheyvens, R. (2018). Linking tourism to the sustainable development goals: A geographical perspective. *Tourism Geographies*, *20*(2), 341–342. doi:10.1080/14616688.2018.1434818
Sen, A. (1999). *Development as Freedom*. New York: Anchor Books.
Sharpley, R. (2000). Tourism and sustainable development: Exploring the theoretical divide. *Journal of Sustainable Tourism*, *8*(1), 1–19. doi:10.1080/09669580008667346
Sharpley, R. (2002). Rural tourism and the challenge of tourism diversification: The case of Cyprus. *Tourism Management*, *23*(3), 233–244. doi:10.1016/S0261-5177(01)00078-4
Sharpley, R. (2009). *Tourism Development and the Environment: Beyond Sustainability?* London: Earthscan.
Sharpley, R. (2020). Tourism, sustainable development and the theoretical divide: 20 years on. *Journal of Sustainable Tourism*, *28*(11), 1932–1946. doi:10.1080/09669582.2020.1779732
Shaw, B., & Shaw, G. (1999). "Sun, sand and sales": Enclave tourism and local entrepreneurship in Indonesia. *Current Issues in Tourism*, *2*(1), 68–81. doi:10.1080/13683509908667844
Sparke, M. (2007). Everywhere but always somewhere: Critical geographies of the Global South. *The Global South*, *1*(1), 117–126.

Stansfield, C. A. (1978). Atlantic City and the resort cycle: Background to the legalisation of gambling. *Annals of Tourism Research*, *5*(2), 238–251. doi:10.1016/0160-7383(78)90222-0

Stone, M., Lenao, M., & Moswete, N. (Eds.) (2020). *Natural Resources, Tourism and Community Livelihoods in Southern Africa*. London: Routledge.

Telfer, D. J., & Sharpley, R. (2008). *Tourism and Development in the Developing World*. London: Routledge.

United Nations (2015). *Transforming Our World: The 2030 Agenda for Sustainable Development. Resolution Adopted by the General Assembly on 25 September 2015*. New York: United Nations.

Wall, G., & Mathieson, A. (2006). *Tourism: Change, Impacts, and Opportunities*. London: Pearson Prentice-Hall.

Wieckowski, M., & J. Saarinen (2019). Tourism transitions and changes, and the creation of new spaces and places in Central-Eastern Europe. *Geographia Polonica*, *92*(4), 369–377. doi:10.7163/GPol.0154a

2 Tourism for change

Change management towards sustainable tourism development

Jarkko Saarinen

Introduction

Tourism is a highly dynamic phenomenon that creates change in destinations and the overall tourism system. These changes are complex with different kinds of socio-cultural, political, economic and environmental outcomes taking place in different time dimensions. Indeed, tourism is a multi-scalar and relational phenomenon taking place simultaneously in different socio-spatial and time scales. However, the main focus of change in tourism studies has been based on a destination scale and impacts caused by tourism activities in destinations. While tourism impact studies have mainly focused on the negatively perceived changes caused by tourism growth in destinations, many policymakers, localities and regions are increasingly using the tourism industry as a tool for positive change. Furthermore, many countries, especially in the Global South, utilise tourism for diversifying their economies that are vulnerable to global change and economic globalisation, for example (see Rogerson & Visser, 2020). However, this has turned out to be a challenging task in practice, as here the relationship between tourism growth and local development has often been controversial and based on uneven exchanges between the Global North and South. All this has highlighted the potential connections between tourism development and sustainability thinking (Saarinen, 2020).

Indeed, before the current COVID-19 pandemic and the resulting crisis that has halted global human mobility, many international organisations and policymakers framed the tourism industry and its local and regional development connections as high-potential tools for putting sustainable development into practice (Hall, 2011). This emphasis on sustainable tourism development will likely remain or even be strengthened in the policymaking agenda in the post-COVID-19 crisis world: even though (or especially because) the crisis has reminded us of the vulnerability of global tourism. Thus, there is probably a greater need for change management and better governance and resilience planning in tourism on the sustainable development path in the future (see Gössling et al., 2021; Saarinen & Gill, 2019). Against this background, this chapter will discuss the general nature and scales of change in tourism. Specific focus is given to the ideas of resilience and sustainable tourism as the key change management frameworks for destination development and how the industry could represent a positive change for destinations in the context of the UN's Sustainable Development Goals (SDGs) Agenda for 2030.

Tourism and change

Changing destinations and change management

As argued by Urry (1995, p. 1): "almost all the major social and cultural theories bear upon the explanation of place". This notion is relevant for tourist destinations and their changes, which are the places the impacts of tourism are the most concrete in practice. Destination development processes have been treated theoretically in various ways and scales in tourism studies (see Butler, 1980; Keller, 1987; Miossec, 1977; Saarinen, 2001; Sirgy & Su, 2000; Strapp, 1988). Simply, the changes taking place in destinations can be categorised as short term, medium term and long term. These dimensions are often deeply intertwined in the tourism development process in practice and are also influenced and guided by various external processes (Ateljevic & Doorne, 2002).

Typically, short-term changes refer to seasonality in tourism (Baum, 1999; Hartmann, 1986). Butler (1994) has defined seasonality as a temporal imbalance in tourism, which can be described in terms of diverse elements such as the numbers and expenditures of visitors, traffic flows and employment, for example (Higham & Hinch, 2002). In practice, seasonality means a concentration of tourist flows and activities in relatively short periods of time (e.g. daily, weekly and annually). Elements causing seasonality are regarded as being natural or institutional. The former refers to various natural phenomena that may have an impact on tourism consumption and production patterns, such as sunlight or daylight, snowfall, rainfall and extreme temperatures (Butler, 1994). The latter elements result from socio-cultural, religious, economic and/or other human structured elements, such as school and public holidays (Hartmann, 1986).

Seasonality is usually considered a problem for tourism development (see Baum, 1999; Butler, 1994; Cannas, 2012), but, in some cases, it can also create diversity for the overall attractiveness of a destination (see Saarinen & Varnajot, 2019). From the problem perspective, tourist destination management organisations (DMOs) and other tourism developers and policymakers aim to manage seasonality, especially reducing the length or existence of low season periods (Bonn et al., 1992). According to Ashworth and Thomas (1999), there are three basic strategies for change management in seasonality. The first one involves changing the product mix to lengthen the season by utilising events, festivals and conference tourism development, for example. These strategies have also been used in Global South contexts (see Lew & Chang, 1999; Rogerson, 2005, 2013). In addition, advances in the use of technology have been successfully applied for extending tourism seasons; Afriski in Lesotho, for example, uses artificial snow-making technology to extend the skiing season in an otherwise vulnerable environment (Hoogendoorn et al., 2020). Second, market diversification can be used for offering destination products to new segments that would be interested in lower seasons. Ashworth and Thomas (1999, pp. 735–736) name "attracting retired people to off-season quieter holidays" as an option, which has also created second-home communities in the Global South in places such as Swakopmund in Namibia and Cape Town and its surroundings in South Africa (see Hall, 2014; Hottola, 2009; Visser & Hoogendoorn, 2015). Rogerson and Rogerson (2020) have also noted the potential of caravan parks for lowering seasonality in the coastal province of the Western Cape, South Africa. Third, Ashworth and Thomas (1999) name seasonal pricing as a way to reduce seasonality, an element widely used in the Global South and North.

Medium-term changes are also quite typical in tourism but are relatively little studied and understood forms of change. Temporally, these changes can last for several months to a few years, depending on the specific case and change elements that are often sudden and unpredictable. Due to this unpredictability, medium-term change management is a very challenging task for DMOs and other destination governance bodies, especially in the Global South where adaptive capacity is low but operational vulnerability is high (see Saarinen et al., 2020). Typically, medium-term changes are caused by external economic or political changes or even shocks that have a serious temporal impact on visitor flows and/or the attractiveness of a destination. In addition to the current COVID-19 crisis, which, it is estimated, will have serious impacts on global tourism and different destinations for a few years at least, past medium-term changes have been caused by external events such as oil crises, terrorism, political conflicts and financial crises. With respect to these kinds of changes, which are usually unpredictable by nature, the issues guiding recovery processes and elements explaining differences in that process are highly challenging to pre-analyse and define in detail (Franklin & Crang, 2001; Gordon & Goodall, 2000). These issues and elements relate to a critical question like why are certain destinations more vulnerable or resilient to external shocks? Also, how can we actively create capacity and planning towards better resilience and recovery in tourism development and governance? (See Bramwell, 2011; Bramwell & Lane, 2011; Saarinen & Gill, 2019). These are the burning issues in the current COVID-19 crisis, defining when and how different destinations will recover, and whether they will return to a similar or different growth path as before the crisis.

Long-term changes are probably the most studied and theorised development processes in tourism. Indeed, there is a long tradition of research on the change and evolution of tourist destinations. As early examples, Gilbert (1939, 1949) studied the development of seaside and inland resorts in Great Britain, and Christaller (1964) sketched phases for a typical development process of tourism in peripheral areas. For Christaller, tourism evolved from individual, unorganised and small-scale activity to institutionalised and large-scale operations with intensive marketing, competition and finally package tours. Stansfield (1978) further developed destination development modelling by analysing the resort cycle of Atlantic City, USA. From a community change perspective, one of the best-known early frameworks was based on Doxey's (1975) Irridex (irritation index) model. The model focuses on changes in a local community's attitudes towards growing tourism and its impacts from initial “euphoria” to “apathy”, followed by “annoyance” and finally “antagonism”. This changing pattern of attitudes towards tourism reflects community responses to the changing impacts of the tourism industry and its negative externalities at a local scale (see Nunkoo & Ramkissoon, 2011; Saarinen, 2019).

Probably the best-known destination development model is based on Butler’s (1980) seminal work on the tourism area life cycle (TALC), which combined a product life cycle model from marketing studies and carrying capacity thinking in natural resource management. For Butler, tourism growth represents a dynamic force that creates changes through specific stages that are characterised by different kinds of tourism impacts in destination environments and host–guest relations. The model depicts a growth process of tourism destinations from small-scale operations and individual travellers (exploration and involvement stages) to a development stage characterised by rapid and massive changes in environment and tourism infrastructure. The development stage is followed by growth towards a mass-scale tourism destination (consolidation/stagnation stages). After the stagnation stage, the options for the future are that a destination cycle can start again (i.e. rejuvenation) or the growth can decline in different degrees (Butler, 1980; Strapp, 1988). This future post-stagnation situation depends, for example, on how the change management has been planned and implemented during the previous stages, especially in the development stage when the impacts are the most visible and often driven mainly by strong market demands, which can potentially sideline host community needs and environmental issues (Keller, 1987). Thus, based on the model, the nature of destination development and the changes taking place are strongly seen as being based on internal processes and results of decisions, planning and governance practices in a destination (Saarinen & Kask, 2008).

The tourism area life cycle process and its specific stages and the overall applicability in theory and practice have been widely debated in the tourism literature (see Agarwal, 2002; Haywood, 1986; Hovinen, 1981, 1982; Keller, 1987; Strapp, 1988). Like the original product marketing cycle, the destination life cycle follows an asymptotic curve resembling the letter “S”. In contrast to the product life cycle, however, the elements turning the growth in tourism towards consolidation and eventually stagnation are based on the various aspects of destination carrying capacity (see Getz, 1983). Although the model indicates that competition with other tourism destinations has an impact on the development of a destination, most relevant indicators are based on local scale changes (Butler, 1980). Instead of this territorial “container” idea, however, tourism destinations are relational spaces that are deeply embedded in wider regional, national and global scale processes, changes and contexts (see Saarinen & Kask, 2008). This is evident with some medium-term changes that may in practice redefine the overall development course of an individual destination, despite being temporal external change factors.

However, in the context of long-term changes, globalisation processes with deepening relations and interdependence of distant places, economies and actors (Bianchi, 2018; Storper, 1997) strongly mould tourism destination structures and courses of development (Saarinen & Kask, 2008). This is largely based on external impacts that destinations can sometimes only react to but not coordinate or even locally manage. While non-locally driven development is not a new issue in modern tourism (see Featherstone & Lash, 1995), contemporary tourism destinations are largely being transformed on a non-local basis and more rapidly than ever before (see Saarinen, 2007; Shaw & Shaw, 1999). Obviously, the capacity to influence the local impacts and outcomes of globalisation and global change, in general, vary contextually. While some destinations in the Global North can locally respond and work with the globalisation process from below, destinations in the Global South are more vulnerable to external impacts and forces of change (see Rogerson & Baum, 2020; Saarinen et al., 2020).

On a theoretical level, this has been effectively outlined by Britton (1982, 1991) in his enclave model for Third World tourism that is based on the long-term change perspective characterised by the structures and processes of unequal political and economic dependency in destination development (see Brohman, 1996). According to Britton (1982), destinations in the Global South are underprivileged in that they have been integrated into the global "world system" ruled by wealthy regions and actors from the Global North (Saarinen, 2017). The model mainly focuses on the macro-economics of tourism development that have a hierarchical structure in the global-local nexus. The top level is controlled by international tourism corporations and other actors located in the core economies. Destination-based branch (e.g. national) offices of core businesses are located in the mid-level of the system model. At the bottom level of the model are those small, locally rooted businesses that are usually highly dependent upon the chain of businesses and actors located at the higher levels of the global (and national) tourism supply chain. Typically, this system leads to a marginalisation of the local economy and hosts actors with major economic leakages in the local tourism economy (see Brohman, 1996; Mbaiwa, 2005). However, as recent studies have shown, there can be other contextual and historical elements beyond the tourism economy and its global value chain taking place in destination regions (see Buzinde & Manuel-Navarrete, 2013; Carlisle & Jones, 2012; Manuel-Navarrete, 2016). These can have a significant impact on the tourism development process and how externally driven changes take place locally. In this respect, one of the key ideas and terms relate to destination resilience and governance.

Change and resilience in tourism

The core idea of resilience refers to change management. Indeed, according to Berkes and Ross (2013, p. 6), resilience "is one of the major conceptual tools [...] to deal with change". Initially, resilience was linked with the stability of ecological systems and how these natural systems reacted with change, disturbance, stress and other random events. Based on this, Holling (1973, p. 17) defined resilience as "the ability of these [natural] systems to absorb changes of state variable, driving variable, and parameters, and still persist". However, nowadays, resilience thinking is widely used outside ecological analyses (see Davidson, 2010;

Davoudi, 2012; Folke, 2006; Magis, 2010), including terms such as community resilience, tourism resilience or resilient tourism destinations (see Calgaro et al., 2014; Saarinen & Gill, 2019; Strickland-Munro et al., 2010).

Among the different understandings of resilience thinking, most tourism-focused research has adopted a so-called socio-ecological resilience approach (see Lew et al., 2016). It is based on the integration of ecological and social processes, adaptive capacity and transformability via innovation and learning (Saarinen & Gill, 2019). From the socio-ecological resilience perspective, a change is seen as a constant state in tourism systems, and resilience indicates the capacity of that system (e.g. tourism destination) to absorb disturbance and sustain itself. This takes place by adapting to and transforming with change (see Berkes & Ross, 2013; Folke, 2006). In this context, Davoudi (2012) has developed the idea of evolutionary resilience that connects resilience thinking with regional change and, thus, destination development (see Brouder & Saarinen, 2019). Furthermore, Simmie and Martin (2010) have emphasised the importance of the adaptive capacity of a system for this evolutionary resilience in regional development. The reason adaptive capacity or related management processes are called "adaptive" is that they recognise the conditions of destination development are in a constant state of change (Saarinen & Gill, 2019), as indicated in most tourism development models.

Tourism resilience research has taken two main directions. The first one is based on the impacts of natural disasters and climate change on tourism destinations and their vulnerability (Becken, 2013; Biggs et al., 2012; Calgaro et al., 2014; Hall et al., 2016), where issues have been actively studied in the Global South contexts (see Betzold, 2015), especially in Southern Africa (see Fitchett et al., 2016; Hambira et al., 2013; Hambira et al., 2020; Rogerson, 2016). From this perspective, the interplay and mutual connections between tourism, environment and related changes are the main research focus (Wall & Mathieson, 2006). On the one hand, tourism can be heavily impacted by environmental change and its direct and indirect outcomes for destinations, tourism-dependent communities and the overall tourism system (see Hoogendoorn & Fitchett, 2018; Kaján & Saarinen, 2013). On the other hand, the industry is also seen as a contributor to environmental and climate change, especially if poorly managed (see Gössling & Hall, 2006).

The second direction is based on an economic emphasis on tourism destination resilience with a focus on diversification and restructuring (Lew, 2014). In a general regional development studies context, this evolutionary economic approach has focused on the historical processes that produce an uneven spatial change (Boschma & Frenken, 2006). This perspective is based on an "evolutionary turn" within economic geography over the last two decades (Coe, 2010; Pike et al., 2015) and it has recently resulted in a major step forward in tourism development studies (see Brouder et al., 2017). As an approach, it aims at a deeper understanding of the relationship between the global tourism economy and the wider community economic development of places beyond (territorial) destination thinking alone (Brouder, 2017). Thus, an important aspect of an evolutionary approach is that it takes tourism out of a tourism-centric focus and into broader regional and relational contexts, which are vital for understanding the resilience of places (Brouder & Saarinen, 2019).

In this context, evolutionary resilience refers to a capacity of a destination to "sustain long-term development" and "respond positively to short-term shocks" (Boschma, 2015, p. 735). With this ability to maintain and create new development relates to the adaptive capacity (Adger, 2000) of destination systems, the evolutionary approach resonates well with the idea of sustainable destination development (see Gill & Williams, 2014). From this perspective, resilient tourism can be interpreted as a sustainable tourism system at a destination scale that has an adaptive capacity and economic, socio-cultural and ecological resources to absorb change and external disturbance and sustain itself by transforming via learning and innovation. Therefore, resilience thinking and sustainable destination governance are highly connected but characterised by complex relations.

Tourism for change towards sustainable development

Sustainable tourism

While there are many similarities between resilience and sustainability thinking in tourism and change (Holladay & Powell, 2013; Lew et al., 2016), there are also some potential major differences. First, sustainable development is not necessarily about adapting to or persisting with change (Saarinen & Gill, 2019). Obviously, sustainable tourism can definitely involve adaptation, but the core idea of sustainable development is to actively change the course of (unsustainable) development, not to adapt to it. Therefore, some scholars connect resilience with adaptation and sustainability with mitigation (Adger, 2000). Second, resilient tourism does not mean that it is sustainable per se: a destination development process can be resilient to changes and external shocks but that state can be based on a negative lock-in, operating against the long-term well-being of local communities and/or environmental integrity, for example (see Espiner et al., 2017; Hambira et al., 2013; Simmie & Martin, 2010). In addition, sustainable development is based on a very long-term, holistic and inter-generational scalar context, while resilience often operates or is placed in more local and shorter time scales (see Lew, 2014).

The idea of sustainable tourism and its relation to changes caused by the tourism industry is highly debated and contested in tourism literature (Liu, 2003; Sharpley, 2000, 2009). As a result, there are many definitions, approaches, principles and criteria for sustainable tourism (see Butler, 1999). One of the most commonly used definitions of sustainable tourism originates from the early 1990s by the World Tourism Organization (UNWTO, 1993, p. 7) stating that sustainable tourism aims to meet "the needs of current tourists and host regions while protecting and enhancing opportunities for the future". Since then, sustainability has been a major focus for tourism policymakers, planners and researchers (Clarke, 1997; Hall, 2011). This led Hall (2011, p. 649) to state that sustainability is "one of the great success stories of tourism research".

In the early phases of research and policymaking, sustainability was perceived as a realistic and optimal target for small-scale tourism operations with limited changes to the environment (see Clarke, 1997). In contrast, mass-scale tourism was considered negatively as something that needs a full transformation towards a smaller scale and, thus, sustainable operations. Nowadays, sustainability is not linked to a specific scale of tourism operation, as all kinds of tourism activities are considered to be able to progress towards sustainable development in tourism (see Butler, 1999). This flexibility has partly created tough criticism of the idea and practices of the current market-oriented and industry-driven interpretation of sustainability (Liu, 2003; Sharpley, 2000). In particular, the relationship between tourism as a global scale growth industry and the idea of sustainable development has been critically debated (Gössling et al., 2021). From another perspective, however, the massive scale of the tourism industry and its potential to have a global impact has resulted in an opposite emphasis that highlights the positive change the industry could create, especially in the Global South, if it were better governed. Indeed, recently, the capacity of tourism to work towards sustainable development has been highlighted in connection with the United Nations SDGs of the 2030 Agenda for Sustainable Development (see Hughes & Scheyvens, 2016; Rogerson & Baum, 2020; Saarinen, 2020).

Tourism and sustainable development goals

In 2015, United Nations member states ratified the SDGs (United Nations, 2015). They defined the agenda for global development by addressing major issues in global development and well-being challenges, such as poverty, inequality, climate, environmental degradation, prosperity and peace and justice. There are 17 goals (Table 2.1) and 169 specific targets, which are set to transform the world by 2030.

The SDGs and specific targets provide opportunities to think about how global tourism could contribute to sustainable change in destination communities and environments, especially in the Global South. In order to advance such change, the United Nations General Assembly established the "International Year of Sustainable Tourism for Development" in 2017 (UNWTO, 2017). According to the International Year of Sustainable Tourism for Development, the industry was set to work towards three specific SDGs: Promote sustained, inclusive and sustainable economic growth, full and productive employment and decent work for all (SDG 8); Ensure sustainable consumption and production patterns (SDG 12); and Conserve and sustainably use the oceans, seas and marine resources for sustainable development (SDG 14). In addition to these, tourism literature indicates that the industry could have the potential to contribute to many other SDGs and their targets (see Saarinen, 2020). Based on this potential, Scheyvens (2018, p. 341) has called for tourism scholars "to consider how we might utilise the SDGs to analyse the linkages between tourism and sustainable development in a wide range of contexts and at different scales".

Tourism has the potential in different degrees to contribute, directly or indirectly, to each of the 17 SDGs (Rasoolimanesh et al., 2020). Related to this, the

Table 2.1 The UN Sustainable Development Goals (SDGs)

Goal 1: No Poverty: Economic growth must be inclusive to provide sustainable jobs and promote equality.
Goal 2: Zero Hunger: The food and agriculture sectors offer key solutions for development, and are central for hunger and poverty eradication.
Goal 3: Good Health and Well-being: Ensuring healthy lives and promoting well-being for all at all ages is essential to sustainable development.
Goal 4: Quality Education: Obtaining a quality education is the foundation to improving people's lives and sustainable development.
Goal 5: Gender Equality: Gender equality is not only a fundamental human right, but a necessary foundation for a peaceful, prosperous and sustainable world.
Goal 6: Clean Water and Sanitation: Clean, accessible water for all is an essential part of the world we want to live in.
Goal 7: Affordable and Clean Energy: Energy is central to nearly every major challenge and opportunity.
Goal 8: Decent Work and Economic Growth: Sustainable economic growth will require societies to create the conditions that allow people to have quality jobs.
Goal 9: Industry, Innovation and Infrastructure: Investments in infrastructure are crucial to achieving sustainable development.
Goal 10: Reduced Inequality: To reduce inequalities, policies should be universal in principle, paying attention to the needs of disadvantaged and marginalised populations.
Goal 11: Sustainable Cities and Communities: There needs to be a future in which cities provide opportunities for all, with access to basic services, energy, housing, transportation and more.
Goal 12: Responsible Consumption and Production: There needs to be responsible production and consumption.
Goal 13: Climate Action: Climate change is a global challenge that affects everyone, everywhere.
Goal 14: Life Below Water: Careful management of this essential global resource is a key feature of a sustainable future.
Goal 15: Life on Land: Sustainably manage forests, combat desertification, halt and reverse land degradation, halt biodiversity loss.
Goal 16: Peace and Justice Strong Institutions: Access to justice for all, and building effective, accountable institutions at all levels.
Goal 17: Partnerships to Achieve the Goal: Revitalise the global partnership for sustainable development.

Source: United Nations, 2015.

World Bank Group (2017) extended the connections between the tourism industry and the SDGs by listing 20 reasons why tourism works for development. They listed five core pillars: (1) sustainable economic growth; (2) social inclusiveness, employment and poverty reduction; (3) resource efficiency, environmental protection and climate; (4) cultural values, diversity and heritage; and (5) mutual understanding, peace and security. These emphasised pillars expand the potential connections between tourism and the SDGs. The sustainability challenge, however, is the idea and nature of change they involve. For example, the sustainable economic growth pillar indicates that tourism stimulates the growth of gross domestic product (GDP) and the increase of international trade and investments. Instead of development, these stimulated elements refer to growth ideology associated with current neoliberal economic thinking (Hall, 2009). Therefore, the growth in GDP, international trade, investments and infrastructure may not necessarily lead to sustainable development for destination communities end environments where global tourism takes place (see Hollenhorst et al., 2014). Thus, while we acknowledge that one-third of all service exports and every tenth job in the world are based on tourism (UNWTO, 2018; WTTC, 2017), we should also ask what kind of jobs tourism creates (SDGs 8 and 10), for whom and at what social cost (e.g. modern slavery issues), and who is included in development based on tourism growth (SDGs 5 and 10), especially in the Global South (Saarinen, 2020). In this respect, the United Nations Conference and Trade Development has indicated that "tourism employment is not gender neutral in Africa, as women and men do not necessarily have the same opportunities in and benefits from the sector" (UNCTAD, 2017, p. 90). This kind of notion may well apply to many other issues than the gender dimension in tourism development in the Global South.

Conclusion

Tourism is a dynamic phenomenon that constantly transforms and creates change in its destinations. There is a rich research tradition describing and modelling the changing characteristics of tourism destinations and economic, socio-cultural and ecological impacts that take place during the tourism development process. While impact studies have mainly focused on negatively perceived changes caused by tourism growth in tourism destinations, many policymakers and regional developers and authorities have increasingly planned to use the tourism industry as a tool for positive change, i.e. development. While policy documents indicate the great potential that tourism has to contribute to the global development agenda, such as the SDGs, the connection between tourism growth and local development and community well-being, it has turned out to be a challenging task in practice, especially in Global South development contexts. This is largely based on uneven political and economic relations between the Global South and the Global North, and higher vulnerability and lower resilience towards external negative changes and shocks in the Global South.

Despite these critical questions and structural issues in global tourism, tourism scholars need to try to understand the change in tourism and the conditions of change for sustainable development that is not solely measured by growth-oriented indicators. Instead, sustainable tourism research should aim to reflect the quality of life and the well-being of people and the environment in destination regions. As noted by Bianchi (2009), this calls for action and practices beyond policies and well-meaning statements towards structural analysis and changes that consider "the material inequalities generated in the tourism economy" (p. 486).

In the tourism economy, the change element works in multi-scalar ways by taking place at different geographical scales at the same time and by operating in destinations in various but intertwined timeframes. Currently, the change aspects of tourism destination development and related discussions and research are dominated by the impacts the COVID-19 pandemic have created for global tourism in general, and literally all tourism destinations in the Global South and North. While sudden and unexpected changes and even shocks are typical for tourism and destination development processes, calling for adaptive capacity and resilience, the current crisis may also create transformational needs towards sustainability for the tourism industry and its international mobilities. Indeed, the COVID-19 crisis may offer an opportunity to rethink growth in tourism and question the volume growth tourism model advocated by powerful tourism policymaking organisations (Gössling et al., 2021; Brouder, 2020). In this respect, the question is whether global tourism should return to the same growth path and pattern as before the crisis or whether we should try to critically rethink and reset some of the premises of the production and consumption modes of global tourism.

At the moment, however, it seems that the tourism industry and overall economic systems will aim to return to the pre-crisis growth mode as soon as possible. This is highly understandable in the Global South, where human development is largely based on the need for economic growth (Carmody, 2019; Moyo, 2018). The International Monetary Fund (IMF, 2020), for example, estimates that COVID-19 will lead to approximately 90 million more people living in extreme poverty in 2020 in the Global South. This contrasts deeply with the aims of SDGs, emphasising the target of no extreme poverty in 2030 and the key idea of the goal that we should "leave no one behind" in global development. In this respect, the tourism industry and the overall global economy had no real opportunities to be prepared for the seriousness of the impacts of the COVID-19 crisis and proactively build greater resilience to take care of the communities and people living close to the poverty line. However, we have no such excuse for the forthcoming climate change crisis. This, together with the current crisis, emphasises the urgent need to develop better change management and resilience planning in tourism (Gössling et al., 2021; Prideaux et al., 2020; Sigala, 2020).

This is especially important in Global South contexts, where Rogerson and Baum (2020, p. 733) have highlighted the need for post-COVID-19 tourism development research "to become more aligned with SDGs". Indeed, tourism destination development should be managed for change, and there is a clear need for better destination governance towards SDGs, given the lesson the current crisis has provided us – not to mention the state of relatively near future affairs characterised by increasing globalisation, related economic insecurity and global climate change (Gössling et al., 2021; Saarinen & Gill, 2019). This better governance would serve the change management needs of tourism and strengthen resilience and adaptive capacity and lead tourism development towards sustainability by utilising the industry as a tool for sustainable development. All this we already know very well in theory and policymaking. However, the core elements and ideas of sustainability "often got lost in translation" when we are doing development (Utting, 2015, p. 1). Thus, the next real challenge – the litmus test – is its implementation.

References

Adger, W. N. (2000). Social and ecological resilience: Are they related? *Progress in Human Geography, 24*(3), 347–364. doi:10.1191/030913200701540465

Agarwal, S. (2002). Restructuring seaside tourism: The resort lifecycle. *Annals of Tourism Research, 29*(1), 25–55. doi:10.1016/s0160-7383(01)00002-0

Ashworth, J., & Thomas, B. (1999). Patterns of seasonality in employment in tourism in the UK. *Applied Economics Letters, 6*(11), 735–739. doi:10.1080/135048599352312

Ateljevic, I., & Doorne, S. (2002). Representing New Zealand: Tourism imaginary and ideology. *Annals of Tourism Research, 29*(3), 648–667. doi:10.1016/s0160-7383(01)00077-9

Baum, T. (1999). Seasonality in tourism: Understanding the challenges. *Tourism Economics, 5*(1), 5–8. doi:10.1177/135481669900500101

Becken, S. (2013). Developing a framework for assessing resilience of tourism subsystems to climatic factors. *Annals of Tourism Research, 43*, 506–528. doi:10.1016/j.annals.2013.06.002

Berkes, F., & Ross, H. (2013). Community resilience: Toward an integrated approach. *Society and Natural Resources, 26*, 5–20. doi:10.1080/08941920.2012.736605

Betzold, C. (2015). Adapting to climate change in small island developing states. *Climatic Change, 133*(3), 481–489. doi:10.1007/s10584-015-1408-0

Bianchi, R. (2009). The "critical turn" in tourism studies: A radical critique. *Tourism Geographies, 11*(4), 484–504. doi:10.1080/14616680903262653

Bianchi, R. (2018). The political economy of tourism development: A critical review. *Annals of Tourism Research, 70*, 88–102. doi:10.1016/j.annals.2017.08.005

Biggs, D., Hall, C. M., & Stoeckl, N. (2012). The resilience of formal and informal tourism enterprises to disasters: Reef tourism in Phuket, Thailand. *Journal of Sustainable Tourism, 20*(5), 645–665. doi:10.1080/09669582.2011.630080

Bonn, M. A., Furr, H. L., & Uysal, M. (1992). Seasonal variation of coastal resort visitors: Hilton Head Island. *Journal of Travel Research, 31*(1), 50–56. doi:10.1177/004728759203100110

Boschma, R. (2015). Towards an evolutionary perspective on regional resilience. *Regional Studies, 49*(5), 733–751. doi:10.1080/00343404.2014.959481

Boschma, R., & Frenken, K. (2006). Why is economic geography not an evolutionary science? *Journal of Economic Geography, 6*(3), 273–302. doi:10.1093/jeg/lbi022

Bramwell, B. (2011). Governance, the state and sustainable tourism: A political economy approach. *Journal of Sustainable Tourism, 19*(4–5), 459–477. doi:10.1080/09669582.2011.576765

Bramwell, B., & Lane, B. (2011). Critical research on the governance of tourism and sustainability. *Journal of Sustainable Tourism, 19*(4–5), 411–421. doi:10.1080/09669582.2011.580586

Britton, S. G. (1982). The political economy of tourism in the Third World. *Annals of Tourism Research, 9*, 331–338. doi:10.1016/0160-7383(82)90018-4

Britton, S. G. (1991). Tourism, capital, and place: Towards a critical geography of tourism. *Environment and Planning D: Society and Space, 9*(4), 451–478. doi:10.1068/d090451

Brohman, J. (1996). New directions in tourism for third world development. *Annals of Tourism Research, 23*(1), 48–70. doi:10.1016/0160-7383(95)00043-7

Brouder, P. (2017). Evolutionary economic geography: Reflections from a sustainable tourism perspective. *Tourism Geographies, 19*(3), 438–447. doi:10.1080/14616688.2016.1274774

Brouder, P. (2020). Reset redux: Possible evolutionary pathways towards the transformation of tourism in a COVID-19 world. *Tourism Geographies*. doi:10.1080/14616688.2020.1760928

Brouder, P., & Saarinen, J. (2019). Co-evolution and resilient regions: Moving towards sustainable tourism development. In J. Saarinen, & A. M. Gill (Eds.), *Resilient Destinations: Governance Strategies in the Transition Towards Sustainability in Tourism* (pp. 67–76). London: Routledge.

Brouder, P., Anton Clavé, S., Gill, A., & Ioannides, D. (Eds.) (2017). *Tourism Destination Evolution*. London: Routledge.

Butler, R. (1980). The concepts of a tourist area cycle of evolution: Implications for management of resources. *Canadian Geographer*, *24*(1), 5–12. doi:10.1111/j.1541-0064.1980.tb00970.x

Butler, R. W. (1994). Seasonality in tourism: Issues and problems. In A. V. Seaton (Ed.), *Tourism: The State of the Art* (pp. 332–339). Chichester: John Wiley & Sons Ltd.

Butler, R. W. (1999). Sustainable tourism: A state-of-the-art review. *Tourism Geographies*, *1*(1), 7–25. doi:10.1080/14616689908721291

Buzinde, C. N., & Manuel-Navarrete, D. (2013). The social production of space in tourism enclaves: Mayan children's perceptions of tourism boundaries. *Annals of Tourism Research*, *43*, 482–505. doi:10.1016/j.annals.2013.06.003

Calgaro, E., Lloyd, K., & Dominey-Howes, D. (2014). From vulnerability to transformation: A framework for assessing the vulnerability and resilience of tourism destinations. *Journal of Sustainable Tourism*, *22*(3), 341–360. doi:10.1080/09669582.2013.826229

Cannas, R. (2012). An overview of tourism seasonality: Key concepts and policies. *Almatourism - Journal of Tourism, Culture and Territorial Development*, *3*(5), 40–58. doi:10.6092/issn.2036-5195/3120

Carlisle, S., & Jones, E. (2012). The beach enclave: A landscape of power. *Tourism Management Perspectives*, *1*, 9–16. doi:10.1016/j.tmp.2011.12.003

Carmody, P. (2019). *Development Theory and Practice in a Changing World*. London: Routledge.

Christaller, W. (1964). Some considerations of tourism location in Europe: The peripheral regions-under development countries-recreation areas. *Regional Science Association; Papers XII, Lund Congress*, pp. 95–105.

Clarke, J. (1997). A framework of approaches to sustainable tourism. *Journal of Sustainable Tourism*, *5*(3), 224–233. doi:10.1080/09669589708667287

Coe, N. M. (2010). Geographies of production I: An evolutionary revolution. *Progress in Human Geography*, *35*(1), 81–91. doi:10.1177/0309132510364281

Davidson, D. J. (2010). The applicability of the concept of resilience to social systems: Some sources of optimism and nagging doubts. *Society and Natural Resources*, *23*(12), 1135–1149. doi:10.1080/08941921003652940

Davoudi, S. (2012). Resilience: A bridging concept or a dead end? *Planning Theory & Practice*, *13*(2), 299–307. doi:10.1080/14649357.2012.677124

Doxey, G. V. (1975). A causation theory of visitor-resident irritants: Methodology and research inferences. *The Impact of Tourism, Sixth Annual Conference Proceedings of the Travel Research Association*, pp. 195–198.

Espiner, S., Orchiston, C., & Higham, J. (2017). Resilience and sustainability: A complementary relationship? Towards a practical conceptual model for the sustainability-resilience nexus in tourism. *Journal of Sustainable Tourism*, *25*(10), 1385–1400. doi:10.1080/09669582.2017.1281929

Featherstone, M., & Lash, S. (1995). Globalisation, modernisation and the spatialisation of social theory. In M. Featherstone & R. Robertson (Eds.), *Global Modernity* (pp. 1–24). London: SAGE.

Fitchett, J., Grant, B., & Hoogendoorn, G. (2016). Climate change threats to two low-lying South African coastal towns: Risks versus perceptions. *South African Journal of Science*, *112*(5/6), 86–94. doi:10.17159/sajs.2016/20150262

Folke, C. (2006). Resilience: The emergence of a perspective for social-ecological systems analyses. *Global Environmental Change*, *16*(3), 253–267. doi:10.1016/j.gloenvcha.2006.04.002

Franklin, A., & Crang, M. (2001). The trouble with tourism and travel theory. *Tourist Studies*, *1*(1), 5–22. doi:10.1177/146879760100100101

Getz, D. (1983). Capacity to absorb tourism: Concepts and implications for strategic planning. *Annals of Tourism Research*, *10*(2), 239–263. doi:10.1016/0160-7383(83)90028-2

Gilbert, E. W. (1939). The growth of island and seaside health resorts in England. *Scottish Geographical Magazine*, *55*(1), 16–35. doi:10.1080/00369223908735100

Gilbert, E. W. (1949). The growth of Brighton. *The Geographical Journal*, *114*(1–3), 30–52. doi:10.2307/1789987

Gill, A. M., & Williams, P. W. (2014). Mindful deviation in creating a governance path towards sustainability in resort destinations. *Tourism Geographies*, *16*(4), 546–562. doi:10.1080/14616688.2014.925964

Gordon, I., & Goodall, B. (2000). Localities and tourism. *Tourism Geographies 2*(3), 290–311. doi:10.1080/14616680050082535

Gössling, S., & Hall, C. M. (Eds.) (2006). *Tourism and Global Environmental Change: Ecological, Social, Economic and Political Interrelationships*. London: Routledge.

Gössling, S, Scott, D., & Hall, C. M. (2021). Pandemics, tourism and global change: A rapid assessment of COVID-19. *Journal of Sustainable Tourism*, *29*(1), 1–20. doi:10.1080/09669582.2020.1758708

Hall, C. M. (2009). Degrowing tourism: Décroissance, sustainable consumption and steady-state tourism. *Anatolia*, *20*(1), 46–61. doi:10.1080/13032917.2009.10518894

Hall, C. M. (2011). Policy learning and policy failure in sustainable tourism governance: From first- and second-order to third-order change? *Journal of Sustainable Tourism*, *19*(4–5), 649–671. doi:10.1080/09669582.2011.555555

Hall, C. M. (2014). Second home tourism: An international review. *Tourism Review International*, *18*(3), 115–135. doi:10.3727/154427214X14101901317039

Hall, C. M., Malinen, S., Vosslamber, R., & Wordsworth, R. (2016). *Business and Post-disaster Management: Business, Organisational and Consumer Resilience and the Christchurch Earthquakes*. London: Routledge.

Hambira, W. L., Saarinen, J., Atlhopheng, J., & Manwa, H. (2013). Climate change adaptation practices in nature-based tourism in Maun in the Okavango Delta area, Botswana: How prepared are the tourism businesses? *Tourism Review International*, *17*(2), 19–29. doi:10.3727/154427213X13649094288025

Hambira, W. L., Saarinen, J., & Moses, O. (2020). Climate change policy in a world of uncertainty: Changing environment, knowledge and tourism in Botswana. *African Geographical Review*, *29*(3), 252–266. doi:10.1080/19376812.2020.1719366

Hartmann, R. (1986). Tourism, seasonality and social change. *Leisure Studies*, *5*(1), 25–33. doi:10.1080/02614368600390021

Haywood, K. M. (1986). Can the tourist-area life cycle be made operational? *Tourism Management*, *7*(3), 154–167. doi:10.1016/0261-5177(86)90002-6

Higham, J., & Hinch, T. (2002). Tourism, sport and seasons: The challenges and potential of overcoming seasonality in the sport and tourism sectors. *Tourism Management*, *23*(2), 175–185. doi:10.1016/S0261-5177(01)00046-2

Holladay, P. J., & Powell, R. B. (2013). Resident perceptions of socio-ecological resilience and the sustainability of community-based tourism development in the commonwealth of Dominica. *Journal of Sustainable Tourism*, *21*(8), 1188–1211. doi:10.1080/09669582.2013.776059

Hollenhorst, S. J., Houge-MacKenzie, S., & Ostergren, D. M. (2014). The trouble with tourism. *Tourism Recreation Research*, *39*(3), 305–319. doi:10.1080/02508281.2014.11087003

Holling, C. S. (1973). Resilience and stability of ecological systems. *Annual Review of Ecology and Systematics*, 1–23. doi:10.1146/annurev.es.04.110173.000245

Hoogendoorn, G., & Fitchett, J. M. (2018). Tourism and climate change: A review of threats and adaptation strategies for Africa. *Current Issues in Tourism*, *21*(7), 742–759. doi:10.1080/13683500.2016.1188893

Hoogendoorn, G., Stockigt, L., Saarinen, J., & Fitchett, J. M. (2020). Adapting to climate change: The case of snow-based tourism in Afriski, Lesotho. *African Geographical Review*. doi:10.1080/19376812.2020.1773878

Hottola, P. (2009). Coastal bird tourism in Namibia: Postcolonial resources and restrains. In P. Hottola (Ed.), *Tourism Strategies and Local Responses in Southern Africa* (pp. 105–126). Wallingford: CABI Publishing.

Hovinen, G. R. (1981). A tourist cycle in Lancaster County, Pennsylvania. *Canadian Geographer*, *15*(3), 283–286. doi:10.1111/j.1541-0064.1981.tb01649.x

Hovinen, G. R. (1982). Visitor cycles – outlook for tourism in Lancaster County. *Annals of Tourism Research*, *9*(4), 565–583. doi:10.1016/0160-7383(82)90073-1

Hughes, E., & Scheyvens, R. (2016). Corporate social responsibility in tourism post-2015: A development first approach. *Tourism Geographies*, *18*(5), 469–482. doi:10.1080/14616688.2016.1208678

IMF (International Monetary Fund) (2020). *World Economic Outlook, October 2020: A Long and Difficult Ascent*. Retrieved from https://www.imf.org/en/Publications/WEO/Issues/2020/09/30/world-economic-outlook-october-2020

Kaján, E., & Saarinen, J. (2013). Tourism, climate change and adaptation: A review. *Current Issues in Tourism*, *16*(2), 167–195. doi:10.1080/13683500.2013.774323

Keller, C. P. (1987). Stages in peripheral tourism development – Canada's Northwest territories. *Tourism Management*, *8*(1), 20–32. doi:10.1016/0261-5177(87)90036-7

Lew, A. A. (2014). Scale, change and resilience in community tourism planning. *Tourism Geographies*, *16*(1), 14–22. doi:10.1080/14616688.2013.864325

Lew, A. A., & Chang, T. C. (1999). Where the world meets: Regionalism and globalization in Singapore's convention industry. *Journal of Convention and Exhibition Management*, *1*(4), 17–36. doi:10.1300/J143v01n04_03

Lew, A. A., Ng, P. T., Ni, C., & Wu, T. (2016). Community sustainability and resilience: Similarities, differences and indicators. *Tourism Geographies*, *18*(1), 18–27. doi:10.1080/14616688.2015.1122664

Liu, Z. (2003). Sustainable tourism development: A critique. *Journal of Sustainable Tourism*, *11*(6), 459–475. doi:10.1080/09669580308667216

Magis, K. (2010). Community resilience: An indicator of social sustainability. *Society and Natural Resources*, *23*(5), 401–416. doi:10.1080/08941920903305674

Manuel-Navarrete, D. (2016). Boundary-work and sustainability in tourism enclaves. *Journal of Sustainable Tourism*, *24*(4), 507–526. doi:10.1080/09669582.2015.1081599

Mbaiwa, J. (2005). Enclave tourism and its socio-economic impacts in the Okavango Delta, Botswana. *Tourism Management, 26*(2), 157–172. doi:10.1016/j.tourman.2003.11.005

Miossec, J. (1977). Un modele de l'espace touristique. *L'Espace Geographique, 6*(1), 4–8.

Moyo, D. (2018). *Edge of Chaos: Why Democracy Is Failing to Deliver Economic Growth and How to Fix it*. New York: Basic Civitas Books.

Nunkoo, R., & Ramkissoon, H. (2011). Developing a community support model for tourism. *Annals of Tourism Research, 38*(3), 964–988. doi:10.1016/j.annals.2011.01.017

Pike, A., MacKinnon, D., & Cumbers, A. (2015). Doing evolution in economic geography. *Economic Geography, 92*(2), 123–144. doi:10.1080/00130095.2015.1108830

Prideaux, B., Thompson, M., & Pabel, A. (2020). Lessons from COVID-19 can prepare global tourism for the economic transformation needed to combat climate change. *Tourism Geographies, 22*(3), 667–678. doi:10.1080/14616688.2020.1762117

Rasoolimanesh, S. M., Ramakrishna, S., Hall, C. M., Esfandiar, K., & Seyfi, S. (2020). A systematic scoping review of sustainable tourism indicators in relation to the sustainable development goals. *Journal of Sustainable Tourism*. doi:10.1080/09669582.2020.1775621

Rogerson, C. M. (2005). Conference and exhibition tourism in the developing world: The South African experience. *Urban Forum, 16*(2–3), 176–195. doi:10.1007/s12132-005-1004-x

Rogerson, C. M. (2013). Urban tourism, economic regeneration and inclusion: Evidence from South Africa. *Local Economy, 28*(2), 188–202. doi:10.1177/0269094212463789

Rogerson, C. M. (2016). Climate change, tourism and local economic development in South Africa. *Local Economy, 31*(1–2), 322–331. doi:10.1177/0269094215624354

Rogerson, C. M., & Baum, T. (2020). COVID-19 and African tourism research agendas. *Development Southern Africa, 37*(5), 727–741. doi:10.1080/0376835X.2020.1818551

Rogerson, C. M., & Rogerson, J. M. (2020). Municipal caravan parks in South Africa: Geography, management and planning. *Bulletin of Geography. Socio-economic Series, 49*(49), 109–124. doi:10.2478/bog-2020-0027

Rogerson, J. M., & Visser, G. (2020). Research trends in South African tourism geographies. In J. M. Rogerson & G. Visser (Eds.), *New Directions in South African Tourism Geographies* (pp. 1–14). Cham: Springer.

Saarinen, J. (2001). The transformation of a tourist destination – theory and case studies on the production of local geographies in tourism in Finnish Lapland. *Nordia Geographical Publications 30*(1), 1–105.

Saarinen, J. (2007). Tourism in peripheries: The role of tourism in regional development in Northern Finland. In B. Jansson & D. Muller (Eds.), *Tourism in High Latitude Peripheries: Space, Place, and Environment* (pp. 41–52). Wallingford: CABI.

Saarinen, J. (2017). Enclavic tourism spaces: Territorialization and bordering in tourism destination development and planning. *Tourism Geographies, 19*(3), 425–437. doi:10.1080/14616688.2016.1258433

Saarinen, J. (2019). Communities and sustainable tourism development: Community impacts and local benefit creation tourism. In S. F. McCool & K. Bosak (Eds.), *A Research Agenda for Sustainable Tourism* (pp. 206–222). Cheltenham: Edward Elgar Publishing.

Saarinen, J. (2020). Tourism and sustainable development goals: Research on sustainable tourism geographies. In J. Saarinen (Ed.), *Tourism and Sustainable Development Goals: Research on Sustainable Tourism Geographies* (pp. 1–10). London: Routledge.

Saarinen, J., & Gill, A. M. (2019). Tourism, resilience and governance strategies in the transition towards sustainability. In J. Saarinen & A. M. Gill (Eds.), *Resilient*

Destinations: Governance Strategies in the Transition Towards Sustainability in Tourism (pp. 15–33). London: Routledge.

Saarinen, J., & Kask, T. (2008). Transforming tourism spaces in changing socio-political contexts: The case of Pärnu, Estonia, as a tourist destination. *Tourism Geographies*, *10*(4), 452–473. doi:10.1080/14616680802434072

Saarinen, J., & Varnajot, A. (2019). The Arctic in tourism: Complementing and contesting perspectives on tourism in the Arctic. *Polar Geography*, *42*(2), 109–124. doi:10.1080/1088937X.2019.1578287

Saarinen, J., Moswete, N., Athlopheng, J., & Hambira, W. (2020). Changing socio-ecologies of Kalahari: Local perceptions towards environmental change and tourism in Kgalagadi, Botswana. *Development Southern Africa*, *37*(5), 855–870. doi:10.1080/0376835X.2020.1809997

Scheyvens, R. (2018). Linking tourism to the sustainable development goals: A geographical perspective. *Tourism Geographies*, *20*(2), 341–342. doi:10.1080/14616688.2018.1434818

Sharpley, R. (2000). Tourism and sustainable development: Exploring the theoretical divide. *Journal of Sustainable Tourism*, *8*(1), 1–19. doi:10.1080/09669580008667346

Sharpley, R. (2009). *Tourism Development and the Environment: Beyond Sustainability?* London: Earthscan.

Shaw, B., & Shaw, G. (1999). "Sun, sand and sales": Enclave tourism and local entrepreneurship in Indonesia. *Current Issues in Tourism*, *2*(1), 68–81. doi:10.1080/13683509908667844

Sigala, M. (2020). Tourism and COVID-19: Impacts and implications for advancing and resetting industry and research. *Journal of Business Research*, *117*, 312–321. doi:10.1016/j.jbusres.2020.06.015

Simmie, J., & Martin, R. (2010). The economic resilience of regions: Towards an evolutionary approach. *Cambridge Journal of Regions, Economy and Society*, *3*(1), 27–43. doi:10.1093/cjres/rsp029

Sirgy, J. M., & Su, C. (2000). Destination image, self-cognity, and travel behaviour: Toward an integration model. *Journal of Travel Research*, *38*(4), 340–352. doi:10.1177/004728750003800402

Stansfield, C. A. (1978). Atlantic City and the resort cycle: Background to the legalisation of gambling. *Annals of Tourism Research*, *5*(2), 238–251. doi:10.1016/0160-7383(78)90222-0

Storper, M. (1997). Territories, flows, and hierarchies in the global economy. In K. R. Cox (Ed.), *Spaces of Globalisation: Reasserting the Power of the Local* (pp. 19–44). New York: Guilford Press.

Strapp, J. D. (1988). The resort cycle and second homes. *Annals of Tourism Research*, *15*(4), 504–516. doi:10.1016/0160-7383(88)90046-1

Strickland-Munro, J., Allison, H., & Moore, S. (2010). Using resilience concepts to investigate the impacts of protected area tourism on communities. *Annals of Tourism Research*, *37*(2), 499–519. doi:10.1016/j.annals.2009.11.001

Tyrell, T.J., & Johnston, R.J. (2008). Tourism sustainability, resiliency and dynamics: Towards a more comprehensive perspective. *Tourism and Hospitality Research*, 8(1): 14–24.

UNCTAD (United Nations Conference on Trade and Development) (2017). *A Commitment to Inclusive Trade*. Geneva: UNCTAD.

United Nations (2015). *Transforming Our World: The 2030 Agenda for Sustainable Development. Resolution Adopted by the General Assembly on 25 September 2015.* New York: United Nations.

UNWTO (World Tourism Organization) (1993). *Sustainable Tourism Development: Guide for Local Planners*. Madrid: World Tourism Organization.

UNWTO (World Tourism Organization) (2017). *Tourism and the Sustainable Development Goals – Journey to 2030*. Madrid: UNWTO.

UNWTO (World Tourism Organization) (2018). *Who We Are*. Retrieved from http://www2.unwto.org/content/who-we-are-0

Urry, J. (1995). *Consuming Places*. London: Routledge.

Utting, P. (2015). Foreword. In P. Utting (Ed.), *Revisiting Sustainable Development* (pp. 1–15). Geneva: UNRISD.

Visser, G., & Hoogendoorn, G. (2015). A decade of second home tourism research in South Africa: Research prospects for the developing world? *South African Geographical Journal*, *97*(2), 111–122. doi:10.1080/03736245.2015.1028976

Wall, G., & Mathieson, A. (2006). *Tourism: Change, Impacts, and Opportunities*. Harlow: Pearson Prentice-Hall.

World Bank Group (2017). *World Bank Annual Report 2017*. Washington, DC: World Bank.

WTTC (World Travel and Tourism Council) (2017). *Global Benchmarking Report*. London: WTTC.

3 Democratising the cultural past

Western values, the Global South and cross-cultural perspectives in heritage tourism

Dallen Timothy

Introduction

Heritage tourism is one of the most prevalent types of tourism. Even traditional sun, sea and sand tourists frequently engage with the cultural past during their holidays, even if only briefly on excursions from their resorts or cruise ships. Likewise, most packaged tours overwhelmingly focus on cultural heritage as their primary attractions for mass consumers. These "casual" or "incidental" heritage tourists are often the largest markets for prevalent cultural sites in highly popular (and overcrowded) destinations (e.g. the Sagrada Familia in Barcelona or St Mark's Square in Venice). "Serious" or "purposeful" heritage tourists may also visit the most popular attractions in the world, but they are more inclined to seek off-the-beaten-path sorts of destinations and desire a more profound cultural experience beyond the most iconic locales (McKercher, 2002; Timothy, 2021).

During the past few decades, researchers and industry promoters have identified plethoric tourism niches. These niches, or special interests, originally grew organically but are now being promoted by destination marketing organisations for their unique market characteristics and economic potential. These are often seen as alternatives to mass, packaged tourism with fewer negative impacts and better engagement between destination residents and tourists. The growth of special interest tourism and the needs of purposeful heritage tourists, together with other contemporary factors, have started shifting demand for the most famous and iconic heritage sites to more mundane and local elements of cultural heritage that may not be ancient, tangible or spectacular. Nevertheless, they are an important part of the cultural heritage of people and places and lend an extra level of perceived authenticity to destination experiences. This, however, is an overtly Western perspective on heritage and tourism. People of the Global South or in traditional societies may have different views about small-scale, local and ordinary heritage and harbour different perspectives on authenticity, branding, special interest products and heritage conservation.

This chapter describes some of the changes taking place in the Global North as regards people's perceptions of heritage and heritage tourism from an overwhelming focus on extraordinary, tangible heritage towards a more holistic view of heritage that includes ordinary places and intangible elements of the past. However, not every culture shares this evolving perspective on supply and demand. The chapter goes on to examine some cross-cultural differences in perceptions of heritage from the perspectives of authenticity, branding and conservation. Because the extant literature on cross-cultural or cross-national perspectives on heritage focuses largely on China as a case for non-Western outlooks, this chapter also refers to China several times, as well as other Global South contexts to illustrate its ideas.

Evolving perspectives in heritage tourism

Tourists and the tourism industry have for the past few hundred years focused on the extraordinary elements of the cultural past and largely ignored the majority of heritage throughout the world (Timothy, 2018a; Weidenfeld, 2010), which is small-scale, ordinary and rather mundane. Extraordinary places, such as castles, fortresses, monasteries, legendary archaeological sites, famous churches and ancient monuments have dominated the cultural tourismscapes in most parts of the world, yet these represent a very small portion of the 100 billion or so people who have lived on the earth since the Anthropocene began. Humankind and the tourism industry have traditionally valued the opulent heritage of kings and queens, popes and priests, military leaders, famous musicians and artists and well-regarded politicians, even though these heritages exemplify an infinitesimally small part of the human experience. This is indicative of the traditional European tendency to regard monumental sites as true and honest representations of history (Cleere, 2001) and is rooted in the "modernist notions of antiquity and grandeur" (Podder et al., 2018, p. 335). Indeed, heritage protection, interpretation and marketing have had an elitist bent in terms of what patrimony has been selected to conserve, reveal and sell to the public (Pielesiak, 2015; Timothy, 2018b).

Thus, the demand for tourism has long been skewed in a different direction from the actual archives of heritage assets because tourists and the tourism industry have focused overwhelmingly on the extraordinary and grandiose elements of the distant past (Podder et al., 2018; Timothy, 2018a), while largely ignoring the vernacular heritage of the ordinary populace. In this process of grandeurisation, the most splendid and oldest attractions have become significant magnets for mass tourism. Although folk villages, open-air and folklife museums in the US, Canada, Australia, Japan, the UK and Europe were established to commemorate the lives of ordinary people, these too have now become mass extraordinary tourist attractions since the 1960s and feature prominently in mainstream tour circuits (Atkinson, 1987; Gordon, 2016; Shafernich, 1993).

Breaking from this history, a shift seems to be occurring in the Global North regarding the recognition and valorisation of ordinary, commonplace heritage, although most tourism, especially in the Global South, lags in appreciating alternative heritage narratives beyond the lavish (and often colonial) manifestations of the past. This relatively recent and Western acknowledgement of alternative heritages stems from an increasing dissatisfaction with traditional mass attractions and the unbalanced history that these attractions have communicated. As education levels have risen, there has been notable growth in social justice, which has translated in the heritage realm into more accurate portrayals of the past and more balanced views beyond the colonialist narratives that have dominated historic sites for decades. Concurrently, there is a growing interest in understanding how ordinary people lived their ordinary lives, which may derive from modern society's unease with modernity (Lowenthal, 2015). There is some evidence to support this proposition of change.

Since the first World Heritage Site designations in 1978, UNESCO's World Heritage List has grown every year to 1,121 properties in 2020 (UNESCO, 2020a, 2020b). UNESCO has been one of the biggest promoters of the grandiose built heritage, as its goals include the representation of "masterpieces" of human creative genius and exemplary illustrations of "universal" heritage. Because of these stated goals, most of the sites acknowledged by ICOMOS and inscribed on the UNESCO World Heritage List are examples of extraordinary and iconic places. Thus, the concept of heritage, even as established by ICOMOS and UNESCO, has traditionally ignored elements of the ordinary past, focusing overwhelmingly on masterpieces. However, out of a fear of their disappearance and with a growing interest in the democratisation of heritage to include more elements of the patrimony of ordinary people and places, UNESCO began a programme to focus on cultural landscapes in 1992, many of which have been inscribed on the World Heritage List. Likewise, UNESCO initiated the Convention for the Safeguarding of Intangible Cultural Heritage in 2003, which emphasises the intangible cultures of humankind and recognises the importance of local inheritances that deserve to be protected (Timothy, 2014a, 2014b). At the time of writing, there are 549 elements of intangible heritage in 127 countries on UNESCO's List of Intangible Heritage (UNESCO, 2020a). In 2019, UNESCO led the International Year of Indigenous Languages as part of its recent efforts to emphasise Indigenous cultures and intangible heritage.

Likewise, the everyday heritage of ordinary people, including marginalised people (e.g. some immigrant groups, ethnic and racial minorities), is in danger of being lost or disinherited as the most imposing heritage is selected for protection and valorisation. In response to this, there has been a broadening of the heritage narrative in the past quarter-century as a way of empowering disempowered groups and of telling a more balanced account of history. This is referred to here as the democratisation of heritage. This movement is increasingly obvious in the US South with regard to giving African slaves and their descendants a louder voice and more control over how their stories are shared with outsiders (Alderman et al., 2016; Cook, 2016; Small, 2013). Similarly, Native Americans' stories are now better controlled by them and offer a counter-narrative to traditional accounts told by white historians and interpreters (Pitchford, 2006; Smith, 2019). We are now also seeing similar changes in other developed countries about their Indigenous populations, such as Australia and Japan (Cheung, 2005; Hiwasaki, 2000; Ruhanen & Whitford, 2019) and much earlier in New Zealand (Amoamo & Thompson, 2010; McIntosh et al., 2004).

Technology also plays a role in this wave of change. With the growth of internet-based technology, especially social media and travel websites, we have seen the beginnings of the globalisation of local and ordinary heritage. Through online evaluations, travel blogs and social media, what was once local heritage is becoming global heritage through increased exposure to the point that what might have once been too ordinary for tourism is now finding its way onto people's travel itineraries. Because of the pervasiveness of social media and the internet, these heritagisation and heritage democratisation processes are beginning to mainstream elements of the cultural past that were once run-of-the-mill and ordinary.

From a tourism-specific perspective, the development of niche tourism is illustrative of this trend towards attractions and destinations that appeal to specialised markets (Agarwal et al., 2018; Novelli, 2005; Rittichainuwat, 2018; Weiler & Hall, 1992), many of which coalesce around ordinary heritage. Although certain attractions, events and destinations have fascinated niche markets for decades, these specialised places and market cohorts have become central to tourism planning, management and marketing in earnest since the 1980s and are increasingly popular among tourists from the Global North.

There are subcategories or overlapping types of tourism that fall under the broader umbrella of heritage tourism because they involve the consumption of heritage assets, places and narratives, and many of the motives for visiting are cultural. These include, among others, agritourism, dark tourism, Indigenous tourism, culinary/food tourism, religious tourism, industrial tourism and many elements of sport tourism. These niches focus largely on ordinary types of heritage resources, which would not normally appeal to the global masses, engaging them within the larger tourism industry. Agritourism is based on farming practices, traditional land tenure and agricultural landscapes, planting and harvesting, traditional tools and local environmental knowledge and other elements of the rural heritage of places (Sznajder et al., 2009). Indigenous tourism involves people's experiences with native peoples, their living cultures, religious rituals and traditions and their built heritage (Carr et al., 2016; Ruhanen & Whitford, 2019). Most Indigenous people have much to share with outsiders, but those that are truly empowered in their heritage narratives can also choose which parts of their cultural inheritance will or will not be shared with outsiders. Culinary tourism is an important manifestation of ordinary heritage-based tourism, focusing largely on traditional foods and their preparation methods, recipes, ingredients, flavours, fragrances and social roles (Porananond, 2015; Timothy, 2016). Food may tell important stories that other elements of heritage cannot – migration, struggles with nature and survival, slavery, colonialism, war, poverty, gendered spaces, faith, familial relations and more. National and regional cuisines are frequently distinguished by the stories they tell, the unique senses of place they convey and the social solidary they reinforce (Bessière, 2013; Timothy, 2016).

North–South perspectives

Through Western discontent, growing social media use, increased acknowledgement by international and national heritage branding bodies and the upwards trajectory of special interest tourism, ordinary heritage is finally coming to the forefront of existential tourist experiences. The heightened appreciation for ordinary heritage previously described is blatantly a Western perspective and illustrates change in the Global North. Yet, many commentators have suggested there are still plentiful differences between how heritage is conceptualised and utilised in different cultures, with particular reference to Western versus traditional societies. Whereas the Western perspective of heritage is evolving as noted earlier, according to Podder et al. (2018, p. 335), the Global South tends to hold onto the notion that heritage "is inherently inclined toward an elitist sense of historicity and romantic nostalgia about the past". Even though the heritage-defining organisations have taken an "ordinary/participatory/subjective" turn throughout the postmodern years, it is yet the visual, exotic and monumental that continue to receive the most attention from heritage discourses, local authorities and tourism producers (Podder et al., 2018, p. 345). However, such a blanket perspective is inaccurate, as will be noted later. Some of the differences in heritage perceptions between the North and the South are examined in the following sections in the contexts of authenticity, branding and conservation.

Authenticity and branding

There is a strong and growing body of tourism literature on the experiential authenticity of visiting cultural places. Western tourists frequently claim that authentic places, cultures and experiences are one of their main motives for visiting, and authenticity is one of the most oft-cited elements of a satisfying experience (Chhabra, 2019; Domínguez-Quintero et al., 2019; Nguyen & Cheung, 2016; Wang, 1999). However, authenticity means different things to different people.

The early work of MacCannell (1973) and the later work of Cohen (1988) and many others were influential in understanding the role of perceived authenticity in the tourist experience, but their ideas were also conceived around Western constructs of authenticity. Although objective authenticity (i.e. measurable and verifiable) is less important than entertainment for most Westerners, there is a growing desire for more verifiably authentic places and events as people want more balanced and holistic interpretations of the past. For some purposeful Western tourists, authenticity entails originality, accuracy and truthfulness. In their eyes, the reality of the destination is most important; what people really wear and eat and where they live are important, whereas the reconstructed traditional homes staged for the tourist gaze or ceremonies performed for tourist consumption are not. Thus, a growing number of postmodern, Western travellers want the ordinary, not the staged extraordinary. From a material culture perspective, Western research has shown that authenticity typically requires original materials, methods, designs, knowledge and skills (Littrell et al., 1993; Soukhathammavong & Park, 2019), while authentic intangible culture must be in its original form and performed by the people whose culture is being reproduced for tourists (Caneen, 2014; Douglas & Douglas, 1991; Timothy, 2021).

In other cultural contexts, the Western concept of authenticity may not exist at all or is substantially different. For Samoans, the staged fire knife dance at the Polynesian Cultural Center in Hawaii is authentic because Samoans created it, the knives used derive from a traditional Samoan weapon and it is Samoans who continue to adapt the knife and the dance in their cultural performances. "Authenticity, then, for these indigenous people consists not in the faithful representation of historic practices, but in the ethnic identity of those who create, own, and control it. It is authentically Samoan because authentic Samoans say it is" (Caneen, 2014, p. 113).

Zhu (2018) suggests that Chinese travellers (in her case, backpackers) rarely seek authenticity in the same way Westerners do because Chinese construct authenticity differently. Whereas Westerners tend to focus on scientific validity and rational thinking (objective authenticity), Chinese authenticity is based more on romanticised aesthetics and emotionality (constructed authenticity) (Nguyen & Cheung, 2016; Zhu, 2018). According to several studies, Chinese travellers are more inclined to want to see touristy places, manicured and physically developed localities and touristified landscapes (front stage) (Nguyen & Cheung, 2016; Yang & Wall, 2009; Zhu, 2018), whereas many Westerners who claim to seek authenticity want the backstage experience, or the "real life" of the destination community, including its grittiness, faults and imperfections (Timothy, 2021; Yeoman et al., 2007). This frequently results in natural attractions, such as scenic areas and national parks, being embellished with plastic decorations, light shows, excess services and signage, and entertainment shows, much of which has little to do with an area's natural heritage. While the Western perspective on natural heritage focuses on an area's naturalness with as little human intervention as possible, some traditional societies see nature as boring, as something to be "developed" and physically modified for visitors' entertainment and comfort. In the words of Gao et al. (2018, p. 668), the Chinese have a "preference for cultural interpretations of natural landscapes", whereas Westerners tend to eschew the anthropocentric view of nature.

Likewise, the value of heritage is enhanced among the Chinese when a site is famous, inspires celebrated works of art, conveys China's power and is visited by someone famous (Ballantyne et al., 2014; Zhu, 2012). It is common in developing countries for heritage sites to gain importance once a famous person visits, such as a president, prime minister, member of the royal house or foreign dignitary. This adds a stamp of approval and officialdom, a value-added "brand", and cements the attraction's place in the valued heritage milieu of a region.

Branding is an extremely important draw for many nationalities. It is especially so for the Chinese and other East Asian cultures (Lojo, 2016; Moy & Phongpanichanan, 2014). Whereas Chinese tourists seek opportunities to visit UNESCO-branded heritage sites, in North America, the UNESCO brand is essentially unknown, and few Americans or Canadians are aware of its meaning or even its very existence (Boyd & Timothy, 2006). While the trend in Western countries is for heritage to become more democratised (inclusive of ordinariness), the trend in China focuses on famousness and branding. When a cultural site is inscribed on the UNESCO World Heritage List, this branding lends an additional layer of interest and increases a locality's fame (Shepherd, 2018; Silverman &

Blumenfield, 2013; Su & Li, 2012). A site's famousness and branding are major criteria not only for a decision to visit but also even to be recognised as heritage. In understanding Chinese tourists' demand for heritage, in the words of Ballantyne et al. (2014, p. 719), "It seems that the key factor is the fame and notoriety of the site, rather than the site itself". In the context of Chinese domestic tourism, Ballantyne et al. (2014) also note that besides being famous, Chinese heritage sites should express the country's power and instil patriotism and national pride. In much of the Global South, the lack of magnificence, widespread recognition, national or international branding and attention by famous people frequently relegates local and ordinary heritage into the rubbish heap of miscellany, disinheriting it in the process.

During a 2016 visit to the historic city of Luoyang, China, I was touring the old city centre with Chinese colleagues when we happened upon a traditional, old shop selling housewares. I commented to my hosts what an interesting and unique piece of Chinese heritage the shop was. Astonished, one of the hosts responded, "That's not heritage. It's too simple". I encountered a similar situation during my doctoral research in Indonesia in the 1990s. During surveys about their views of the tourism potential of their villages around Yogyakarta, several participants asked the same question: "Why would tourists want to come to our village? We're just poor farmers; our homes and villages are nothing special". As the discussions progressed, the villagers expressed their understanding of "heritage" to be something far grander than what their ordinary existence and vernacular villagescapes could offer. These two anecdotal examples illustrate cultural differences between perceptions of heritage. From my (outsider) perspective, the homes, villages and historic shops of Indonesia and China are a crucial part of those countries' heritage milieu and contribute to an overall heritage ambience. In many countries of the South or traditional societies, however, there is a pervasive sense that heritage is only extraordinary; it is not associated with daily life or peasantry. Heritage remains, in many parts of the Global South, the domain of the social elites, nobility, colonisers or chieftains.

This also raises questions about different views of authenticity and heritage between traditional societies. Although the perspectives described earlier may be pervasive in some countries, it is not necessarily so throughout the entire Global South. Each culture and society will see the world through different lenses. Many traditional societies, such as the Maasai of Tanzania and Kenya, and the Indigenous Maya of southern Mexico, have found considerable success in packaging some of their ordinary cultural heritage for tourist consumption and have begun to provide authentic experiences, even if that concept means different things to residents and tourists.

Special interest products

North–South differences may also be seen with the development of special interest heritage tourisms. For example, dark tourism is essentially a Western concoction, but other countries have started to jump on the dark tourism bandwagon to the point where several Global South locations have become important dark destinations, such as the Killing Fields in Cambodia and the Kigali Genocide Memorial in Rwanda. While the growth of dark tourism embraces death (thanatourism) and sites of human suffering more ardently today, in some native communities and less-developed regions such a notion is rather absurd. In the Global South, two extremes can be found as regards thanatopsis: utilising death and the dead in celebratory events and avoiding death and the dead entirely. The unique funerary rights of Indonesia's Toraja people involves living with the corpses of dead relatives for months and sometimes years at a time. The annual *Ma'nene* festival entails exhuming the bodies of entombed deceased ancestors for their annual cleaning, dressing, feeding and photographing to celebrate the important role they played in the family. In this process, descendants brush the corpses' hair and teeth, offer them food and drink, hug and kiss them and pose for family portraits with younger generations. This extreme contact with decedents is celebrated locally each year and has become a major spectacle for Western tourists because of their morbid fascination with death (Adams, 2018; Junaid, 2015). For the Toraja, the celebration is not considered morbid at all. Other extremes among Indigenous people eschew close proximities to cemeteries, contact with the dead or even touching artefacts associated with the deceased. Some traditional cultures teach that graveyards are scary and should not be visited and that touching or removing artefacts would incite the decedents to return and retrieve them (Klassen, 2013, p. 94). As these examples show, the ordinariness of death is negotiable and differs between cultural groups and Western tourists. How dark heritage is viewed by different societies is only one perspective on how niche tourism functions between the Global North and the Global South.

Conservation and protection

In many traditional societies of the Global South, it is difficult to incite enthusiasm for heritage protection in communities that are suffering from poverty, illiteracy and hunger. Prioritising conservation is not a luxury many can support when they cannot afford to feed and clothe their children (Timothy, 1999; Timothy & Nyaupane, 2009). Yet, do-gooders from the outside frequently come into their communities in neocolonial fashion and demand that they protect their cultural heritage for its aesthetic and scientific value, sometimes with the promise that they will reap the rewards of protecting the past, through tourism – a promise that only occasionally comes to fruition. Broken promises and unremitting poverty frequently lead to the utilisation of archaeological remains for construction material and the pillaging of ancient sites for relics that can be sold to tourists or on the illicit antiquities market (Timothy, 2020).

Although this is almost universally illegal and strikes horror in the hearts of many Westerners, some Indigenous people are unashamed or unrepentant of these types of behaviours. There exist many conflicts in values between outsiders' (neocolonialists) and Indigenous people's views of heritage assets. For some native peoples, prehistoric remains, ancient burial sites and their accompanying treasures are blessings and gifts from their ancestors who left these behind for the social and economic betterment of their descendant communities and whose spirits reveal the locations of artefacts to their progeny (Greer, 2010). This rationale is frequently used to justify harvesting ancestral riches for sale on the black market (Matsuda, 1998; Timothy, 2020, p. 138). On the other hand, some Indigenous people exercise prohibitions against touching or removing artefacts, because these items belong to the ancestors and were left on the earth for a reason; the artefacts do not belong to modern-day natives and should not be removed or disturbed (Klassen, 2013, p. 94).

As this example of looting shows, Indigenous populations and societies of the South have their own views of heritage that often clash with those of Western preservation ideals. The Western perspective on heritage protection usually considers periods of time in isolation and focuses on the "preservation of things" as static, "unchanging monuments to the past" (Smith, 2006, p. 5). In the Western conservation mindset, historical objects accrue a reverence value by virtue of their age and the stories they can tell. Western society will do almost anything to preserve what is old for its scarcity and aesthetic value, largely because of nostalgic reasons that help people in today's beleaguered world connect with a simpler past and provide a sense of rootedness in an increasingly rootless Global North (Lowenthal, 2015).

On the other hand, many Aboriginal groups see heritage as a living document, to be added to by subsequent generations. Today's Aboriginal people are a part of their past, and the past continues in them in the present (Klassen, 2013). Western heritage tenets abhor modern-day Aboriginals repainting over rock art in Australia, for example. This abhorrence is a "failure to recognize or accept that Indigenous societies may hold views of heritage which may be radically different from those which prevail in dominant institutions" (Thorley, 2002, p. 110). For many native people, historical "things" are not inherently valuable; they are simply objects that may be used today as tools or building materials, added to (e.g. rock art) or discarded (Thorley, 2002).

Despite China's Confucian ethos of respect for the past, including heritage (Huo, 2016), during China's cultural revolution, Mao Zedong taught the doctrine of the "Four Olds". According to the Four Olds dogma, pre-communist customs, cultures, habits and ideas were to be destroyed and replaced with the revolutionary concepts of new customs, new cultures, new habits and new ideas. This replacement campaign began in the 1960s, resulting in many changes throughout the country. Many old records were burned. Place names were changed to reflect revolutionary figures and events. Many historic examples of Chinese architecture were destroyed, temples were defiled and literature and works of art were annihilated (Grasso et al., 2009; Gruber, 2007; Huo, 2016). Later, under Deng Xiaoping's leadership, China began opening to the outside world and the "Four Modernizations" were established in agriculture, industry, science and technology (Grasso et al., 2009). Both of these dogmatic schemes emphasised changes in heritage appreciation and protection. During Mao's time, much of the country's built heritage was destroyed, and some of its intangible heritage was suppressed or abolished. During Deng's reign, the acceleration towards modernisation also brought about the destruction of much of the country's heritage, particularly in rural areas where many traditional agricultural settlements were replaced with concrete-block villages in the name of "progress".

Many traditional societies see the protection of historic buildings as inconsistent with progress or development. Traditional lifestyles and old buildings may be seen as unfashionable and indicators of backwardness and poverty, whereas demolishing old structures to make way for new buildings is frequently equated with modernisation, progress and development (Rezaei et al., 2018; Timothy, 2021). This has empowered governments and development agencies to raze historic areas of ancient cities and rural villages, as well as traditional cultural landscapes, to make way for modern housing units, rural infrastructure, resorts and amusement parks for tourists (Gruber, 2007; Zan, 2014).

Another example of such policies can be found in Beijing with the city's traditional urban neighbourhoods known as *hutongs*. Here, the government has been planning for several years to demolish many of these urban neighbourhoods, some of which date from the thirteenth and fourteenth centuries, to make way for the reconstruction of an idealised and more "prosperous" urban landscape. According to one report, one-third of the city's hutongs have been demolished and another third have been renovated to the point where they no longer resemble the originals (Huo, 2016; Nix, 2018). Despite these neighbourhoods being an important part of Beijing's (ordinary) heritage, where "residents gather to gossip and play cards in the alleyways and narrow lanes, vendors go door-to-door hawking their wares and people use communal toilets" (Nix, 2018, n.p.), the city plans to flatten them to make room for more prosperous-looking neighbourhoods that will no longer resemble their traditional character.

Conclusion

There is a growing movement within heritage studies, including heritage tourism, which aims to democratise heritage in more holistic ways, so that the past of ordinary people may be valued and valorised just as the past of the elites has been valued and valorised. Cultural geographers, anthropologists, archaeologists and historians have long appreciated the vernacular cultural landscapes of ordinary people living ordinary lives – places that lack enormous, extravagant and globally celebrated cultural heritage, yet these have been largely overlooked in the global tourism marketplace until recently.

The disregard for mundane elements of heritage is beginning to change in the Global North and some parts of the Global South, but many traditional, non-Western societies remain unconvinced of the value and importance of ordinary places and ordinary lives, even to the point where these might not be considered "heritage" at all and unworthy of protection. This is disconcerting, for as Podder et al. (2018) argue from a non-Western perspective, these run-of-the-mill elements of human heritage form an extremely important part of the cultural environment, which if not acknowledged and supported will continue to be endangered.

Many destinations are uninclined to salvage the ordinary past if there is no economic rationale for doing so. Many developing countries have extraordinary archaeological heritage, ethnic cultures and built environments that are deemed more urgent and profitable. These almost always take precedent over the protection of elements of the ordinary past. Only when Indigenous people or other residents begin to value their own heritage and fight for it in terms of policy and economic development (i.e. tourism), will the same evolution of heritage valuation and democratisation begin to resemble what we see emerging in the Global North. Thankfully, increasing numbers of people in the Global South are embracing and treasuring elements of their ordinary heritage and are increasingly packaging them for tourism through community-based rural tourism initiatives. If kept at an appropriate scale and locally governed, such efforts have the potential to support the principles of sustainable development. Slowly, such perspectives are beginning to supplement the grandiose views of heritage found in much of the developing world. These emerging efforts appeal to special niche markets that desire a truly rural and authentic experience in the Global South and are indicative of some communities valuing their everyday heritage, even if mass tourism does not.

There are significant differences between countries, Indigenous groups and traditional societies in what is considered heritage and perceived to be worthwhile saving or packaging for tourist consumption. Additional intercultural comparative research is sorely needed beyond Europe, North America and China. It is critical for scholars and industry representatives to understand how different nations and cultures, especially those in the Global South, might envision the meaning, brandability and salability of their cultural past. In accomplishing this, we can better ensure that heritage and its protection become more holistic, sustainable and democratic in scope, which can help empower communities that have long been relegated to the cultural and economic margins of the state (Timothy, 2015).

References

Adams, K. M. (2018). Leisure in the "Land of the Walking Dead": Western mortuary tourism, the internet, and zombie pop culture in Toraja, Indonesia. In A. Kaul & J. Skinner (Eds.), *Leisure and Death: An Anthropological Tour of Risk, Death, and Dying* (pp. 97–120). Louisville, CO: University Press of Colorado.

Agarwal, S., Busby, C., & Huang, R. (Eds.) (2018). *Special Interest Tourism: Concepts, Contexts, and Cases*. Wallingford: CABI.

Alderman, D. H., Butler, D. L., & Hanna, S. P. (2016). Memory, slavery, and plantation museums: The River Road project. *Journal of Heritage Tourism*, *11*(3), 209–218. doi: 10.1080/1743873X.2015.1100629

Amoamo, M., & Thompson, A. (2010). (re)Imaging Māori tourism: Representation and cultural hybridity in postcolonial New Zealand. *Tourist Studies*, *10*(1), 35–55. doi:10.1177%2F1468797610390989

Atkinson, F. (1987). The Beamish open air museum. *Museum International*, *39*(3), 132–138. doi:10.1111/j.1468-0033.1987.tb00683.x

Ballantyne, R., Hughes, K., Ding, P., & Liu, D. (2014). Chinese and international visitor perceptions of interpretation at Beijing built heritage sites. *Journal of Sustainable Tourism, 22*(5), 705–725. doi:10.1080/09669582.2013.861469

Bessière, J. (2013). "Heritagisation", a challenge for tourism promotion and regional development: An example of food heritage. *Journal of Heritage Tourism, 8*(4), 275–291. doi:10.1080/1743873X.2013.770861

Boyd, S. W., & Timothy, D. J. (2006). Marketing issues and World Heritage Sites. In A. Leask & A. Fyall (Eds.), *Managing World Heritage Sites* (pp. 53–66). Oxford: Butterworth Heinemann.

Caneen, J. M. (2014). Tourism and cultural identity: The case of the Polynesian Cultural Center. *Athens Journal of Tourism*, 1(2), 102–120. doi:10.30958/ajt.1-2-1

Carr, A., Ruhanen, L., & Whitford, M. (2016). Indigenous peoples and tourism: The challenges and opportunities for sustainable tourism. *Journal of Sustainable Tourism, 24*(8–9), 1067–1079. doi:10.1080/09669582.2016.1206112

Cheung, S. C. (2005). Rethinking Ainu heritage: A case study of an Ainu settlement in Hokkaido, Japan. *International Journal of Heritage Studies, 11*(3), 197–210. doi:10.1080/13527250500160500

Chhabra, D. (2019). Authenticity and the authentication of heritage: Dialogical perceptiveness. *Journal of Heritage Tourism, 14*(5–6), 389–395. doi:10.1080/1743873X.2019.1644340

Cleere, H. (2001). The uneasy bedfellows: Universality and cultural heritage. In R. Layton, P. G. Stone & J. Thomas (Eds.), *Destruction and Conservation of Cultural Property* (pp. 22–29). London: Routledge.

Cohen, E. (1988). Authenticity and commoditization in tourism. *Annals of Tourism Research, 15*(3), 371–386. doi:10.1016/0160-7383(88)90028-X

Cook, M. R. (2016). Counter-narratives of slavery in the Deep South: The politics of empathy along and beyond River Road. *Journal of Heritage Tourism, 11*(3), 290–308. doi:10.1080/1743873X.2015.1100624

Douglas, N., & Douglas, N. (1991). Where the tiki are wired for sound and poi glow in the dark: A day at the Polynesian Cultural Center. *Islands Business Pacific, 17*(12), 60–64.

Domínguez-Quintero, A. M., González-Rodríguez, M. R., & Roldán, J. L. (2019). The role of authenticity, experience quality, emotions, and satisfaction in a cultural heritage destination. *Journal of Heritage Tourism, 14*(5–6), 491–505. doi:10.1080/1743873X.2018.1554666

Gao, J., Zhang, C., & Huang, Z. (2018). Chinese tourists' views of nature and natural landscape interpretation: A generational perspective. *Journal of Sustainable Tourism, 26*(4), 668–684. doi:10.1080/09669582.2017.1377722

Gordon, A. (2016). *Time Travel: Tourism and the Rise of the Living History Museum in Mid-Twentieth-Century Canada*. Vancouver: University of British Columbia Press.

Grasso, J., Corrin, J., & Kort, M. (2009). *Modernization and Revolution in China: From the Opium Wars to the Olympics* (4th ed.). London: M.E. Sharpe.

Greer, S. (2010). Heritage and empowerment: Community-based indigenous cultural heritage in northern Australia. *International Journal of Heritage Studies, 16*(1–2), 45–58. doi:10.1080/13527250903441754

Gruber, S. (2007). Protecting China's cultural heritage sites in times of rapid change: Current developments, practice and law. *Asia Pacific Journal of Environmental Law, 10*(3/4), 253–301.

Hiwasaki, L. (2000). Ethnic tourism in Hokkaido and the shaping of Ainu identity. *Pacific Affairs*, *73*(3), 393–412. doi:10.2307/2672026

Huo, Z. (2016). Legal protection of cultural heritage in China: A challenge to keep history alive. *International Journal of Cultural Policy*, *22*(4), 497–515. doi:10.1080/10286632.2015.1015534

Junaid, I. (2015). Sustainable tourism in Toraja: Perspective of indigenous people. *ASEAN Journal on Hospitality and Tourism*, *14*(1), 45–55. doi:10.5614%2Fajht.2015.14.1.5

Klassen, M. A. (2013). *Indigenous Heritage Stewardship and the Transformation of Archaeological Practice: Two Case Studies from the Mid-Fraser Region of British Columbia*. Unpublished doctoral dissertation, Simon Frasier University, Burnaby, BC.

Littrell, M. A., Anderson, L. F., & Brown, P. J. (1993). What makes a craft souvenir authentic? *Annals of Tourism Research*, *20*(1), 197–215. doi:10.1016/0160-7383(93)90118-M

Lojo, A. (2016). Chinese tourism in Spain: An analysis of the tourism product, attractions and itineraries offered by Chinese travel agents. *Cuadernos de Turismo*, *37*, 243–268. doi:10.6018/turismo.37.256231

Lowenthal, D. (2015). *The Past is a Foreign Country – Revisited*. Cambridge: Cambridge University Press.

MacCannell, D. (1973). Staged authenticity: Arrangements of social space in tourist settings. *American Journal of Sociology*, *79*(3), 589–603. doi:10.1086/225585

Matsuda, D. (1998). The ethics of archaeology, subsistence digging, and artifact looting in Latin America: Point muted counterpoint. *International Journal of Cultural Property*, *7*, 87–97. doi:10.1017/S0940739198770080

McIntosh, A. J., Zygadlo, F. K., & Matunga, H. (2004). Rethinking Maori tourism. *Asia Pacific Journal of Tourism Research*, *9*(4), 331–352. doi:10.1080/1094166042000311237

McKercher, B. (2002). Towards a classification of cultural tourists. *International Journal of Tourism Research*, *4*(1), 29–38. doi:10.1002/jtr.346

Moy, L. Y. Y., & Phongpanichanan, C. (2014). Does the status of a UNESCO World Heritage City make a destination more attractive to Mainland Chinese tourists? A preliminary study of Melaka. *Procedia: Social and Behavioral Sciences*, *144*, 280–289. doi:10.1016/j.sbspro.2014.07.297

Nguyen, T. H. H., & Cheung, C. (2016). Chinese heritage tourists to heritage sites: What are the effects of heritage motivation and perceived authenticity on satisfaction?. *Asia Pacific Journal of Tourism Research*, *21*(11), 1155–1168. doi:10.1080/10941665.2015.1125377

Nix, E. (2018). In China, it's out with the old, in with the old. *History*. Retrieved from https://www.history.com/news/in-china-its-out-with-the-old-in-with-the-old

Novelli, M. (Ed.) (2005). *Niche Tourism: Contemporary Issues, Trends and Cases*. Oxford: Butterworth-Heinemann.

Pielesiak, I. (2015). Managing "ordinary heritage" in Poland: Łódź and its post-industrial legacy. *European Spatial Research and Policy*, *22*(2), 73–92. doi:10.1515/esrp-2015-0026

Pitchford, S. R. (2006). Identity tourism: A medium for Native American stories. *Tourism Culture & Communication*, *6*(2), 85–105. doi:10.3727/109830406777410616

Podder, A. K., Hakim, S. S., & Bosu, S. P. (2018). Ordinary heritage. *Archnet-IJAR: International Journal of Architectural Research*, *12*(2), 334–346. doi:10.1108/ARCH-01-2019-0012

Porananond, P. (2015). Khun Tok Dinner: The transformation of a Lanna eating style into a tourist attraction in Chiang Mai, Thailand. *International Journal of Asia-Pacific Studies, 11*, 129–144.

Rezaei, N., Rasouli, M., & Azhdari, B. (2018). The attitude of the local community to the impact of building reuse: Three cases in an old neighborhood of Tehran. *Heritage & Society, 11*(2), 105–125. doi:10.1080/2159032X.2019.1583805

Rittichainuwat, B. N. (2018). *Special Interest Tourism*. Newcastle upon Tyne: Cambridge Scholars Publishing.

Ruhanen, L., & Whitford, M. (2019). Cultural heritage and Indigenous tourism. *Journal of Heritage Tourism, 14*(3), 179–191. doi:10.1080/1743873X.2019.1581788

Shafernich, S. M. (1993). On-site museums, open-air museums, museum villages and living history museums: Reconstructions and period rooms in the United States and the United Kingdom. *Museum Management and Curatorship, 12*(1), 43–61. doi:10.1080/09647779309515342

Shepherd, R. (2018). When sacred space becomes a heritage place: Pilgrimage, worship, and tourism in contemporary China. *International Journal of Religious Tourism and Pilgrimage, 6*(1), 34–46. doi:10.21427/D7TM64

Silverman, H., & Blumenfield, T. (2013). Cultural heritage politics in China: An introduction. In T. Blumenfield & H. Silverman (Eds.), *Cultural Heritage Politics in China* (pp. 3–22). New York: Springer.

Small, S. (2013). Still back of the big house: Slave cabins and slavery in southern heritage tourism. *Tourism Geographies, 15*(3), 405–423. doi:10.1080/14616688.2012.723042

Smith, L. (2006). *Uses of Heritage*. London: Routledge.

Smith, S. A. (2019). Heritage tourism and New Western history: A narrative analysis of six Colorado museums. *Journal of Heritage Tourism, 14*(1), 1–18. doi:10.1080/17438 73X.2018.1452245

Soukhathammavong, B., & Park, E. (2019). The authentic souvenir: What does it mean to souvenir suppliers in the heritage destination?. *Tourism Management, 72*, 105–116. doi:10.1016/j.tourman.2018.11.015

Su, M., & Li, B. (2012). Resource management at World Heritage Sites in China. *Procedia Environmental Sciences, 12*, 293–297. doi:10.1016/j.proenv.2012.01.280

Sznajder, M., Przezbórska, L., & Scrimgeour, F. (2009). *Agritourism*. Wallingford: CABI.

Thorley, P. (2002). Current realities, idealized pasts: Archaeology, values and Indigenous heritage management in central Australia. *Oceania, 73*(2), 110–125. doi:10.1002/j.1834-4461.2002.tb01632.x

Timothy, D. J. (1999). Participatory planning: A view of tourism in Indonesia. *Annals of Tourism Research, 26*(2), 371–391. doi:10.1016/S0160-7383(98)00104-2

Timothy, D. J. (2014a). Contemporary cultural heritage and tourism: Development issues and emerging trends. *Public Archaeology, 13*(3), 30–47. doi:10.1179/14655187 14Z.00000000052

Timothy, D. J. (2014b). Views of the vernacular: Tourism and heritage of the ordinary. In J. Kaminski, A. Benson, & D. Arnold (Eds.), *Contemporary Issues in Cultural Heritage Tourism* (pp. 32–44). London: Routledge.

Timothy, D. J. (2015). Cultural heritage, tourism and socio-economic development. In R. Sharpley & D. J. Telfer (Eds.), *Tourism and Development: Concepts and Issues* (2nd ed., pp. 237–249). Bristol: Channel View Publications.

Timothy, D. J. (Ed.) (2016). *Heritage Cuisines: Traditions, Identities and Tourism*. London: Routledge.

Timothy, D. J. (2018a). Making sense of heritage tourism: Research trends in a maturing field of study. *Tourism Management Perspectives*, *25*, 177–180. doi:10.1016/j.tmp.2017.11.018

Timothy, D. J. (2018b). Producing and consuming heritage tourism: Recent trends. In S. Gmelch & A. Kaul (Eds.), *Tourists and Tourism: A Reader* (3rd ed., pp. 167–178). Long Grove, IL: Waveland Press.

Timothy, D. J. (2020). Plundering the past: Tourism and the illicit trade in archaeological remains. In D. J. Timothy & L. G. Tahan (Eds.), *Archaeology and Tourism: Touring the Past* (pp. 134–151). Bristol: Channel View Publications.

Timothy, D. J. (2021). *Cultural Heritage and Tourism: An Introduction* (2nd ed.). Bristol: Channel View Publications.

Timothy, D. J., & Nyaupane, G. (Eds.) (2009). *Cultural Heritage and Tourism in the Developing World: A Regional Perspective*. London: Routledge.

UNESCO (2020a). *Intangible Cultural Heritage*. Retrieved from https://ich.unesco.org/en/lists

UNESCO (2020b). *World Heritage List*. Retrieved from https://whc.unesco.org/en/list/

Wang, N. (1999). Rethinking authenticity in tourism experience. *Annals of Tourism Research*, *26*, 349–370. doi:10.1016/S0160-7383(98)00103-0

Weidenfeld, A. (2010). Iconicity and flagshipness of tourist attractions. *Annals of Tourism Research*, *37*(3), 851–854. doi:10.1016/j.annals.2010.02.007

Weiler, B., & Hall, C. M. (1992). *Special Interest Tourism*. London: Belhaven.

Yang, L., & Wall, G. (2009). Authenticity in ethnic tourism: Domestic tourists' perspectives. *Current Issues in Tourism*, *12*(3), 235–254. doi:10.1080/13683500802406880

Yeoman, I., Brass, D., & McMahon-Beattie, U. (2007). Current issue in tourism: The authentic tourist. *Tourism Management*, *28*(4), 1128–1138. doi:10.1016/j.tourman.2006.09.012

Zan, L. (2014). Cultural heritage in China: Between policies, development, professional discourse, and the issue of managing. *Public Archaeology*, *13*(1–3), 99–112. doi:10.1179/1465518714Z.00000000058

Zhu, G. (2012). China's architectural heritage conservation movement. *Frontiers of Architectural Research*, *1*(1), 10–22. doi:10.1016/j.foar.2012.02.009

Zhu, X. (2018). The comparative study on authentic experience of Western and Chinese backpacker travelers based on narrative analysis: A case study in Yangshuo, China. *Tourism Tribune*, *33*(3), 116–126.

4 Antecedents of tourism vicarious nostalgia

Conceptual model, systematic review and research agenda

Oscar Hengxuan Chi and Christina Geng-Qing Chi

Introduction

Nostalgia, "a wistful or excessively sentimental yearning for return to or of some past period or irrecoverable condition" (Merriam Webster Dictionary), is categorised into personal and vicarious nostalgia (Marchegiani & Phau, 2013). *Personal nostalgia* is associated with a person's autobiographical memory (Neisser, 1988) and reflects the degree to which people yearn for their past such as childhood and family (Marchegiani & Phau, 2013). In contrast, *vicarious nostalgia* (also called *historical nostalgia*) is related to people's collective memory and represents how much people miss a historic era, historic events and ancient culture (Halbwachs, 1950).

Tourism literature defines *nostalgia tourism* as travelling to a destination that allows an evocation of the past (Lee, 2014). Although visiting friends or relatives may trigger an evocation of personal nostalgia, the type of nostalgia involved in most tourism contexts, especially heritage tourism, is often more vicarious than personal (Kim, 2005). For example, travellers who visit *The Door of No Return* and *The House of Slaves*, a world heritage site that is located on Gorée Island, Dakar, Senegal, may revive the past of Africa in which, millions of Africans, both men and women, youths and elders, were chained and forced to be sold into slavery through the Island and were unable to return to their homes (Appiah & Gates, 2005). Travellers today are not able to directly experience this history, instead, through the trip, they recollect a history of the dark times and remember the human toll of African slavery (Fisher, 2013). The nostalgia evoked during the trip is likely to be a vicarious imagination experience drawn from collective memory.

Existing studies have investigated tourists' nostalgic experiences in heritage sites such as Barcelona, Spain (Oscarsson, 2013), Chiangkan, Thailand (Nilnoppakun & Ampavat, 2015), Macau, China (Leong et al., 2015) or Jidong mural alley, Korea (Park et al., 2018), suggesting that the traditional residency houses, local lifestyles, old churches or temples evoke tourists' nostalgia for the traditional culture and lifestyle. Other studies have explored nostalgia in film tourism (e.g. Bandyopadhyay, 2008), food tourism (Renko & Bucar, 2014), sport tourism (Fairley et al., 2018) or theme cruise tourism (Weaver, 2011), indicating that the sense of traditional setting activates travellers' collective memory of an ideal past.

Despite a few studies reporting the effects of nostalgia on travellers' behaviours (e.g. Cho et al., 2019b; Fairley et al., 2018; Kim et al., 2019; Mun et al., 2018; Park et al., 2018; Russell, 2008; Verma & Rajendran, 2017), there is a lack of understanding of nostalgia by the academia and the tourism industry. According to Christou et al. (2018), tourism operators are usually unaware of the potential nostalgia triggers, whereas tourism nostalgia research is dominated by narrative case studies, leaving the antecedents of tourism nostalgia underexplored. The insufficient understanding of tourism nostalgia not only undermines the authentic travelling experience (Park et al., 2018) but also buries a potential opportunity to stimulate tourists' behaviour.

This chapter aims to deepen people's understanding of the antecedents of tourism vicarious nostalgia. The chapter is organised into four sections. First, conceptualisations of nostalgia from different research domains are reviewed and discussed. Afterwards, an integrative conceptual model is proposed. Next, a systematic literature review is conducted to assess the existing works regarding tourism vicarious nostalgia. Last, a research agenda is provided to guide future vicarious nostalgia research in the tourism domain.

Conceptualisations of nostalgia

From a psychopathological disease (Beck, 2013), a type of depression (McCann, 1941) to a positive evocation (Davis, 1979), the definition of nostalgia has undergone significant changes since the term was coined by a Swiss physician in the 1600s. Today, people generally describe nostalgia as "a longing for the past" (Holbrook, 1993, p. 245) and believe that nostalgia is associated with a series of cognitive and emotional reactions (Sedikides & Wildschut, 2018; Stern, 1992). However, different research domains still conceptualise nostalgia in distinct manners.

Conceptualising nostalgia as an *emotion* is a common approach in marketing and psychology domains. Previous studies that examine nostalgia in advertising (e.g. Stern, 1992), brand image (e.g. Shields & Johnson, 2016), social media (e.g. Youn & Jin, 2017), personality (e.g. Cheung et al., 2017) or psychology (e.g. Sedikides & Wildschut, 2018) suggest that nostalgia is an emotion-laden construct. Others believe nostalgia involves both cognitive efforts and ambivalent emotions (Merchant & Rose, 2013). In a nostalgic experience, cognitive activities such as thinking, imagining (Stern, 1992) and fantasising (Merchant & Rose, 2013), are activated to simulate a sense of return to the past. Afterwards, positive emotions (e.g. sentimental) are generated via the evocation of an idealised past (Merchant & Rose, 2013). Whereas negative emotions (e.g. loss) may also appear since the past will not truly revive (Holak & Havlena, 1998).

Another nostalgia research stream conceptualises nostalgia as a preference. According to Holbrook and Schindler (1991), nostalgia is a preference for an object that commonly appeared in the past. Holbrook and Schindler believe that the phenomenon of nostalgia is like imprinting (Holbrook & Schindler, 1991). As observed from animals, a psychological bond between an animal and an object establishes if the animal was frequently exposed to the object in its early life (Bornstein, 1989), leading to a preference towards similar objects in the future (Holbrook & Schindler, 1991).

Some scholars in psychology and personality studies conceptualise nostalgia as a *coping strategy* (Garrido, 2018). When people are experiencing distress, nostalgic experiences can yield positive impacts on people's self-regard, social connectedness and existential meaning (Wildschut et al., 2006). Scholars also argue that a negative mood leads to the intention of escaping from the present and returning to the past in which people seek a sense of security (Merchant & Ford, 2008).

An integrative conceptual model

Based on different conceptualisations of nostalgia, various antecedents of nostalgia can be found to stimulate travellers' vicarious nostalgia. Nostalgia-as-emotion suggests that nostalgia tends to be driven by *travellers' characteristics*. Nostalgia-as-preference points to nostalgia as associated with tourism *motivation*. Nostalgia-as-coping-strategy implies that *negative moods* are likely to trigger nostalgia. Moreover, in the tourism context, travellers' *awareness of nostalgia triggers* is also a critical antecedent of nostalgia. Hence, this chapter proposes an integrative conceptual model of tourism vicarious nostalgia (Figure 4.1), which is discussed in the following sections.

Traveller's characteristics

In the tourism context, vicarious nostalgia is believed to provide travellers with a recurrence of a "golden age" or a historical event (Merchant & Ford, 2008). Hence, nostalgia has been conceptualised as an emotion. Nostalgic travellers are likely to exhibit both positive emotions such as appreciation and negative emotions such as missing the past, simultaneously. Studies that conceptualise nostalgia as an emotion suggest that nostalgia is driven by different personal characteristics (Cheung et al., 2017) such as nostalgia proneness and social identities.

Nostalgia proneness

Vicarious nostalgia can be fostered by nostalgia proneness (Cheung et al., 2017; Holbrook, 1993). Nostalgia proneness refers to a past-oriented attitude and is an important trait that associates with a person's psychological condition, life experiences or general characteristics (Holbrook, 1993). People have different levels of nostalgia proneness, which determines to what extent a person desires to pursue an object in the past (Holbrook, 1993). In the tourism context, travellers with higher levels of nostalgia proneness are more likely to experience tourism nostalgia.

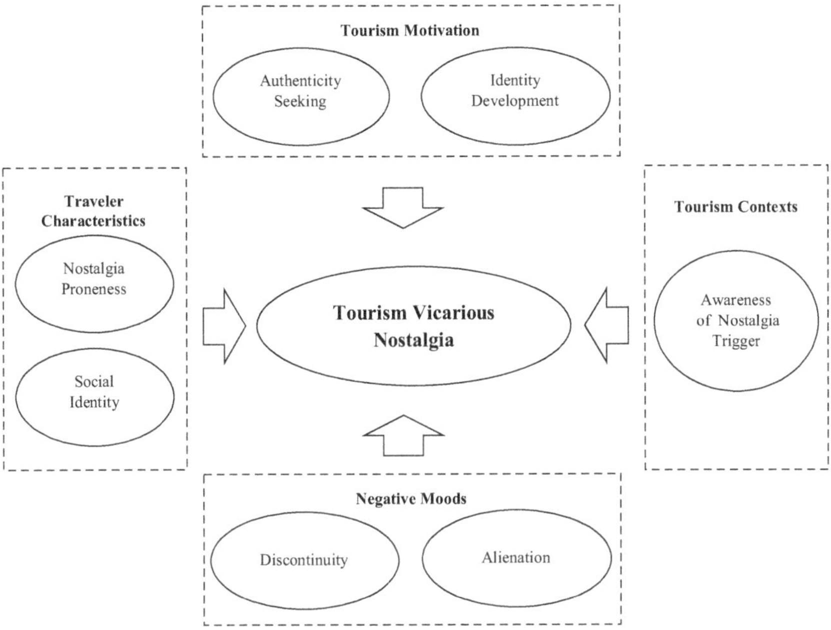

Figure 4.1 An integrative conceptual model of tourism vicarious nostalgia. Source: Authors.

Social identity

Social identity is described as a self-concept that values the belonging of a group (Tajfel, 1978) and is significantly associated with collective knowledge and memory (Roediger & Abel, 2015). Based on the theory of collective memory (Halbwachs, 1950; Marchegiani & Phau, 2013), the shared knowledge and memory within a social group is a fundamental driver of vicarious nostalgia. Scholars believe that collective memory alters and controls people's understanding of the past or even imbues a new meaning of a historical event (Larson & Lizardo, 2007). Thus, whether and how travellers place themselves in a fantasised past are influenced by the connection between their social identity and the tourism destination (Marchegiani & Phau, 2013).

Tourism motivation

In the consumer behaviour domain, studies find that nostalgia acts as a motivational force of purchasing (Leong et al., 2015; Sedikides & Wildschut, 2016). Conceptualising nostalgia as a preference, scholars suggest that consumers have certain psychological needs that lead them to seek products that can evoke nostalgia to satisfy those needs (Holbrook & Schindler, 1991). Drawing on this conceptualisation, in the tourism context, the evocation of vicarious nostalgia is likely to be driven by tourists' psychological demands such as authenticity seeking and identity development.

Authenticity seeking

Authenticity is described as perceived genuineness (Trilling, 1972) and results in a feeling of "purity and sacredness" (Merchant & Ford, 2008). The complex and stressful modern life leads some people to be dissatisfied with and distrustful of current surroundings and events (Merchant & Ford, 2008). As a result, they are looking for an authentic past that can only be revived in a nostalgic experience (Chase & Shaw, 1989). Travellers who seek authenticity are likely to be attracted by the heritage elements in a tourism destination since these elements can evoke their vicarious nostalgia and make them want to re-live the past (Wang, 1999). Based on these arguments, it is expected that a traveller who seeks authenticity is likely to experience tourism vicarious nostalgia.

Identity development

Based on social identity theory, people continuously develop and reinforce their group identifications to enhance a sense of belonging to a social group (Stets & Burke, 2000) through self-categorisation and social comparison (Hogg & Abrams, 1988). Self-categorisation refers to the reinforcement of the similarities between a person and a social group. Social comparison is the comparison between a person's social group (in-group) and other social groups (out-groups) leading to a perception that the in-group is better (Stets & Burke, 2000).

Tourism offers people opportunities to develop and recover the sense of self (Muller & O' Cass, 2001). Scholars find that travellers choose certain tourism experiences to reinforce their belonging to a place or a culture (Bond & Falk, 2013). On the other hand, travellers are also likely to utilise tourism activities to distinguish themselves from out-groups (Bond & Falk, 2013). In this situation, nostalgia promotes the connection between an individual and a social group by activating collective memories, values and beliefs (Brown & Humphreys, 2002). Therefore, nostalgia is likely to act as a means of identity development (Pourtova, 2013). In the marketing domain, Zhou et al. (2013) report that consumers prefer nostalgic products because these products reinforce their social identities. Based on the argument earlier, travellers who pursue identity development tend to experience vicarious nostalgia, which allows them to reinforce or redefine identification in a fantasised past.

Negative moods

Scholars suggest that negative feelings tend to trigger nostalgia as a result of conceptualising nostalgia as a mean of coping (e.g. Garrido, 2018). Drawing on this conceptualisation, traveller's negative moods such as discontinuity and alienation are likely to prompt them to escape from the unpleasant and stressful present through nostalgic tourism experiences.

Discontinuity

Discontinuity refers to the disconnection between an individual's past and present (Davis, 1979). When people are experiencing changes such as unemployment, the death of a family member or moving away from an old house, the feeling of discontinuity occurs (Davis, 1979; Merchant & Ford, 2008). In this case, people could mitigate the negative mood by recalling the past since nostalgic experiences provide a sense of security (Davis, 1979). Sedikides and Wildschut (2018) find that nostalgia plays a counteraction role of discontinuity and can effectively improve perceived self-continuity. Based on the arguments earlier, travellers who suffer a sense of discontinuity are more likely to experience vicarious nostalgia (Merchant & Ford, 2008).

Alienation

Alienation is another factor that impacts people's nostalgia (Merchant & Rose, 2013). Alienation is defined as the increase in the psychological and physical distance between a person and his/her fellows (Kanungo, 1979). The rapid development of society and the fast pace of life exert tremendous pressures on people and lead to a lack of interpersonal relationships (Bewes, 1997; Lears, 1998). For people who lack but desire social involvement, their feeling of alienation could be reduced by imagining the past when people paid more attention to community value, social groups and families (Merchant & Ford, 2008). For this reason, travellers who have a high level of alienation are more likely to experience vicarious nostalgia (Merchant & Rose, 2013).

Awareness of nostalgia triggers

Despite the distinctive conceptualisations of nostalgia, scholars generally believe that nostalgic experience is activated by the presence of nostalgia triggers, which can be in different formats. Existing studies find that many objects can evoke nostalgia, such as music (Barrett et al., 2010), films (Kim et al., 2019), advertising (Stern, 1992) or heritage sites (Ali, 2015). Wildschut et al. (2006) found that sensory inputs, tangible items, similar events and certain dates and time could all trigger nostalgia. Similarly, in the tourism context, historical buildings, ancient cultures or traditional customs may trigger vicarious nostalgia. However, travellers are not always aware of these nostalgia triggers (Christou et al., 2018). Therefore, the awareness of nostalgia triggers can be a factor that influences tourism vicarious nostalgia. Based on the discussions in this section, an integrative conceptual model of nostalgia is illustrated in Figure 4.1.

A systematic review of tourism vicarious nostalgia literature

To understand the current stage of tourism vicarious nostalgia research, a systematic review of the literature was performed. This review included studies that focused on vicarious nostalgia or historical nostalgia in the tourism context.

Articles that explored other types of nostalgia (e.g. personal nostalgia) were excluded. Since many studies do not distinguish vicarious nostalgia from other types of nostalgia, a broad set of keywords was used in the article searching process. Keywords, such as *nostalgia, nostalgic, tourism, travel, traveler, tourist* and *destination* were used to search research papers and book chapters that were written in English and published in the most recent two decades (2000–2020) from four databases: *Science Direct*, *Sage*, *Springer Link* and *Taylor & Francis*. This process created an initial pool of 131 articles. Afterwards, three researchers were instructed to individually refine the article pool based on the inclusion and exclusion criteria. Their results were discussed and combined, yielding a refined pool of 31 articles. Next, researchers were asked to categorise the articles based on the factors included in the proposed conceptual model (Figure 4.1). The final review results are discussed in the following sections.

Via the systematic review, tourism vicarious nostalgia was found to predict various positive outcomes, including visit intention (Cho et al., 2019b), tourism involvement (Kim et al., 2019), satisfaction (Park et al., 2018; Russell, 2008; Verma & Rajendran, 2017), attitude towards destinations (Mun et al., 2018) and memento collecting (Fairley et al., 2018). While few studies explored nostalgic proneness and negative moods (e.g. discontinuity and alienation), the results of existing tourism literature supported other proposed antecedents such as social identity, authenticity seeking, identity development and awareness of nostalgia triggers.

The systematic review reveals that a traveller's social identity and collective memory promote nostalgia. For example, in the sport tourism context, Cho et al. (2019a) found that travellers who had more experiences (versus less) with football games were more likely to experience nostalgia. The frequent exposure of football games allows travellers to build strong football-related identities, leading to a high level of nostalgia. Similarly, Fairley et al. (2018) explored travellers' nostalgia in the 2011 Tour de France and found that travellers' collective memory from watching TV or reading books resulted in vicarious nostalgia. Cho et al. (2020) further report a positive association between travellers' team identity and evoked nostalgia in professional baseball games, further suggesting that social identity may drive tourism vicarious nostalgia.

Previous tourism research has demonstrated that traveller's authenticity seeking plays a salient role in promoting nostalgia. For example, Yeh and Lin (2017) showed that the appreciation of local culture and the recurrence of history were critical drivers of travellers' nostalgia in a cultural festival in Taiwan. In a study conducted in a heritage site located in southern China, Gao et al. (2020) explored how travellers' perceived authenticity, involvement and nostalgia influence tourism satisfaction. The results revealed a positive relationship between perceived authenticity and evoked nostalgia. Leong et al. (2015) studied travellers' nostalgia in Macau and found that heritage and historical elements of the destination induced travellers' nostalgia, in which travellers established an emotional connection between history and themselves. Focusing on film tourism, Kim and King (2019) argue that the nostalgic experience during film-watching is partially formed from the audience's need for vicariously revisiting the "good old days". Moreover, by developing a scale of Vicarious Nostalgia Evoked by Heritage Tourism (VNEHT), Chi and Chi's (in press) study is the first to measure tourism vicarious nostalgia and has provided empirical evidence of the positive relationship between authenticity seeking and the level of evoked nostalgia using cross-cultural samples. All these findings suggest that travellers' vicarious nostalgia is facilitated by their motivation for seeking authenticity.

For this reason, a branch of studies focuses on the preservation and development of destination authenticity. For example, to investigate the factors that make New Orleans an attractive destination, Vesey and Dimanche (2003) argue that the conservation of authentic historic context is critical for this city to develop nostalgic images that allow travellers to fantasise about the past. Focusing on museum tourism, Arnold-de-Simine (2013) found that different museums evoked different nostalgic experiences. War museums revive the "glorious" past of a country, while folk museums reproduce the simple traditional lifestyle. Therefore, tourism destinations should carefully consider and design evoked nostalgic experiences since travellers seek various authentic pasts.

Another battery of studies highlights that people's need for identity development is the antecedent of seeking and experiencing vicarious nostalgia. For instance, Cho et al. (2019c) report that nostalgic European football club fans are motivated to visit their favourite football stadiums because they could enhance their team identification through the nostalgic experience evoked during the visit. In the film tourism context, Bandyopadhyay (2008) has investigated Indian diasporas' vicarious nostalgic experiences evoked by Bollywood films and found that Indian diasporas are motivated to watch Bollywood films out of the need for promoting and developing social identities. The history and lifestyles presented in these movies allow people to fantasise about an ideal past of India. All these findings indicate that identity development is likely to be a significant determinant of tourism vicarious nostalgia.

Previous studies have identified a variety of tourism elements that trigger travellers' nostalgia in different tourism contexts. These nostalgia triggers are often tangible objects such as Buddhist temples in Chiang Khan, Thailand (Nilnoppakun & Ampavat, 2015), emigrant-built houses in Brazil (Neto et al., 2016) or different themes in theme cruise tourism (Weaver, 2011). Among these studies, scholars mainly examine how effective these tangible objects are to trigger travellers' nostalgia.

Besides tangible objects, other studies aim at intangible tourism elements such as senses, experiences or services. For example, Fairley et al. (2018) suggest that experiences collected during sports tourism contribute to the evocation of travellers' nostalgia. Christou et al. (2018) argue that a nostalgic atmosphere (e.g. warm services, the involvement of traditional tourism activities or social interaction) may also trigger nostalgia in addition to tangible objects (e.g. traditional food or old pictures). D'Arcens (2011) examined the effectiveness of the simulated medieval odour used by medievalist tourism operators in triggering travellers' vicarious nostalgia. All studies emphasised that travellers should be aware that tourism operators need to provide effective nostalgia triggers.

Future research agenda

Based on the review of tourism nostalgia research, this chapter identified several research areas that could be further explored in order to deepen people's understanding of tourism vicarious nostalgia, especially in the area of African tourism.

Traveller characteristics

Personal characteristics are salient determinants of evoked nostalgia (Cheung et al., 2017). However, based on the results of the systematic review, traveller characteristics are generally neglected by tourism researchers. Using the proposed vicarious nostalgia framework, future research can explore how tourism nostalgia is influenced by nostalgia proneness, which is believed to have a significant impact on nostalgia in other research domains (Cheung et al., 2017; Holbrook, 1993) including hospitality (Hunt & Johns, 2013). In addition, the existing tourism studies that explore travellers' social identity are mainly through the lens of sport tourism. This calls for studies of travellers' social identity in other tourism contexts (e.g. heritage tourism).

Negative moods

Surprisingly, this chapter has found few tourism studies that examine travellers' negative moods, although negative moods (e.g. discontinuance and alienation) have been reported as important (Garrido, 2018) or even major (Wildschut et al., 2006) antecedents of nostalgia in other research filed. This might be explained by the fact that tourism being primarily hedonic is not usually associated with negative moods. However, given that the relationship between negative moods and nostalgia has been well discussed in other research domains (Merchant & Ford, 2008; Merchant & Rose, 2013), future research is encouraged to examine the relationship in different tourism contexts.

Nostalgia in different tourism stages

According to Fairley et al. (2018), nostalgia can be experienced before, during and after a trip. The associated memory regarding a destination is evoked in the travel planning stage. While at the destination site, the nostalgia is greatly enhanced by the presence of nostalgia triggers. After the trip, nostalgia becomes a mixture of personal and historical nostalgia due to the established connection between historical memory and personal travel experience. Still, as a result of a lack of understanding on this topic, there is a need for future studies to further explore nostalgia in different tourism stages.

Nostalgia type conversion

Previous studies imply a possible conversion of nostalgia types (e.g. vicarious to personal nostalgia) during tourism activities (Russell, 2008). For example, travellers who visit a heritage site may have vicarious nostalgia. After the visit, they memorise not only the heritage site, but also make personal memories, resulting in a transformation from vicarious to personal nostalgia. Thus, future studies can examine different types of nostalgia evoked during the trip and the possible conversion after the trip.

Africa nostalgia tourism

Africa is an ideal place for tourism nostalgia research. Destinations such as *the door of no return*, Senegal, carry generations of collective memories and become top choices of pilgrimages for African Americans hoping to reconnect with their heritage. In destinations such as *Fez*, Morocco, traditional culture and lifestyles are well preserved, enabling travellers to revive the value of community, social groups and families. Destinations such as the *Serengeti National Park*, Tanzania, boast spectacular natural resources that trigger travellers' nostalgia for the harmony of man and the environment. Destinations such as Robben Island bring back memories of South Africa's sad history but is also a reminder of the power of the human spirit. There is also growing village tourism where tourists visit local rural villages to understand the development of African identify. For this reason, future studies should consider exploring nostalgia in the context of African tourism.

Conclusion

Vicarious nostalgia is a critical traveller experience that exists under multiple tourism contexts. Concentrating on the antecedents of nostalgia, this chapter developed a conceptual model and assessed current tourism research regarding vicarious nostalgia. A research agenda was provided to identify research gaps and guide future research directions. This chapter pointed to the insufficiency of current tourism literature and advocated for more generalisable and heterogeneous tourism nostalgia research.

References

Ali, F. (2015). Heritage tourist experience, nostalgia, and behavioural intentions. *Anatolia: An International Journal of Tourism and Hospitality Research, 26*(3), 472–475. doi:10.1080/13032917.2015.1013477

Appiah, A., & Gates, H. L. (2005). *Africana: The Encyclopedia of the African and African American Experience* (2nd ed.). Oxford: Oxford University Press.

Arnold-de-Simine, S. (2013). Nostalgia and post-nostalgia in heritage sites. In S. Arnold-de-Simine (Ed.), *Mediating Memory in the Museum* (pp. 54–67). London: Palgrave Macmillan.

Bandyopadhyay, R. (2008). Nostalgia, identity and tourism: Bollywood in the Indian diaspora. *Journal of Tourism and Cultural Change, 6*(2), 79–100. doi:10.1080/14766820802140463

Barrett, F., Grimm, K., Robins, R., Wildschut, T., Sedikides, C., Janata, P., & Phelps, E. A. (2010). Music-evoked nostalgia: Affect, memory and personality. *Emotion, 10*(3), 390–403. doi:10.1037/a0019006

Beck, J. (2013, August 14). When nostalgia was a disease. *Atlantic*. Retrieved from https://www.theatlantic.com/health/archive/2013/08/when-nostalgia-was-a-disease/278648/

Bewes, T. (1997). *Cynicism and Postmodernity*. London: Verso.

Bond, N., & Falk, J. (2013). Tourism and identity-related motivations: Why am I here (and not there)? *International Journal of Tourism Research, 15*(5), 430–442. doi:10.1002/jtr.1886

Bornstein, M. H. (1989). Sensitive periods in development: Structural characteristics and causal interpretations. *Psychological Bulletin, 105*, 179–197. doi:10.1037/0033-2909.105.2.179

Brown, A. D., & Humphreys, M. (2002). Nostalgia and the narrativization of identity: A Turkish case study. *British Journal of Management, 13*(2), 141–159. doi:10.1111/1467-8551.00228

Chase, M., & Shaw, C. (1989). *The Imagined Past: History and Nostalgia*. Manchester: Manchester University Press.

Cheung, W., Sedikides, C., & Wildschut, T. (2017). Nostalgia proneness and reduced prejudice. *Personality and Individual Differences, 109*(C), 89–97. doi:10.1016/j.paid.2016.12.045

Chi, O. H., & Chi, C. G. (2020). Reminiscing other people's memories: Conceptualizing and measuring vicarious Nostalgia Evoked by Heritage tourism. *Journal of Travel Research*. doi:10.1177/0047287520969904

Cho, H., Joo, D., & Chi, C. G. (2019a). Examining nostalgia in sport tourism: The case of US college football fans. *Tourism Management Perspectives, 29*, 97–104. doi:10.1016/j.tmp.2018.11.002

Cho, H., Joo, D., Moore, D., & Norman, W. C. (2019b). Sport tourists' nostalgia and its effect on attitude and intentions: A multilevel approach. *Tourism Management Perspectives, 32*, 100563. doi:10.1016/j.tmp.2019.100563

Cho, H., Khoo, E. C., & Lee, H.-W. (2019c). Nostalgia, motivation, and intention for international football stadium tourism. *Asia Pacific Journal of Tourism Research, 24*(9), 912–923. doi:10.1080/10941665.2019.1653339

Cho, H., Joo, D., & Woosnam, K. M. (2020). Sport tourists' team identification and revisit intention: Looking at the relationship through a nostalgic lens. *Journal of Hospitality & Tourism Research*. doi:10.1177/1096348020926550

Christou, P., Farmaki, A., & Evangelou, G. (2018). Nurturing nostalgia?: A response from rural tourism stakeholders. *Tourism Management, 69*, 42–51. doi:10.1016/j.tourman.2018.05.010

D'Arcens, L. (2011). Laughing in the face of the past: Satire and nostalgia in medieval heritage tourism. *Postmedieval: A Journal of Medieval Cultural Studies, 2*(2), 155–170. doi:10.1057/pmed.2011.9

Davis, F. (1979). *Yearning for Yesterday: A Sociology of Nostalgia*. New York: Free Press.

Fairley, S., Gibson, H., & Lamont, M. (2018). Temporal manifestations of nostalgia: Le Tour de France. *Annals of Tourism Research, 70*, 120–130. doi:10.1016/j.annals.2017.09.004

Fisher, M. (2013, June 28). What Obama really saw at the "Door of No Return", a disputed memorial to the slave trade. *The Washington Post*. Retrieved from https://www.washingtonpost.com/news/worldviews/wp/2013/06/28/what-obama-really-saw-at-the-door-of-no-return-a-debunked-memorial-to-the-slave-trade/

Gao, J., Lin, S., & Zhang, C. (2020). Authenticity, involvement, and nostalgia: Understanding visitor satisfaction with an adaptive reuse heritage site in urban

China. *Journal of Destination Marketing & Management, 15*, 100404. doi:10.1016/j.jdmm.2019.100404

Garrido, S. (2018). The influence of personality and coping style on the affective outcomes of nostalgia: Is nostalgia a healthy coping mechanism or rumination?. *Personality and Individual Differences, 120*, 259–264. doi:10.1016/j.paid.2016.07.021

Halbwachs, M. (1950). *The Collective Memory*. New York: Harper Colophon Books.

Hogg, M., & Abrams, D. (1988). *Social Identifications - A Social Psychology of Intergroup Relations and Group Processes*. New York: Routledge.

Holak, S. L., & Havlena, W. J. (1998). Feelings, fantasies, and memories: An examination of the emotional components of nostalgia. *Journal of Business Research, 42*(3), 217–226. doi:10.1016/S0148-2963(97)00119-7

Holbrook, M. B. (1993). Nostalgia and consumption preferences: Some emerging patterns of consumer tastes. *Journal of Consumer Research, 20*(2), 245–256. doi:10.1086/209346

Holbrook, M. B., & Schindler, R. (1991). Echoes of the dear departed past: Some work in progress on nostalgia. *Advances in Consumer Research, 18*, 330.

Hunt, L., & Johns, N. (2013). Image, place and nostalgia in hospitality branding and marketing. *Worldwide Hospitality and Tourism Themes, 5*(1), 14–26. doi:10.1108/17554211311292411

Kanungo, R. N. (1979). The concepts of alienation and involvement revisited. *Psychological Bulletin, 86*(1), 119–138. doi:10.1037/0033-2909.86.1.119

Kim, H. (2005). Research note: Nostalgia and tourism. *Tourism Analysis, 10* (1), 85–88.

Kim, S., & King, S. (2019). Nostalgia film tourism and its potential for destination development. *Journal of Travel & Tourism Marketing, 36*(2), 236–252. doi:10.1080/10548408.2018.1527272

Kim, S., Kim, S., & Petrick, S. (2019). The effect of film nostalgia on involvement, familiarity, and behavioral intentions. *Journal of Travel Research, 58*(2), 283–297. doi:10.1177/0047287517746015

Larson, J. A., & Lizardo, O. (2007). Generations, identities, and the collective memory of Che Guevara 1. *Sociological Forum, 22*(4), 425–451. doi:10.1111/j.1573-7861.2007.00045.x

Lears, J. (1998). Looking backward: In defense of nostalgia. *Lingua Franca: The Review of Academic Life, 7*(10), 59–66.

Lee, B. (2014). Nostalgia tourism. In J. Jafari & H. Xiao (Eds.), *Encyclopedia of Tourism*. Cham: Springer.

Leong, A. M. W., Yeh, S., Hsiao, Y., & Huan, T. C. (2015). Nostalgia as travel motivation and its impact on tourists' loyalty. *Journal of Business Research, 68*(1), 81–86. doi:10.1016/j.jbusres.2014.05.003

Marchegiani, C., & Phau, I. (2013). Personal and historical nostalgia—A comparison of common emotions. *Journal of Global Marketing, 26*(3), 137–146. doi:10.1080/08911762.2013.804617

McCann, W. H. (1941). Nostalgia: A review of the literature. *Psychological Bulletin, 38*, 165–182. doi:10.1037/h0057354

Merchant, A., & Ford, J. (2008). Nostalgia and giving to charity: A conceptual framework for discussion and research. *International Journal of Nonprofit and Voluntary Sector Marketing, 13*(1), 13–30. doi:10.1002/nvsm.300

Merchant, A., & Rose, G. M. (2013). Effects of advertising-evoked vicarious nostalgia on brand heritage. *Journal of Business Research, 66*(12), 2619–2625. doi:10.1016/j.jbusres.2012.05.021

Muller, T., & O' Cass, A. (2001). Targeting the young at heart: Seeing senior vacationers the way they see themselves. *Journal of Vacation Marketing, 7*, 285–301. doi:10.1177/135676670100700401

Mun, N. Y., Lee, W., & Jeong, C. (2018). Traveling from South to North: The relationships between historical nostalgia, novelty seeking, and attitudes to visit North Korea. *International Journal of Tourism Sciences, 18*(3), 170–191. doi:10.1080/15980634.2018.1517934

Neisser, U. (1988). What is ordinary memory the memory of? In U. Neisser & E. Winograd (Eds.), *Remembering Reconsidered: Ecological and Traditional Approaches to the Study of Memory* (pp. 356–373). Cambridge: Cambridge University Press.

Neto, A. H., Ahmad, R., & Hertzog, A. (2016). "Brazilians'1 houses": An example of nostalgia and a proposal of touristic empowerment. *Tourism and Hospitality Research, 16*(3), 294–307. doi:10.1177/1467358415602956

Nilnoppakun, A., & Ampavat, K. (2015). Integrating cultural and nostalgia tourism to initiate a quality tourism experiences at Chiangkan, Leuy Province, Thailand. *Procedia Economics and Finance, 23*, 763–771. doi:10.1016/S2212-5671(15)00545-6

Oscarsson, O. M.-R. (2013). Archaeology, nostalgia, and tourism in post-Civil War Barcelona (1939–1959). *Journal of Urban History, 39*(3), 478–494. doi:10.1177/0096144212443709

Park, S., Hwang, D., Lee, W., & Heo, J. (2018). Influence of nostalgia on authenticity, satisfaction, and revisit intention: The case of Jidong mural alley in Korea. *International Journal of Hospitality & Tourism Administration*, 1–17. doi:10.1080/15256480.2018.1511497

Pourtova, E. (2013). Nostalgia and lost identity. *Journal of Analytical Psychology, 58*(1), 34–51. doi:10.1111/j.1468-5922.2013.02016.x

Renko, S., & Bucar, K. (2014). Sensing nostalgia through traditional food: An insight from Croatia. *British Food Journal, 116*(11), 1672–1691. doi:10.1108/BFJ-02-2014-0089

Roediger, H. L., & Abel, M. (2015). Collective memory: A new arena of cognitive study. *Trends in Cognitive Sciences, 19*(7), 359–361. doi:10.1016/j.tics.2015.04.003

Russell, D. W. (2008). Nostalgic tourism. *Journal of Travel & Tourism Marketing, 25*(2), 103–116. doi:10.1080/10548400802402271

Sedikides, C., & Wildschut, T. (2016). Past forward: Nostalgia as a motivational force. *Trends in Cognitive Sciences, 20*(5), 319–321. doi:10.1016/j.tics.2016.01.008

Sedikides, C., & Wildschut, T. (2018). Finding meaning in nostalgia. *Review of General Psychology, 22*(1), 48–61. doi:10.1037/gpr0000109

Shields, A., & Johnson, J. (2016). Childhood brand nostalgia: A new conceptualization and scale development. *Journal of Consumer Behaviour, 15*(4), 359–369. doi:10.1002/cb.1578

Stern, B. (1992). Historical and personal nostalgia in advertising text: The fin de siècle effect. *Journal of Advertising, 21*(4), 11–22. doi:10.1080/00913367.1992.10673382

Stets, J., & Burke, P. (2000). Identity theory and social identity theory. *Social Psychology Quarterly, 63*(3), 224–237. doi:10.2307/2695870

Tajfel, H. (Ed.) (1978). *Differentiation Between Social Groups: Studies in the Social Psychology of Intergroup Relations*. London: Academic Press.

Trilling, L. (1972). *Sincerity and Authenticity*. Cambridge, MA: Harvard University Press.

Verma, A., & Rajendran, G. (2017). The effect of historical nostalgia on tourists' destination loyalty intention: An empirical study of the world cultural heritage site – Mahabalipuram, India. *Asia Pacific Journal of Tourism Research, 22*(9), 977–990. doi:10.1080/10941665.2017.1357639

Vesey, C., & Dimanche, F. (2003). From Storyville to Bourbon Street: Vice, nostalgia and tourism. *Journal of Tourism and Cultural Change*, *1*(1), 54–70. doi:10.1080/14766820308668159

Wang, N. (1999). Rethinking authenticity in tourism experience. *Annals of Tourism Research*, *26*(2), 349–370. doi:10.1016/s0160-7383(98)00103-0

Weaver, A. (2011). The fragmentation of markets, neo-tribes, nostalgia, and the culture of celebrity: The rise of themed cruises. *Journal of Hospitality and Tourism Management*, *18*, 54. doi:10.1375/jhtm.18.1.54

Wildschut, T., Sedikides, C., Arndt, J., & Routledge, C. (2006). Nostalgia: Content, triggers, functions. *Journal of Personality and Social Psychology*, *91*(5), 975–993. doi:10.1037/0022-3514.91.5.975

Yeh, H. R., & Lin, L. Z. (2017). Exploring tourists' nostalgic experiences during culture festivals: The case of the Sung Chiang Battle Array. *Current Issues in Tourism*, *20*(4), 391–424. doi:10.1080/13683500.2014.983055

Youn, S., & Jin, S. (2017). Reconnecting with the past in social media: The moderating role of social influence in nostalgia marketing on Pinterest. *Journal of Consumer Behaviour*, *16*(6), 565–576. doi:10.1002/cb.1655

Zhou, L., Wang, T., Zhang, Q., & Mou, Y. (2013). Consumer insecurity and preference for nostalgic products: Evidence from China. *Journal of Business Research*, *66*(12), 2406–2411. doi:10.1016/j.jbusres.2013.05.027

Part II

Change in tourism

5 Images, instruments and the governance capacity of local governments in tourism development planning

Evidence from Ghana[1]

Emmanuel Akwasi Adu-Ampong

Introduction

An outcome of past and ongoing decentralisation and other public sector reforms in many countries has been a downward shift in a number of key socio-economic development responsibilities from national governments to local governments. It was anticipated that such reforms would strengthen the capacity of local governments, reduce central government control and increase the managerial autonomy of local government authorities and personnel in the development process (Awortwi, 2010; Crook, 2003; Manning, 2001; Smoke, 2003). Governance activities such as local revenue generation and mobilisation, public service provision and participatory planning processes are some of the elements that have been shifted to local governments. In addition, local governments are responsible for maintaining local infrastructure such as markets, public toilets and sewage and environmental waste management services. Specifically, decentralisation reforms have sought to induce effective local economic development strategies devised and implemented by local governments (Helmsing, 2003; Nel, 2001; Rogerson & Rogerson, 2010). While responsibilities have definitely increased for local governments, the required inputs for action have not always been decentralised and/or easily available. These issues militate against the work of local governments in driving development agendas, including in utilising tourism as a tool for local economic development and poverty reduction (Rogerson, 2020). In light of these issues, it is important to understand the governance capacity of local governments in planning for and utilising tourism for local economic development.

The tourism sector is highly fragmented and therefore requires the collaboration of a range of stakeholders from the public, private and civil society sectors for effective management (Adu-Ampong, 2017a). When it comes to the governance and management of the tourism sector, the state through its public sector agencies has been rightly identified as a key actor with enduring power (Adu-Ampong, 2017b; Jeffries, 2001). The state public sector is normally constituted by the three tiers of national, regional and local levels. Much of the existing focus has been on national and regional level government involvement in destination development.

The essential role played by local governments in this process often tends to go unnoticed even though local governments have the biggest responsibility of maintaining destination competitiveness and diversification of the tourism product (McKercher & Ritchie, 1997; Ruhanen, 2013). Indeed, research has shown that achieving sustainable tourism development is a key role of local governments (Dredge, 2006; Ruhanen, 2013). This notwithstanding, studies on assessing the governance capacity of local government in tourism is relatively rare (Churugsa et al., 2007; Rogerson, 2020; Slocum & Backman, 2011).

This chapter presents the case of the Komenda-Edina-Eguafo-Abrem (KEAA) local government vis-à-vis the Elmina 2015 Strategy in Ghana. This chapter makes a unique contribution by developing the concept of stakeholder governance capacity and using it to analyse the ways in which the extent of stakeholder governance capacity shapes the development, implementation and ultimate outcomes of tourism-led local economic development initiatives. After this short introduction, I explore the evolving understandings around government, governing and interactive governance before outlining the concept of stakeholder governance capacity. This is followed by an overview of the case study and research methods. The findings and discussion are then presented before conclusions drawn in the final section.

Government, governing and the interactive governance of societal opportunities

The concept of governance is used in describing public administration and management, public–private partnerships and the coordination of mechanisms aimed at realising collective interests. In this sense, governance is both a theory and a science, a structure and a process (Pierre & Peters, 2000). Governance is commonly used in terms of: (i) the "empirical manifestations of state adaptation to its external environment as it emerges in the late twentieth century" and (ii) a "conceptual or theoretical representation of co-ordination of social systems and … the role of the state in that process" (Pierre, 2000, p. 3). This role of the state and the way of governing society has changed over the years and hence governance is no longer used as a synonym for government but rather to signify this change (Rhodes, 1996). The state is increasingly seen as adapting to contemporary governance structures of multiple actors with a shifting of roles (Kooiman, 2000) rather than a shrinking "hallowing out of the state" (Rhodes, 1994). A nuanced understanding of contemporary governance structures posits a transformation of the state in the interactive processes through which policy decisions are made and implemented for governing different sectors of society (Bell & Hindmoor, 2009; Kjær, 2004; Peters & Pierre, 2016; Torfing et al., 2012).

Increasingly, the mutual interdependence of a variety of stakeholders in the governance process is being appreciated in terms of interactive governance. A stakeholder, following Gray (1985), can be understood as one who is affected by the actions of others and has the right capacity and skillset to participate in the decision-making process. Thus, interactive governance is achieved through the interaction of a variety of stakeholders in the public, private and civil society sectors. Definitionally, interactive governance is seen as "the whole of public as well as private interactions taken to solve societal problems and create societal opportunities; including the formulation and application of principles guiding those interactions and care for institutions that enable them" (Kooiman et al., 2005, p. 17). However, since what constitutes a societal problem or societal opportunity is context specific, it is only through public and private sector interactions that the potential use of tourism as a tool for poverty reduction through tourism-led local economic development planning can be considered as a societal opportunity that can be created, enhanced and realised.

It has been acknowledged that tourism is a socio-economic and political activity within a highly complex system made up of a variety of actors and stakeholders (Adu-Ampong, 2017a; Britton, 1982; Ioannides & Debbage, 1998; Tosun, 2000). Consequently, the tourism policy and planning process is not simply a technocratic process but rather a social and political process that is open to interactive contestations of actors with the requisite governing capacities (Dredge & Jenkins, 2007, 2011). Given that tourism is place based, Rogerson (2014) has argued that it is only when tourism policy and planning is embedded in local and regional economic development planning that it can have a greater impact on poverty reduction. This is where the governing issues of coordination, steering and goal-oriented interventions (Kooiman, 1993) become important for local governments. It is within this context that local governments are seen as key actors in leading this coordination and collaborative process in relation to embedding tourism within local economic development planning. Yet, there is very little understanding of the capacity of local governments to effectively lead such interactive processes in utilising resources and assets (Rogerson, 2020).

Stakeholder governance capacity for tourism-led local economic development planning

The concept of stakeholder governance capacity was developed from the interactive governance framework drawing on the work of Kooiman (2000, 2003) and Kooiman et al. (2008). Stakeholder governance capacity comprises of the notion of governance capacities and stakeholder interactions that operate at both an intentional/action level and a structural/conditional level (see Table 5.1). On the one hand, stakeholder interactions are seen as the actions emanating from the values, goals, visions and purposes of stakeholders at the intentional level. This

Table 5.1 Stakeholder governance capacity

Governance capacity	*Intentional/action level*	*Structural/conditional level*
	Images	Culture, data collection
	Instruments	Resources
	Action potential	Institutions, power relations

Source: Author's conceptualisation based on the literature.

intentional/action level simply refers to the choice or agency available to stakeholders to choose how they act or interact with others, while at the structural level, stakeholder interactions refer to the institutions, rules and patterns of communication that enable and/or constrain actions at the intentional level. Stakeholder interaction essentially states that the less constraining the structures, the more flexibility and freedom available to stakeholders to choose their goals, interests and how to act (Caffyn & Jobbins, 2003; Kooiman, 2000). On the other hand, governance capacities are the capabilities stakeholders are able to derive from within the governance system and governing arrangements. Governance capacities are constituted by three components operating at the intentional level but embedded within structures: images (as part of culture), instruments (as part of resources) and action potential (as part of institutional power relations) (Caffyn & Jobbins, 2003; Kooiman, 1999, 2000; Slocum & Backman, 2011).

Stakeholder governance capacity therefore concerns the extent to which a given stakeholder or set of stakeholders have at their disposal the vision, tools and ability to effect desired change in a given sector. This concept is operationalised through the subconcepts of images, instruments and action potential as described in the following sections.

Images at the intentional level represent the mental constructs held by stakeholders regarding contemporary and future governance situations. Images are the guiding lights for governing activities. At the structural level, images typically consist of visions, judgements, ways of data collection and analysis, presuppositions and convictions through which images held by stakeholders get filtered. Images speak to specific issues at hand and may involve fundamental assumptions about the role of the state vis-à-vis other actors and of the relationship between tourism and poverty reduction, for instance (Kooiman, 2008; Kooiman & Bavinck, 2005). Given that each stakeholder possesses an image of some sort, the key question is whether stakeholders make them explicit in their governing interactions at both intentional and structural levels.

Instruments are the tools available to enable stakeholders to link images to tangible action. These range from "soft" to "hard" instruments such as licences, standards, taxes, educational forums, peer pressure and physical force. The choice of instruments available to and deployed by stakeholders is not neutral but dependent on power positions within the governance system. Consequently, the choice, design and use of instruments frequently involves competing interests and tensions (Kooiman, 2008). At the structural level, instruments are seen as resources – material, human, technical, social capital – that need to be developed and implemented. Given the range of possible resources and instruments, the critical questions to ask are, which instruments are chosen, how are they developed and do the deployed instruments achieve their stated goals?

Action potential refers to the socio-political scope for action within which stakeholders operate and the extent to which they are able to put an instrument into effect in order to translate images into societal outcomes. At the structural level, action potential is embedded in power relations. These can be political, as in the willingness of a government to encourage or constrain tourism revenue sharing between stakeholders. Thus, an action potential for a simple task of implementing a set of policy guidelines often requires the mobilisation of support and adequate socio-political will among other stakeholders. Central to action potential are questions and issues of legitimacy and authority and the ability of stakeholders to achieve their goals through the enabling and constraining structural contexts of power relations with others (Kooiman, 2000). The action potential available to stakeholders is often a matter of power between different tiers of government.

Images, instruments and action potential are not always distinguishable, nor are they always linear in conceptualisation. These elements are highly interconnected within an inherently interactive process of governance (Chuenpagdee et al., 2008; Kooiman & Bavinck, 2005). The recognition and understanding of these interconnections are important in assessing the extent to which a given sector such as tourism can be effectively governed towards desired goals. This is because the structural aspects of these elements and processes may have important implications on outcomes (Kooiman, 2008). Consequently, in the context of this chapter, I deploy the concept of stakeholder governance capacity as *the capabilities available to stakeholders to interact with others, form images and select appropriate tools for expressing their values and goals while being embedded in the institutional structures that shape their socio-political room for action.* In this conceptualisation of stakeholder governance capacity, the issue of power (Bachrach & Baratz, 1962) is acknowledged as being an important element.

The issue of power within interactive governance and the shaping of stakeholder governance capacity are important to bear in mind. In particular, an imbalance of power among stakeholders has the potential for driving tourism development towards the particular interests of certain stakeholders at the expense of others and communities at large. In light of the foregoing, some of the insights on power are embedded within the conceptualisation of stakeholder governance capacity used here. This is particularly in terms of using the restrictive face of power (Bachrach & Baratz, 1962) to understand the dynamics of non-decision-making in this tourism-led local economic development planning. Besides, where appropriate, the analysis will also highlight the relation between power and capacity. However, the focus of this chapter is not to analyse power structures per se but to primarily outline the stakeholder governance capacity of the Komenda-Edina-Eguafo-Abrem (KEEA) Municipal Assembly as the local government actor involved in the Elmina 2015 Strategy.

In the context of this chapter and case study on the Elmina 2015 Strategy, as a tourism-led local economic development planning, the elements and issues to consider include (a) the images held by stakeholders with regard to the role of tourism in local economic development and poverty reduction; (b) the instruments available to translate such visions into achievable objectives and the actions taken to meet these objectives; (c) the sources of the images held by tourism stakeholders; (d) how the images travel across governance scales; (e) the mechanisms through which images get filtered; (f) the resources out of which instruments are shaped; (g) the politico-cultural context of action; and (h) the power relations that constrain and/or enable actions.

Case study, methodology and methods

The tourism sector in Ghana continues to receive significant state priority as a key driver for job creation, economic growth and sustainable development for the country. The sector is seen as one of seven pillars for transforming the national economy and development trajectory (Adu-Ampong, 2019a). This state backing and prioritising has been demonstrated through public sector infrastructure development and investment with the 2019 budget allocating GHS75 m ($16.2 m) to the tourism ministry. Tourism is currently the fourth-highest foreign exchange earner for the country (Adu-Ampong, 2018). It is estimated that the sector contributed to some 5% of national GDP with about 623,000 jobs (5.2% of total national employment) in the year ending 2019 (WTTC, 2020). Tourism governance in Ghana involves a myriad set of public, private, third-sector and traditional authority institutions and steering mechanisms through which development, planning, marketing and management are accomplished (Yankholmes, 2018). Currently, the Ministry of Tourism, Arts and Culture (MoTAC) is the highest public sector institution for tourism planning, development and management and primarily operates through its main implementation agencies of the Ghana Tourism Authority (GTA), the Ghana Tourism Development Company (GTDC) and the Ghana Museums and Monuments Board (GMMB). The GTA and GMMB have regional offices across the country as part of decentralisation efforts of local level management for tourism resources and activities. In addition to these major public sector agencies, the Ghana Tourism Federation (GHATOF) as the apex body of the private sector represents the interests of all private sector associations involved in the tourism sector of Ghana in governing and managing the sector.

The multiplicity of additional state institutions in allied sectors at the regional and local levels means that tourism governance at these levels is often fraught with coordination challenges. This is compounded by the dispersal of both governance capacity and resources across these multiple institutions. Adu-Ampong (2017b) in his study of institutional collaboration in tourism planning and development in Ghana's Central Region argued that one of the most pertinent issues impeding the development of tourism is the high level of fragmentation in tourism governance, administration and management from the national to the local level. Notwithstanding these, the past five years have seen efforts by the state to streamline tourism governance institutions, structures and processes.

The research reported in this chapter derives from the PhD thesis project of the author carried out between 2013 and 2016. This PhD project sought to understand, explain and analyse how governance interactions shape the planning and use of tourism for local economic development in Ghana. This meant a careful study of how the three tiers of government related to each other within the tourism sector of Ghana and especially in the context of the Elmina 2015 Strategy, which was carried out in the KEEA Municipal Assembly in the Central Region of Ghana (Adu-Ampong, 2016).

Elmina is the capital settlement of the KEEA Municipal Assembly in the Central Region of Ghana, a region considered as the tourism hub of the country

(Adu-Ampong, 2017a). Imbued with over 700 years of well-documented extraordinary history, there is no space here to outline this complex history (for details, see Adu-Ampong, 2021; Arthur, 2012; Essah, 2001; Holden et al., 2011; Sonne, 2010). Suffice to say that Elmina is endowed with a number of cultural heritage tourism resources and assets including the UNESCO World Heritage Site of Elmina Castle and Fort St. Jago, 18th and 19th century traditionally built Asafo Posts, Old Dutch Merchant houses, the Old Dutch Cemetery, the Bakatue and Edina Bronya traditional festivals and a number of historical landmarks in addition to Benya Lagoon and a long stretch of coastal beach land. It is in this context that the Elmina 2015 Strategy emerged as an outcome of the Elmina Cultural Heritage Management Programme (ECHMP) that began in 2000/2001.

The ECHMP was largely funded by the Dutch government to focus on urban heritage conservation and the revitalisation of identified mutual heritage between Ghana and the Netherlands. This was led by the KEEA Municipal Assembly in collaboration with the Institute of Housing and Urban Studies of the Erasmus University of Rotterdam in the Netherlands, the GMMB, Ghana Institute of Local Government Studies (ILGS) and the Department of International Relations and International Organisations at the University of Groningen in the Netherlands. A series of public and stakeholder consultations lead to the production of the Elmina 2015 Strategy ('Building on the Past to Create a Better Future') developed around five priority areas: (1) drainage and waste management; (2) fishing and the fishing harbour; (3) tourism and local economic development; (4) health; and (5) education. The overall success of the Elmina 2015 Strategy was predicated on the effective implementation of priority three – tourism and local economic development. A range of projects including the restoration of Elmina Castle, Fort St. Jago, the Dutch Cemetery, old merchant houses and the dredging of the fishing harbour, as well as training of artisans and the construction of the Elmina craft market were to be implemented. The ultimate goal being that by the year 2015, Elmina's socio-economic development fortunes would be flourishing through increased tourism and the city would become "the engine of equitable socio-economic development that impacts on the wealth creation, poverty reduction and improved local governance on a sustainable basis" (Arthur & Mensah, 2006 p. 310). However, at the end of 2015, this vision has not fully materialised as originally envisaged. The much-anticipated hope of tourism catalysing local economic development and poverty reduction came undone (see Adu-Ampong, 2016, 2021). For this chapter, I focus on explaining this outcome by examining the inadequacy of stakeholder governance capacity held by the KEEA Municipal Assembly vis-à-vis other stakeholders.

The full details of the methods of data collection and analysis have been reported in Adu-Ampong (2016). While the data collection period appears dated, the findings and implications have contemporary relevance in Ghana as well as in other countries in the Global South seeking to utilise tourism for development. In Ghana, for instance, there are new attempts in the context of the "Year of Return 2019" and "Beyond the Return" tourism marketing campaigns in utilising slavery heritage tourism as an avenue for local and national economic development.

In summary, semi-structured in-depth interviews, policy document analysis and observations were used as the basis of data collection. Fieldwork was carried out on two separate occasions (August–November 2014 and July–September 2015). A total of 59 key informant interviews were conducted in addition to 9 community member interviews, 3 of which ended up becoming a form of focus group with between 2 and 5 people. The interviewees included past and present KEEA officers, Directors at the Ministry of Tourism, Culture and Creative Arts (currently Ministry of Tourism, Arts and Culture), Ghana Museums and Monuments Board, Ghana Heritage Conservation Trust, community members and homeowners who benefitted from the building projects under the Elmina 2015 Strategy, as well as members of the private sector umbrella organisation of the Ghana Tourism Federation. In addition, there were a number of informal conversations with community members and extensive field notes taken that provided background information for this research. These were triangulated with the analysis of national, regional and local level tourism-related development plans, especially the Elmina 2015 Strategy documents including progress reports on project implementation. The interviews were used to confirm evidence from other sources while also opening up new lines of inquiry to the (re)analysis of documents and other pieces of evidence. The iterative process of using document analysis and interviews to confirm, corroborate and highlight differences ensured a robust data collection and analysis process.

Data analysis followed the framework approach for qualitative analysis as developed by Ritchie and Spencer (1994). The five-step process of the approach consisted of familiarisation, identifying a thematic framework, indexing, charting, and mapping and interpretation. An iterative process through these steps enabled the data to be scrutinised for meaning, salience and connections between what interviewees perceived in their governing interactions and what actually pertained in their day-to-day activities. Significantly, this approach to analysis made it easier to identify salience and connections between the governing capacities, governing (in)activities and the challenges of implementing the Elmina 2015 Strategy process. Data analysis focused on building up broad subthemes of commonalities and differences between actors in terms of the images held about tourism development, the instruments available to implement these visions and the available governance capacity (see Adu-Ampong, 2016, for a detailed explanation of data analysis).

Findings

The development of the Elmina 2015 Strategy plan was funded by the Dutch Culture Fund of the Dutch Ministry of Foreign Affairs and the Dutch Ministry of Education, Culture and Society. The main stakeholders were the KEEA Municipal Assembly, the GMMB, the ILGS in close collaboration with the Institute of Housing Studies and Urban Development (IHS) of the Erasmus University of Rotterdam and University of Groningen (RUG) both in the Netherlands. The initial idea for renovating the Elmina Castle came from the GMMB, who entered into discussions with the IHS and RUG. The discussion led to the acknowledgement that to secure funding, the project scope needed to go beyond simply restoring the castle. Findings from the interviews show that the Dutch counterparts had a much stronger and decisive role in shaping the initial ideas as highlighted in the following two quotes.

> let me put it this way, the people of Urban Solutions were originally employees of IHS – Institute of Housing Studies in Rotterdam – and to make themselves functionally and also to maintain themselves in jobs that they wanted, they identified places which need help and together with the locals brought up strategic plans with action plans, with budgets and in collaboration with the locals, sought funds from institutions and it is from the funding that they also carve out their remuneration.
>
> (Interview with Dr Anthony Annah-Prah, Head of Implementation of Elmina Cultural Heritage Management Programme, 3 November 2014)

Consequently, building on the experiences of the IHS, the decision was made to link the restoration of the castle to the revitalisation of the whole Elmina township based on the recognition that the community needed to see a direct benefit from any restoration work on the castle. The ambition was as follows:

> The Elmina 2015 strategy goes beyond restoring monuments. It believes that with improved tourist infrastructure, the revitalisation of the existing monuments and the development of other important cultural sites, tourism can become a major economic activity in Elmina and serve to improve the general standard of living in the town. The strategy identifies tourism as an entry point in facilitating sustainable development through an integrated development approach.
>
> (KEEA, 2002, p. 10)

For this reason, the KEEA Municipal Assembly was brought on board as the leading agency for the project proposal rather than the GMMB. This was because the KEEA Municipal Assembly is the local government body charged with the responsibility for overall planning and development of the district of which Elmina is the district capital. It is noteworthy that although the project was basically proposing a vision of heritage tourism-led local economic development planning process, neither the GTA nor the Ministry of Tourism, Culture and Creative Arts were involved in shaping the strategy at the initial stages. Funding for the final proposal was secured from the Culture Fund of the Ministry of Foreign Affairs and the Ministry of Education, Culture and Science in the Netherlands. Additional support was also secured through European Union funding schemes. For the remainder of this section, I focus on assessing the governance capacity of the KEEA Municipal Assembly and how this affected the realisation of the Elmina 2015 Strategy.

Image of tourism's role in local economic development held by stakeholders

The original image held by key stakeholders shifted from a project of castle restoration to one of an integrated tourism-led local economic development project. In essence, this image was that an integrated approach to cultural heritage asset restoration and management could lead to an improvement in the tourist, social and economic conditions of Elmina. Thus, stakeholders held that diversifying tourist attractions and maintaining existing assets would lead to increased visits and duration of stay of tourists coming to Elmina. This would then feed into the local economy through the multiplier effects of tourist spending in town, particularly at a purpose-built art and craft market, which later fell into disuse (see Figures 5.1 and 5.2).

Figure 5.1 The deserted and empty Elmina arts and craft market. Source: Author.

Figure 5.2 The deserted and empty Elmina arts and craft market. Source: Author.

The Elmina 2015 Strategy document developed by the KEEA Municipal Assembly presented the image held by stakeholders regarding tourism's role in local economic development. This document made reference to information from the Ghana Tourism Authority concerning the average length of tourist stay in Elmina but offered no systematic review of tourist behaviour in Elmina. The facts about the number of visitors to the castle and their short stay and lack of interaction within the Elmina Township therefore shaped stakeholders' views. Stakeholders saw that given a diversified and wide range of attraction sites, visitors were more likely to spend more time and money within the community, which would then catalyse a number of local economic development opportunities. While it is not clear that a detailed market research of tourist behaviour was undertaken, the strategy projected an annual visit of 150,000 tourists in Elmina. The Elmina 2015 Strategy was therefore underlined by a number of assumptions regarding tourist behaviour, a tourism multiplier effect and the likelihood of increased tourism leading to a rejuvenation of other sectors of the local economy. For instance, these assumptions are made evident in this interview quote:

> You see, I don't know how to, I, I, am not find the words to really describe, because the moment you tackle tourism, you tackle all other sectors of the economy. We say that people produce let's say agriculture, people are producing they are not processing so people lose their harvest and all their produce and all that, do you think that if we have another million or 2 million people visiting Ghana and spend a day or two, our demand for foodstuff wouldn't go up, that whatever is produced in the village will get bought.
>
> (Interview with Francis Cobbinah, Executive Director of Ghana Heritage Conservation Trust, 29 September 2014)

There was a broadly shared consensus among stakeholders about what they perceived to be the relationship between tourism and local economic development. Stakeholders held the image summarised in the aspiration of the Elmina 2015 Strategy, which was to "restore and manage mutual heritage existing in Elmina; for it to become the forefront of an integrated programme to improve the tourist, social and economic condition of Elmina" (KEEA, 2002, p. 86)

Instruments available to stakeholders

A key aspect of stakeholder governance capacity is the availability of instruments that can be selected to translate images into tangible actions and outcomes. Here, I mainly focus on the politico-legal and administrative and financial instruments.

Politico-legal and administrative instruments.

The initiation by the GMMB of what became the Elmina 2015 Strategy relied on the legal and administrative prerogative that made them the legal custodian of

Elmina Castle with management responsibilities for it. However, once the project had been expanded to cover overall local economic development of Elmina, there was a need to involve the KEEA Municipal Assembly. The politico-legal backing of the KEEA Municipal Assembly rested on the provisions in the Ghana Local Government Act 462 of 1993 and Section 5 of the National Development Planning System Act 480 of 1994. Under these legislations, the KEEA Municipal Assembly is a planning authority and the only institution legally mandated to prepare local and/or subdistrict development plans either directly or through delegation to lower tiers of government like town and urban councils. Therefore, the proposals for the ECHMP fell into the category of a subdistrict local economic development plan of Elmina and hence the KEEA Municipal Assembly had to be the leading institution for the project in order to provide the requisite politico-legal instruments. It must also be noted that at this point in time, a new political party had just come in into office for the first time and was eager to show that they were bringing the promised development projects to the community. This political reality was instrumental in shaping the enthusiasm and support given to the project in the beginning (see Adu-Ampong, 2021, for details on this).

Notwithstanding its legal mandate for the overall development of the municipality, the remit of the KEEA Municipal Assembly did not extend to the management of Elmina Castle or many of the other tourist assets of the municipality. These areas fell under the responsibility of the GMMB. Thus, while the KEEA Municipal Assembly came to lead the project, it was the GMMB that had the ultimate governing authority over the golden goose upon whose golden eggs all other components of the project hinged. In addition, given the dual governance structure, it was imperative for the traditional council to deploy their traditional authority instruments and give backing to the project (Yankholmes, 2018). These traditional instruments consisted largely of the mobilisation of community support for and participation in the projects. Thus, the politico-legal instruments that made the ECHMP process possible were dispersed across many stakeholders.

Financial instruments.

At the structural level, financial resources have the propensity to enable and/or constrain the way in which images are translated into tangible actions. The availability of financial resources makes it possible to bring together the material, human, technical and social capital assets required to develop and implement projects. Within the Elmina 2015 Strategy, the source and conditions of the financial resources available served as both an avenue of constraint and enablement in shaping the final outcomes. Initial funding for the preparation of the Elmina Profile and the Elmina 2015 Strategy in addition to the numerous project-specific proposals was through the Culture Fund of the Netherlands Ministry of Foreign Affairs and funds from the Dutch Ministry of Education, Culture and Science. Project implementation funding came from the Culture Fund of the Ministry of Foreign Affairs. This funding was used to complete five main projects (see Table 5.2). The remaining projects were completed with funding from the 9th

Table 5.2 Financing of associated projects implemented under the Elmina 2015 Strategy

Financial instruments	*Projects implemented*
Dutch Culture Fund	Restoration of Nana Etsiapa Hall Renovation of Nana Kobina Gyan (formerly Victoria/ Trafalgar) Square Restoration of Van Dyke House Restoration of the Dutch Cemetery and surrounding area (including Catholic Museum)
9th European Union Development Fund under the Elmina Cultural Heritage Local Economic Development Programme (CHLDEP)	Restoration of selected parts of St. George's (Elmina) Castle and Fort Jago Restoration of St. Jago Hill Renovation of selected historic buildings through a home-owner scheme (20 in total) Establishment of tourist information and tourist facilities Training for local economic development (tourist guides, artisans, financial management etc)
Royal Netherlands Embassy in Ghana Funding	Renovation of four Asafo Posts in Elmina

Source: Author's compilation based on project documents and the Elmina 2015 Strategy document.

European Union Development Fund under the so-called Cultural Heritage and Local Development in Elmina Programme (CHLDEP).

These main sources of financial instruments being project-specific meant that after completion, there was no additional financial provision for continued maintenance of some of the projects. Some projects, like the renovation of historic buildings, were to be partly used for income-generating ventures like guest rooms and commercial stores. However, not all these income-generating activities were successful since the image held by stakeholders regarding increased tourist numbers and length of stay in the community were not realised. An Elmina Development Fund (EDF) was also established, envisaged as a source for finance for ongoing maintenance and improvement in facilities. In addition, the EDF was to be used to train young people as tour guides as well as paving city streets in parts of Elmina.

The EDF was to be financed from donations and a daily levy of $1 on all tourists to Elmina – to be collected by owners of tourist facilities. In the 2003 document, the EDF was expected to initially generate $25,000 annually and to increase afterwards. However, the expected increase in tourist numbers and length of stay did not happen in part because a visit to Elmina Castle and the nearby Cape Coast Castle is seen as a return day trip. Thus, the EDF was poorly developed and deployed as a financial instrument. In addition to the fact that the fund did not accrue due to poor marketing and collection, there was no clear institutional arrangement as to the management of the fund. Indeed, in the draft final project report by Urban Solutions as project consultants made no mention of the EDF. This absence of a pot of funding to carry on project implementation, especially the challenging aspect of linking increased tourist numbers to local economic development, meant that the ultimate objective of the Elmina 2015 Strategy remained unfulfilled.

The lack of financial sustainability being planned into the project from the start constrained the governance capacity of stakeholders. In the absence of additional financial instruments, stakeholders were unable to maintain and further improve upon completed projects in pursuit of the ultimate objective of using tourism to catalyse local economic development. The GMMB, for instance, should have been made to set aside a percentage of visitor fees for continued maintenance. As this was not planned for, all the visitor fees that accrue to the GMMB from Elmina Castle went straight into the national consolidated fund. Access to this fund for maintenance work had to go through the national office of the GMMB.

Action potential for project implementation

Action potential is about the socio-political room for action in which stakeholders operate. It is here that stakeholders' images of current and future situations are translated into tangible actions through the selection and application of appropriate instruments. This is a process that usually takes place within the intentional level of action potential. At the structural level, action potential is embedded within institutional power relations that serve to enable and/or constrain the ability of stakeholders to meet their governing needs. The Elmina 2015 Strategy required mobilising support and generating adequate socio-political will among all stakeholders. To a large extent, the action potential available is shaped by how stakeholders are able to answer questions about their source of legitimacy and authority.

The most notable observation of the Elmina 2015 Strategy is that the greatest action potential was aggregated in state institutions, and, to an extent, in the structures of political parties. The (incomplete) decentralised governance and planning system provided the source of legitimacy and authority for the KEEA Municipal Assembly in leading the planning process (Arthur, 2012). The KEEA Municipal Assembly was therefore able to derive considerable authority and legitimacy from its control of the technical planning knowledge production system. This decentralised governance system also served as a structural constraint to the governing role of the KEEA vis-à-vis the GMMB. The GMMB is the only state institution with the administrative and legal mandate for the day-to-day management and governing activities of the Elmina Castle and Fort St. Jago. This was a source of action potential for them to translate their image of the castle into reality, which sometimes differed from the image of other stakeholders in terms of the commercial use of these sites.

The fragmented action potential did not paralyse the project in the early phases because of high levels of cross-sectoral coordination and mobilisation of support. For instance, the issue of revenue sharing of the visitor fees at Elmina Castle was resolved informally through a national cabinet-level arrangement between the GMMB, the Elmina Traditional Authority and the KEEA Municipal Assembly.

The traditional authority and the KEEA Municipal Assembly were therefore given about 5–10% of the total revenue from Elmina Castle on a yearly basis (interviews with Justice Amoah, Former Municipal Planning Officer of KEEA/Former National Project Coordinator of Elmina 2015 Strategy, 17 July 2015; Ebo Dadzie, Head of Tourism Sub-Committee, KEEA, 16 July 2015; Clifford Ashun, Elmina Castle Manager, 10 August 2015). This arrangement facilitated and strengthened the socio-political room for action and the capacity of key stakeholders to meet the governing needs of the Elmina 2015 Strategy. However, this arrangement was stopped in 2009 after a change of government on the basis that there was no legislative instrument backing it. Thus, after 2009, the new socio-political environment – in which there was no revenue sharing – constrained the action potential of stakeholders in meeting their governance needs. This led to the decline in the level of communication, collaboration and coordination among stakeholders and ultimately in a decrease in governance capacity for the realisation of the objectives of the Elmina 2015 Strategy.

Discussion and conclusion

> In 2015 we envisage Elmina to be a thriving town in which fishing is still the main economic activity but has significant other sectors of employment in the salt mining, boat building but most importantly the tourist industry. The unique heritage assets and the picturesque harbour make the town Ghana's prime tourist destination. The excellent road conditions allow tourists to reach Elmina by car within 2 hours from Accra, while the building of airstrip is underway. The town is to be visited by over 150,000 tourists annually that spend an average of two nights in the town choosing from a wide range of choices for accommodation. The local population is also benefitting from the tourist boom and consequent economic development.
>
> (KEEA, 2002, p. 85)

At the end of the year 2015, the reality on the ground showed a remarkable difference from the objectives of the Elmina 2015 Strategy as outlined in the above quote. This divergence of objectives and outcomes while common to many development plans and projects raises a number of important questions about tourism-led local economic development planning processes more generally and the particular case of the Elmina 2015 Strategy specifically. In this chapter, I have outlined an argument to the effect that this divergence can be explained in terms of the extent of stakeholder governance capacity within the planning process.

Stakeholder governance capacity is about the alignment of images, instruments and action potentials as outlined earlier. A process tracing of the Elmina 2015 Strategy shows that there were many instances of misalignment of these three elements within and between stakeholders. A prime example of such misalignment can be seen in the relationship between the KEEA Municipal Assembly and the GMMB, who were two key stakeholders. The dynamics between these two institutions, which could have offered the potential for positive outcomes, however, resulted in a negative outcome due to competing interests that were not reconciled or aligned. The KEEA Municipal Assembly was keen on getting access to part of the revenue accruing from visitor fees at Elmina Castle. The GMMB on the other hand sought to keep these revenues away from other stakeholders. Part of the reason for this was due to the complexity and diversity in their internal procedures as well as the different instruments available to them. The legislative instrument that guided how revenue from Elmina Castle was to be disbursed stipulated that all funds should go to the GMMB and not to any other body. Thus, while there was a shared image between the KEEA Municipal Assembly and the GMMB in their efforts to utilise tourism for local economic development and poverty reduction, the legislative, as well as the financial, instruments to make this happen were dominated by the GMMB. The GMMB was unwilling to match its image with available instruments leading to a lack of alignment. This misalignment affected the governability of the Elmina 2015 Strategy – where governability refers to the ability of governors to meet governance needs given their capacities (Jobbins, 2003; Kooiman, 1993, 2008). The Elmina 2015 Strategy brought about governances needs that required the pooling together of not only shared visions, but critically, also instruments, especially financial ones, in order to meet those needs. This pooling of instruments and coordination activities did not effectively take place.

The fragmentation of instruments among stakeholders contributed to undermining the vision of how to make the transition to the stage of full implementation of the Elmina 2015 Strategy. This was further compounded by the change in action potential when the semi-formally brokered revenue-sharing arrangement was overturned. The end of revenue sharing between the Ghana Museums and Monuments Board, the KEEA Municipal Assembly and the Elmina Traditional Authority shaped the socio-political action potential of stakeholders. This constrained their ability to meet the governance needs of the Elmina 2015 Strategy. Overall, therefore, there was a low level of governance capacity to ensure that the "software" aspects of the Strategy were fully implemented so that tourism could indeed catalyse local economic development and poverty reduction in Elmina.

This chapter highlighted how the extent of stakeholder governance capacity in developing and implementing the Elmina 2015 Strategy shaped the expected outcomes. The initial stages of the project provided stakeholders with a shared vision of tourism's potential contribution to local economic development. This led to the collaborative pooling together of individual capacities in establishing a strong stakeholder governance capacity base. This pooling together of resources, however, appears to have been facilitated by the fact that the resources out of which financial instruments were fashioned came from an external source. Consequently, once this external funding had ended, old fractures in stakeholder relationships opened up, leading to a reduction in governance capacity.

The low level of governance capacity surrounding the implementation of the Elmina 2015 Strategy is best represented in the (in)actions of the KEEA Municipal Assembly. The KEEA Municipal Assembly remains the key governing stakeholder when it comes to the management and administration of local economic development and poverty alleviation in Elmina and its immediate environs. The KEEA Municipal Assembly as the local government authority was expected to take on a number of responsibilities. However, these were not always accompanied by corresponding access to the resources from which to fashion governing instruments. For instance, while the KEEA Municipal Assembly understood the potential of tourism to catalyse local economic development and poverty reduction, the legal and, importantly, financial instruments for managing the tourism assets in Elmina were held by the GMMB. This limited the ability of the KEEA Municipal Assembly to marshal and align the images, instruments and action potentials needed to develop full governance capacity.

With increasing attention on local governments to facilitate sustainable tourism development (Churugsa et al., 2007; Dredge, 2006; Rogerson, 2020; Ruhanen, 2013), the analysis in this chapter raises important questions about the expectations surrounding local governments and their role in tourism development. Given the multiplicity of other state and non-state institutions involved in tourism, what role is there for local government? How do local governments reconcile their day-to-day management of public service provision with the long-term planning required of tourism-led local economic development initiatives? How are local governments able to develop the required governance capacity for such initiatives when instruments are dispersed among stakeholders who do not collaborate? How can stakeholders collaborate in pooling resources and instruments together to develop and sustain high levels of governance capacity? These are questions that future research needs to take up (Adu-Ampong, 2019b).

In this chapter, I have shown how the mere presence of local governments and other state institutions in tourism development, while a necessary condition, is not sufficient for effective development of stakeholder governance capacity. The findings outline the proximate causes of the inadequacy of stakeholder governance capacity and the resulting effect on the Elmina 2015 Strategy outcomes. In the midst of a plethora of institutions involved in tourism development and management, local governments need to be provided with adequate resources and instruments in order to lead the collaborative process required to develop stakeholder governance capacity. Given the ongoing spread of decentralisation and other public sector administration reforms in many countries, it is crucial that we fully understand the capacity of local governments in taking up these additional responsibilities.

In terms of the local government role in tourism development planning, two main recommendations can be made based on the discussions from this chapter. First, the increasing calls for local governments to take a more active role in sustainable tourism development need to go hand in hand with attention paid to how local governments are being equipped to engage in this process. It is essential to bear in mind that the state and internal power dynamics between key institutional stakeholders within it remain crucial in setting the policy agenda and thus must be brought back into how we conceptualise the tourism governance process. This is particularly pertinent in terms of the use of tourism for local economic development and poverty reduction goals. This is an important yet often ignored aspect of local government policy processes and needs to be considered when it comes to the tourism–poverty nexus. A second recommendation is to embed and integrate tourism planning within the local economic development planning process while involving the private sector and regional and local tourism operators. Given that the embedding and integration process exposes tourism planning to existing local government politics and planning there is a need for active involvement of private sector agencies. Tourism private sector actors and operators provide the needed safeguards that ensure the continuity of planning and project implementation because their business objectives depend on successful implementation.

Note

1 This chapter is based on the author's PhD thesis which was completed at the Department of Urban Studies and Planning, University of Sheffield, UK, in 2017.

References

Adu-Ampong, E. A. (2016). *Governing Tourism-Led Local Economic Development Planning: An Interactive Tourism Governance Perspective on the Elmina 2015 Strategy in Ghana*. Unpublished doctoral dissertation. University of Sheffield, Sheffield, UK.

Adu-Ampong, E. A. (2017a). Divided we stand: Institutional collaboration in tourism planning and development in the Central Region of Ghana. *Current Issues in Tourism*, *20*(3), 295–314. doi:10.1080/13683500.2014.915795

Adu-Ampong, E. A. (2017b). State of the nation address and tourism priorities in Ghana – A contextual analysis. *Tourism Planning and Development*, *14*(1), 135–138. doi:10.1080/21568316.2015.1101392

Adu-Ampong, E. A. (2018). Tourism and national economic development planning in Ghana, 1964–2014. *International Development Planning Review*, *40*(1), 75–95. doi:10.3828/idpr.2018.2

Adu-Ampong, E. A. (2019a). Historical trajectories of tourism development policies and planning in Ghana, 1957–2017. *Tourism Planning & Development*, *16*(2), 124–141. doi:10.1080/21568316.2018.1537002

Adu-Ampong, E. A. (2019b). The tourism-development nexus from a governance perspective: A research agenda. In R. Sharpley & D. Harrison (Eds.), *A Research Agenda for Tourism and Development* (pp. 53–70). Cheltenham: Edward Elgar Publishing.

Adu-Ampong, E. A. (2021). "Today is party A, tomorrow is party B": The Elmina 2015 Strategy and the politics of tourism-led local economic development planning in Ghana. In M. Novelli, E. Adu-Ampong, & M. A. Riberio (Eds.), *The Routledge Handbook of Tourism in Africa*. Abingdon: Routledge.

Arthur, S. N. A. (2012). *The Unfinished Business of Decentralisation: Political Accountability of Local Government in Ghana: A Case Study of the Komenda-Edina-Eguafo-Abrem (KEEA) Municipality*. PhD thesis. Rheinische Friedrich-Wilhelms-Universität, Bonn.

Arthur, S. N. A., & Mensah, J. V. (2006). Urban management and heritage tourism for sustainable development: The case of Elmina cultural heritage and management programme in Ghana. *Management of Environmental Quality: An International Journal*, *17*(3), 299–312. doi:10.1108/14777830610658719

Awortwi, N. (2010). The past, present, and future of decentralisation in Africa: A comparative case study of local government development trajectories of Ghana and Uganda. *International Journal of Public Administration*, *33*(12), 620–634. doi:10.1080/01900692.2010.514451

Bachrach, P., & Baratz, M. S. (1962). Two faces of power. *American Political Science Review*, *56*(4), 947–952. doi:10.2307/1952796

Bell, S., & Hindmoor, A. (2009). *Rethinking Governance: The Centrality of the State in Modern Society*. New York: Cambridge University Press.

Britton, S. G. (1982). The political economy of tourism in the third world. *Annals of Tourism Research*, *9*(3), 331–358. doi:10.1016/0160-7383(82)90018-4

Caffyn, A., & Jobbins, G. (2003). Governance capacity and stakeholder interactions in the development and management of coastal tourism: Examples from Morocco and Tunisia. *Journal of Sustainable Tourism*, *11*(2–3), 224–245. doi:10.1080/09669580308667204

Chuenpagdee, R., Kooiman, J., & Pullin, R. (2008). Assessing governability in capture fisheries, aquaculture and coastal zones. *Journal of Transdisciplinary Environmental Studies*, *7*(1), 1–20.

Churugsa, W., McIntosh, A. J., & Simmons, D. (2007). Sustainable tourism planning and development: Understanding the capacity of local government. *Leisure/Loisir*, *31*(2), 453–473. doi:10.1080/14927713.2007.9651391

Crook, R. C. (2003). Decentralisation and poverty reduction in Africa: The politics of local–central relations. *Public Administration and Development*, *23*(1), 77–88. doi:10.1002/pad.261

Dredge, D. (2006). Policy networks and the local organisation of tourism. *Tourism Management*, *27*(2), 269–280. doi:10.1016/j.tourman.2004.10.003

Dredge, D., & Jenkins, J. J. M. (2007). *Tourism Planning and Policy*. Milton, QLD: John Wiley & Sons Australia, Limited.

Dredge, D., & Jenkins, J. J. M. (Eds.) (2011). *Stories of Practice: Tourism Policy and Planning*. Farnham: Ashgate Publishing, Ltd.

Essah, P. (2001). Slavery, heritage and tourism in Ghana. *International Journal of Hospitality & Tourism Administration*, *2*(3–4), 31–49. doi:10.1300/J149v02n03_02

Gray, B. (1985). Conditions facilitating interorganisational collaboration. *Human Relations*, *38*(10), 911–936. doi:10.1177%2F001872678503801001

Helmsing, A. (2003). Local economic development: New generations of actors, policies and instruments for Africa. *Public Administration and Development*, *23*(1), 67–76. doi:10.1002/pad.260

Holden, A., Sonne, J., & Novelli, M. (2011). Tourism and poverty reduction: An interpretation by the poor of Elmina, Ghana. *Tourism Planning & Development*, *8*(3), 317–334. doi:10.1080/21568316.2011.591160

Ioannides, D., & Debbage, K. G. (1998). *The Economic Geography of the Tourist Industry: A Supply-Side Analysis*. New York: Routledge.

Jeffries, D. (2001). *Governments and Tourism*. Oxford: Butterworth-Heinemann.

Jobbins, G. (2003). The effects of stakeholder interactions on capacity for integrated coastal governance in Morocco and Tunisia. *Aquatic Ecosystem Health & Management*, *6*(4), 455–464. doi:10.1080/714044175

KEEA (Komenda-Edina-Eguafo-Abrem Municipal Assembly) (2002). *The Elmina 2015 Strategy: Building on the Past to Create a Better Future*. Elmina: Komenda-Edina-Eguafo-Abrem Municipal Assembly.

Kjær, A. M. (2004). *Governance*. Malden, MA: Polity Press.

Kooiman, J. (1993). *Modern Governance: New Government – Society Interactions*. London: SAGE.

Kooiman, J. (1999). Social-political governance. *Public Management Review*, *1*(1), 67–92. doi:10.1080/14719037800000005

Kooiman, J. (2000). Societal governance: Levels, modes and orders of social political interaction. In J. Pierre (Ed.), *Debating Governance: Authority, Steering and Democracy* (pp. 138–164). Oxford: Oxford University Press.

Kooiman, J. (2003). *Governing as Governance*. London: SAGE.

Kooiman, J. (2008). Exploring the concept of governability. *Journal of Comparative Policy Analysis: Research and Practice*, *10*(2), 171–190. doi:10.1080/13876980802028107

Kooiman, J., & Bavinck, M. (2005). The governance perspective. In J. Kooiman, M. Bavinck, S. Jentoft, & R. Pullin (Eds.), *Fish for Life: Interactive Governance for Fisheries* (pp. 11–24). Amsterdam: Amsterdam University Press.

Kooiman, J., Bavinck, M., Jentoft, S., & Pullin, R. (2005). *Fish for Life: Interactive Governance for Fisheries*. Amsterdam: Amsterdam University Press.

Kooiman, J., Bavinck, M., Chuenpagdee, R., Mahon, R., & Pullin, R. (2008). Interactive governance and governability: An introduction. *Journal of Transdisciplinary Environmental Studies*, *7*(1), 1–11.

Manning, N. (2001). The legacy of new public management in developing countries. *International Review of Administrative Sciences*, *67*(2), 297–312. doi:10.1177%2F0020852301672009

Mckercher, B., & Ritchie, M. (1997). The third tier of public sector tourism: A profile of local government tourism officers in Australia. *Journal of Travel Research*, *36*(1), 66–72. doi:10.1177%2F004728759703600110

Nel, E. (2001). Local economic development: A review and assessment of its current status in South Africa. *Urban Studies*, *38*(7), 1003–1024. doi:10.1080%2F00420980120051611

Peters, B. G., & Pierre, J. (2016). *Comparative Governance: Rediscovering the Functional Dimension of Governing*. Cambridge: Cambridge University Press.

Pierre, J. (2000). *Debating Governance: Authority, Steering and Democracy*. Oxford: Oxford University Press.

Pierre, J., & Peters, G. B. (2000). *Governance, Politics, and the State*. New York: St. Martin's Press.

Rhodes, R. A. W. (1994). The hollowing out of the state: The changing nature of the public service in Britain. *The Political Quarterly*, *65*(2), 138–151. doi:10.1111/j.1467-923X.1994.tb00441.x

Rhodes, R. A. W. (1996). The new governance: Governing without government. *Political Studies*, *44*(4), 652–667. doi:10.1111%2Fj.1467-9248.1996.tb01747.x

Ritchie, J., & Spencer, L. (1994). Qualitative data analysis for applied policy-research. In A. Bryman & R. G. ,Burguess (Eds.), *Analysing Qualitative Data* (pp. 173–194). London: Routledge.

Rogerson, C. M. (2014). Strengthening tourism-poverty linkages. In A. A. Lew, C. M. Hall, & A. M. Williams (Eds.), *The Wiley Blackwell Companion to Tourism* (pp. 600–610). Oxford: Wiley Blackwell.

Rogerson, C. M. (2020). Using municipal tourism assets for leveraging local economic development in South Africa. *Bulletin of Geography. Socio-Economic Series*, *48*(48), 47–63. doi:10.2478/bog-2020-0013

Rogerson, C. M., & Rogerson, J. M. (2010). Local economic development in Africa: Global context and research directions. *Development Southern Africa*, *27*(4), 465–480. doi:10.1080/0376835X.2010.508577

Ruhanen, L. (2013). Local government: Facilitator or inhibitor of sustainable tourism development? *Journal of Sustainable Tourism*, *21*(1), 80–98. doi:10.1080/09669582.2012.680463

Slocum, S. L., & Backman, K. F. (2011). Understanding government capacity in tourism development as a poverty alleviation tool: A case study of Tanzanian policy-makers. *Tourism Planning & Development*, *8*(3), 281–296. doi:10.1080/21568316.2011.591157

Smoke, P. (2003). Decentralisation in Africa: Goals, dimensions, myths and challenges. *Public Administration and Development*, *23*(1), 7–16. doi:10.1002/pad.255

Sonne, J. (2010). *The Role of Tourism in Poverty Reduction in Elmina, Ghana*. PhD Dissertation, University of Bedfordshire, Bedfordshire, UK.

Torfing, J., Peters, B. G., Pierre, J., & Sorensen, E. (2012). *Interactive Governance: Advancing the Paradigm*. Oxford: Oxford University Press.

Tosun, C. (2000). Limits to community participation in the tourism development process in developing countries. *Tourism Management*, *21*(6), 613–633. doi:10.1016/S0261-5177(00)00009-1

WTTC (World Travel and Tourism Council) (2020). *Ghana, 2020 Annual Research: Key Highlights*. London: World Travel and Tourism Council

Yankholmes, A. (2018). Tourism as an exercise in three-dimensional power: Evidence from Ghana. *Tourism Management Perspectives*, *25*, 1–12. doi:10.1016/j.tmp.2017.11.001

6 From small island developing states to large ocean states

Tourism in the changing periphery of island states in the Global South

C. Michael Hall

Introduction

For many people, the small island developing states (SIDS) are the epitome of the Global South. The total resident population of all the SIDS is only 65 million people, slightly less than 1% of the world's population, yet this group faces unique social, economic and environmental issues as a result of their small size, resource base and, in many cases, distance to markets. Yet, at the same time, these countries, geographically concentrated primarily in the Caribbean, Indian Ocean and the Pacific, are often portrayed as paradises in tourism promotions to their markets in the developed world (Hall, 2015).

Tourism is usually a central element of SIDS economies. Travel and tourism services account for more than 50% of least developed countries (LDC) services exports (UN Office of the High Representative for the Least Developed Countries, Landlocked Developing Countries and Small Island Developing States [UN-OHRLLS], 2018). However, at the same time, because of their specific geographies, SIDS are greatly at risk from global environmental change, and particularly global heating and the climate crisis, but are also substantially affected by biodiversity loss and the drawdown of other parts of their natural capital. As a result of these factors, SIDS are at the forefront of the impacts of disasters arising from hurricanes, cyclones, flooding and drought, as the world's climate shifts in response to anthropogenic climate change. This combination of factors means that SIDS have become a focal point for sustainable development thinking and policy making, including for tourism. UN-OHRLLS (2018), for example, notes that with respect to vulnerability to climate change, "The tourism sector and its related infrastructure play a critical role, especially in coastal LDCs and SIDS" (2018, p. 39).

This chapter argues that sustainability approaches concerning SIDS have shifted dramatically in recent years and present a changed perspective on how SIDS are framed in sustainable development terms. These changes have substantial long-term implications for policy making and the role of tourism in island economies and the wider Global South. They also reflect a broader appreciation of the changing vulnerabilities of islands, including in the social sciences (Moore, 2010). This chapter is broadly framed into three main parts. First, an outline of

SIDS and the challenges that they face. Second, the importance of islands in sustainability thinking. Third, the dramatic changes that have occurred in reframing SIDS as large ocean states, particularly in the Pacific, and the possible policy implications for tourism. Finally, the chapter concludes by commenting on the changing positioning of SIDS within the Global South.

Small island developing states

SIDS were first formally recognised by the international community at the 1992 United Nations Conference on Environment and Development (UNCED) held in Rio de Janeiro (commonly known as Agenda 21). Chapter 17, paragraph 124 of Agenda 21 states:

> [SIDS] and islands supporting small communities are a special case both for environment and development. They are ecologically fragile and vulnerable. Their small size, limited resources, geographic dispersion and isolation from markets, place them at a disadvantage economically and prevent economies of scale.
>
> (UN DESA, 2010, p. iii)

However, there is no agreed definition of SIDS, even within the United Nations Community. UNCTAD (2012) recognised 29 SIDS of which eight are included in the group of 48 LDCs, while the Department of Economic and Social Affairs, Division for Sustainable Development, identified 39, two of which also qualify as LDCs (Table 6.1). UN-OHRLLS (2020) recognises 58 SIDS as of September 2020, 38 of which are UN members, with the remaining 20 states being either non-UN members or associate members of regional commissions.

The modern development challenges facing SIDS first began to emerge during decolonisation. However, the sustainable development context was arguably first formally recognised by the international community at the 1992 United Nations Conference on Environment and Development (UNCED) held in Rio de Janeiro. Chapter 17, paragraph 124 of Agenda 21 provides a good summary of the problems facing SIDS, given that they

> and islands supporting small communities are a special case both for environment and development. They are ecologically fragile and vulnerable. Their small size, limited resources, geographic dispersion and isolation from markets, place them at a disadvantage economically and prevent economies of scale.
>
> (UN DESA, 2010, p. iii)

The majority of SIDS lag behind the global averages for human development (UNDP, 2019) (see Table 6.1). Undoubtedly, the characteristics of SIDS make them vulnerable to exposure to global change (Hall, 2015). The Intergovernmental Panel on Climate Change (IPCC) forecast that global sea levels will rise on

Table 6.1 Characteristics of SIDS

SIDS (UN, 2015)	*SIDS (UNCTAD, 2012)*	*SIDS UN-OHRLLS (2020)*[1]	*LDC*[2]	*High Environmental Vulnerability*[3]	*Human Development Index (rank) (2012)*[4]	*Human Development Index (rank) (2019)*[5]
American Samoa		N		NA	NA	NA
Anguilla		N		NA	NA	NA
Antigua and Barbuda	×	UN		×	0.760 (67)	0.776 (74)
Aruba		N		NA	NA	NA
Bahamas	×	UN		NA	0.794 (49)	0.805 (60)
Bahrain		UN		×	0.796 (48)	0.838 (45)
Barbados	×	UN		×	0.825 (38)	0.813 (56)
Belize		UN			0.702 (96)	0.720 (103)
Bermuda		N		NA	NA	NA
British Virgin Islands		N		NA	NA	NA
Cape Verde	×	UN		×	0.586 (132)	0.651 (126)
Cayman Islands		N		NA	NA	NA
Commonwealth of Northern Marianas		N		NA	NA	NA
Comoros	×	UN	×	×	0.429 (169)	0.538 (156)
Cook Islands		N		NA	NA	NA
Cuba		UN		×	0.780 (59)	0.778 (72)
Curacao		N			NA	NA
Dominica	×	UN		NA	0.745 (72)	0.724 (98)
Dominican Republic		UN		×	0.702 (96)	0.745 (89)
Federated States of Micronesia	×	UN		NA	0.645 (117)	0.614 (135)
Fiji	×	UN		×	0.702 (96)	0.724 (98)
French Polynesia				NA	NA	NA
Grenada	×	UN		×	0.770 (63)	0.763 (78)
Guadeloupe		N		NA	NA	NA

Guam		N		NA	NA	NA
Guinea-Bissau		UN	×		0.364 (176)	0.461 (178)
Guyana		UN			0.636 (118)	0.670 (123)
Haiti		UN	×	×	0.456 (161)	0.503 (169)
Jamaica	×	UN		×	0.730 (85)	0.726 (96)
Kiribati	×	UN	×	×	0.629 (121)	0.623 (132)
Maldives	×	UN		×	0.688 (104)	0.719 (104)
Marshall Islands	×	UN		×	NA	0.698 (117)
Martinique		N		NA	NA	NA
Mauritius	×	UN		×	0.737 (80)	0.796 (66)
Montserrat		N		NA	NA	NA
Nauru	×	UN		×	NA	NA
New Caledonia		N		NA	NA	NA
Niue		N		NA	NA	NA
Palau	×	UN		×	0.791 (52)	0.814 (55)
Papua New Guinea	×	UN		×	0.466 (156)	0.543 (155)
Puerto Rico		N		NA	NA	
Samoa	×	UN	×	×	0.702 (96)	0.707 (111)
Sao Tome and Principe	×	UN	×		0.525 (144)	0.609 (137)
Seychelles	×	UN		×	0.806 (46)	0.801 (62)
Singapore		UN		NA	0.895 (18)	0.935 (9)
Sint Martin		N		NA	NA	NA
Solomon Islands	×		×	×	0.530 (143)	0.557 (153)
St. Kitts and Nevis	×	UN		×	0.745 (72)	0.776 (73)
St. Lucia	×	UN		NA	0.725 (88)	0.745 (89)

(Continued)

Table 6.1 Continued

SIDS (UN, 2015)	*SIDS (UNCTAD, 2012)*	*SIDS UN-OHRLLS (2020)*[1]	*LDC*[2]	*High Environmental Vulnerability*[3]	*Human Development Index (rank) (2012)*[4]	*Human Development Index (rank) (2019)*[5]
St. Vincent and the Grenadines	×	UN		×	0.733 (83)	0.728 (94)
Suriname		UN			0.684 (105)	0.724 (98)
Timor-Leste	×	UN	×	NA	0.806 (134)	0.626 (131)
Tonga	×	UN		×	0.702 (95)	0.717 (105)
Trinidad and Tobago	×	UN		×	0.760 (67)	0.799)63)
Turks and Caicos Islands		N		NA	NA	NA
Tuvalu	×	UN	×	×	NA	NA
US Virgin Islands				NA	NA	NA
Vanuatu	×	UN	×	×	0.626 (124)	0.597 (141)
LDCs					0.449	0.528
SIDS					0.648	0.723
World					0.694	0.731

1. The United Nations Office of the High Representative for the Least Developed Countries, Landlocked Developing Countries and the Small Island Developing States (UN-OHRLLS) (2020). UN, UN members; N, Non-UN Members/Associate Members of the Regional Commissions. Retrieved from: https://www.un.org/ohrlls/content/list-sids.
2. Least developed country category according to UNCTAD (2012).
3. Above average environmental vulnerability on UNEP environmental vulnerability index compared with the average for all LDCs (UN 2010). The Environmental Vulnerability Index is based on 50 indicators covering natural/anthropogenic risks, resilience and ecosystem integrity, and covers issues related to climate change, biodiversity, water, agriculture and fisheries, human health, desertification, and exposure to natural disasters (UN, 2010).
4. The Human Development Index (HDI) is a composite index measuring average achievement in three basic dimensions of human development – a long and healthy life, knowledge and a decent standard of living (UNDP, 2013).
5. Figures from UNDP (2019).

average between 24 and 30 cm by 2046–2065 and 40 and 63 cm by 2081–2100 (IPCC, 2013). For example, it is estimated that 29% of the population of all SIDS live within 5 m above sea level, making them highly susceptible to sea-level rise and high-magnitude extreme weather events (UN-OHRLLS, 2015). The situation for SIDS is further complicated by a high coastline-to-land-area ratio. This means that many settlements and critical infrastructure are increasingly subject to erosion, storms and tidal surges, saline intrusion and the intersection of groundwater with the surface, all of which can lead to inundation of low-lying areas (Nunn, 2013). Such vulnerabilities are also exacerbated by population increase and growing urbanisation. For example, Ebeye Atoll, the capital of the Marshall Islands, has a population density of over 40,000 people/km^2 (Chui & Terry, 2013). As SIDS populations grow and move into more at-risk areas in search of livelihoods, the number of people exposed to risk stemming from severe weather events also increases, while the cost of such events also affects development capacity (UN-OHRLLS, 2015).

The capacity of SIDS to adapt to climate change is affected by their overall level and rate of economic development, and possibly distribution of wealth, as well as their propensity to be affected by sea-level rise, climate variability and severe weather events (Hall, 2015). The IPCC (2014) states that the barriers to adaptation strategies in island settings include "inadequate access to financial, technological and human resources, issues related to cultural and social acceptability of measures, constraints imposed by the existing political and legal framework", and significantly given the role of tourism in development processes in the Global South, an "emphasis on island development as opposed to sustainability" (IPCC, 2014, p. 27). It should therefore be no surprise that the Alliance of Small Island States' (AOSIS) Declaration on Climate Change, states "climate change poses the most serious threat to our survival and viability, and … undermines our efforts to achieve sustainable development goals and threatens our very existence" (AOSIS, 2009, p. 1). Such an existential risk may therefore be expected to figure significantly in considerations of sustainability in an islands and SIDS context.

Islands and SIDS in sustainability thinking

Islands have an important place in the development of sustainability thinking. According to Hall (2015, p. 55):

> Islands are significant to helping understand the problem of transitioning to sustainable development because their finite space represents an analogue with that of the Earth with respect to issues of managing resource use and waste within a relatively bounded system.

Ideas of carrying capacity, for example, are grounded in ideas of "islands of sustainability" (IOS) (Bebbington, 1997; Burnett et al., 2014; Deschenes & Chertow, 2004; Péti, 2012; Wallner et al., 1996), whereby sustainability is achieved at different spatial scales, ranging from the local, when those areas live according to their carrying capacity, i.e. ability to live and develop without running down natural capital. According to Wallner et al. (1996, p. 1764), this means that:

> In order to reach regional sustainability the area balance—taking into account the appropriated area from other regions (imports), the area actually occupied for a region's own purposes, and the area made available for other regions (exports)—should not be negative in such a way that the appropriated area exceeds the others.

As well as being bounded entities offering "islands of sustainability" that may provide a hierarchy of solutions at ever greater scale (Connell, 2018), IOS are also regarded as important "innovative disturbances" (Wallner & Narodoslawsky, 1996) that are able to jeopardise the structural stability of unsustainable systems whether they be at a macro-regional, national or global scale, and may provide opportunities to introduce elements of sustainable development into the wider system. Such potential shifts as a result of IOS, Wallner et al. (1996) argue, would therefore contribute to the transition towards sustainable development as a result of the accompanying paradigm change from a mechanistic to a more holistic (synergetic, network) or integrated sustainability paradigm (Hall, 2015).

The bounded systems of the IOS approach appear to be a useful analogue to examine issues of island transition (Hall, 2015). However, real islands are never completely bounded (Kerr, 2005). Physical flows of matter and energy extend over products, processes and firms, and local, regional or national boundaries and borders, as well as flows of capital and people (Hall, 2015). The latter being especially important for SIDS given the critical economic role of migrant flows and remittances (Grydehøj & Kelman, 2020). Moreover, different islands and SIDS can have substantially different political jurisdictions and governance capacities and socio-cultural structures. Nevertheless, issues of scale and relative isolation remain key characteristics of islands. The relatively "simpler ecologies" of islands are also matched by their economies which tend to have a narrow base. Nevertheless, islands are systems that are closed and bounded in many respects and thus provide a manageable unit of research and a "living laboratory" on the realities of sustainable development, including with respect to scales of application and analysis (Hall, 2010; Pungetti, 2012). Of course, these same properties "present island populations with the challenges of limited resource availability, tenuous resource security and limited natural carrying capacity" (Deschenes & Chertow, 2004, p. 202).

Hall (2015) used an island biogeographic approach to examine whether it was possible for SIDS to become islands of sustainability. He concluded that islands that are located close to mainland areas in the developed regions of the world with ample resources and relatively low population levels may have the potential to develop dense self-sustaining network economies within existing resource and human capacities, but for many SIDS this outcome was seen as a highly doubtful prospect. Hall (2015) argued that many island businesses had to occupy a specialised niche in order to survive. The limited resource base of most islands therefore means that if a relative advantage exists, it only does so in a small number of sectors, usually fisheries, tourism, financial services and traded agricultural products (Kerr, 2005). However, such specialisation not only leads to a lack of diversity in the business base but also makes the economy extremely vulnerable to external economic and environmental change and crisis and even more dependent on remittances and aid payments (Benhamou & Cassin, 2020; McGillivary et al., 2010; Pelling & Uitto, 2001).

The degree of specialisation in island economies, and to tourism in particular, makes them highly vulnerable to competition as well as changes in demand and accessibility. For example, in the Caribbean, around 12% of the total labour force, approximately two million people, are estimated to work in the tourist sector, generating around 14% of regional GDP and 15% of total exports (UN-OHRLLS, 2015). The combination of social and economic factors that characterise most SIDS means that the capacity of the majority of SIDS to undertake effective environmental and climate change adaptation measures are highly constrained. Such a situation affects the urgent need of many SIDS to mainstream or integrate climate change adaptation and sustainable planning strategies into development plans and policies (IPCC, 2014). For example, in the case of tourism in the Caribbean:

> [A]s much as 29 percent of major resort properties in the CARICOM countries would be partially or fully inundated by a one-meter sea-level rise. Furthermore, about 49 percent are estimated to be damaged or destroyed by combined sea-level rise and storm surge and sea-level rise-enhanced erosion given their lack extensive coastal protection in order to preserve aesthetics of natural beach areas and views to the sea.
>
> (UN-OHRLLS, 2015, p. 16)

The issue of climate change is therefore clearly a substantial influence in the sustainable development agenda for SIDS in a policy context

Sustainable development and SIDS policy settings

Since the 1990s, the sustainable development focus of SIDS has moved through three main periods which broadly follow on from UN conferences on sustainable development and their implementation. These stages reflect the outputs of the various UN International Conferences on Small Island Developing States. The first, from 1994, is the Barbados Programme of Action (UN, 1994) which is an annex to Agenda 21. The preamble to the Annex states:

> Small island developing States have valuable resources, including oceans, coastal environments, biodiversity and, most importantly, human resources. Their potential is recognized, but the challenge for small island developing States is to ensure that they are used in a sustainable way for the well-being of present and future generations. Although they are afflicted by economic difficulties and confronted by development imperatives similar to those of

> developing countries generally, small island developing States also have their own peculiar vulnerabilities and characteristics, so that the difficulties they face in the pursuit of sustainable development are particularly severe and complex.
>
> (UN, 1994, para. 4)

The Annex prescribed specific actions that would enable SIDS to achieve sustainable development within the context laid out in Agenda 21, and set forth specific policies, actions and measures to be taken at the national, regional and international levels to achieve these goals. The Barbados Programme of Action is also notable because the first major section of the annex is focussed on climate change, stating:

> Small island developing States are particularly vulnerable to global climate change, climate variability and sealevel rise. As their population, agricultural land and infrastructure tend to be concentrated in the coastal zone, any rise in sealevel will have significant and profound effects on their economies and living conditions; the very survival of certain low-lying countries will be threatened. Inundation of outlying islands and loss of land above the high-tide mark may result in loss of exclusive economic rights over extensive areas and in the destruction of existing economic infrastructure as well as of existing human settlements. Global climate change may damage coral reefs, alter the distribution of zones of upwelling and affect both subsistence and commercial fisheries production. Furthermore, it may affect vegetation and saline intrusion may adversely affect freshwater resources. The increased frequency and intensity of the storm events that may result from climate change will also have profound effects on both the economies and the environments of small island developing States.
>
> (UN, 1994, para. 18)

The Annex therefore clearly lays out the challenge for SIDS within the UN Framework Convention for Climate Change, some 20 years before the Paris Agreement, and which remain absolutely relevant to the present day. Tourism is also given special attention in the Annex with its own section (Sec. VIII). The "Basis for Action" on tourism notes that the sector is "one of only a few development options for those small States, will continue to be very important for their future growth" (UN, 1994, para. 39). Although, it goes on to state, "However, if not properly planned and managed, tourism could significantly degrade the environment on which it is so dependent. The fragility and interdependence of coastal zones and the unspoilt areas on which eco-tourism depends calls for careful management" (UN, 1994, para. 39). Substantial attention is given in the section to ecotourism and cultural aspects of tourism and although national, regional and international actions suggested note the desire for tourism and sustainable development to be mutually supportive there is no specific suggestion of an interrelationship between tourism and climate change.

The second major period commenced with the Mauritius Strategy of 2005 (UN, 2005). The Mauritius Strategy was designed to encourage the continued implementation of the 1994 programme. As with the 1994 strategy, climate change and sea-level rise was the first section of the strategy, noting that "small island developing States believe that they are already experiencing major adverse effects of climate change. Adaptation to adverse impacts of climate change and sea-level rise remains a major priority for small island developing States" (UN, 2005, para. 16). Coastal and marine resources received considerably more coverage in the Mauritius Strategy although the tourism section was much smaller and covered only three paragraphs. There was no mention of ecotourism and no explicit connection of tourism development to climate change, although links were made to biodiversity conservation, instead, emphasis was placed more on sustainable tourism. However, again, there were warnings about potential impacts: "it is recognized that the sector is open to many exogenous shocks. In addition, if tourism is not developed sustainably, it can damage or even destroy the natural environment that attracts tourism in the first place" (UN, 2005, para. 50).

The third period is from 2014 and is marked by the SIDS Accelerated Modalities of Action (SAMOA) Pathway (UN, 2014). The Pathway reaffirms the special circumstances of SIDS and emphasises the risk of climate change even more by noting in the preamble:

> We recognize that sea-level rise and other adverse impacts of climate change continue to pose a significant risk to small island developing States and their efforts to achieve sustainable development and, for many, represent the gravest of threats to their survival and viability, including, for some, through the loss of territory.
>
> (UN, 2014, para. 11)

Also of significance with respect to the larger picture of sustainable development is that the Pathway acknowledges the emergence of the Sustainable Development Goals (SDGs), which were under development at the time in the suggestion, "support of the sustainable development of small island developing States would require appropriate consideration in the post-2015 development agenda" (UN, 2014, para. 13). However, while climate change is a major theme in the Pathway and an agreed priority area, it follows "Sustained and sustainable, inclusive and equitable economic growth with decent work", which was the first post-preamble section in the Pathway, reflecting a different set of priorities than previous SIDS conferences.

Tourism is also now included under the heading of "sustainable tourism" as a subsection of "Sustained and sustainable, inclusive and equitable economic growth with decent work" with sustainable tourism representing "an important driver of sustainable economic growth and decent job creation" (UN, 2014, para. 30), with states being encouraged to develop and implement "policies that promote responsive, responsible, resilient and sustainable tourism, inclusive of all peoples" (UN, 2014, para. 30(a)). Again, however, there is no indication of the interrelationships between tourism and climate change, although mention is made in the substantial section on Oceans and Seas. Sustainable transport, which is clearly of significance to SIDS given their relative isolation from major markets, only notes that SIDS "are committed to continuing and enhancing support for the efforts of small island developing States … To increase energy efficiency in the transport sector" (UN, 2014, para. 67). There is no statement with respect to emissions reduction, although sustainable tourism is mentioned under a heading regarding the promotion of sustainable consumption and production (UN, 2014, para. 69).

From SIDS to large ocean states

As noted in the previous section, thinking with respect to sustainable development context has not remained stagnant. While climate change is a consistent issue, the position of economic development, and tourism as a contributor, has shifted over time. Significantly, such changes are symptomatic of wider movements in the politics of the Global South. In the case of SIDS, this is represented in the growth of new regionalism (Hall & Amore, 2021). This situation is important from a tourism perspective because the governance of tourism relationships and connectivities cannot readily be seen as being isolated from broader issues of institutional architecture (Tarte, 2014). In the case of the Pacific, for example, tourism and aviation bodies are part of the Council of Regional Organisations in the Pacific that also includes the Pacific Islands Forum Secretariat (PIFS), the Secretariat of the Pacific Community (SPC), the University of the South Pacific (USP), the Secretariat of the Pacific Regional Environment Program (SPREP) and the Forum Fisheries Agency (FFA). Significantly, countries such as Fiji have argued for a new "island-centred" regionalism (Maclellan, 2015). As Roch Wamytan, a leader of the Kanak independence movement in New Caledonia states:

> All of us want a certain autonomy and to cease constantly being under the influence of the colonial powers … We want to have a space where we can talk amongst ourselves without each time having to refer to the big countries, each of which has their own interests.
>
> (Quoted in Maclellan, 2015, p. 4)

Assertions of a new Pacific regionalism can be seen in the strengthening of the Pacific Small Island Developing States (PSIDS) caucus at the United Nations, which has created a significant platform for Pacific states to promote their interests internationally. This is matched by the promotion of a new Framework on Pacific Regionalism by the Pacific Island Forum Secretariat (2014). Rather than a "sinking nation paradigm", as is often portrayed in Western media, Pacific island countries are increasingly placing the oceans at the centre of regional frameworks, and increasingly identifying themselves as "large ocean countries" (Maclellan, 2015) rather than small island states. For example, a Pacific Oceans Alliance was launched by the Forum Secretariat and Secretariat of the Pacific Regional

Environment Program in 2014 to address concerns over ocean conservation, climate change and sustainable development.

The rise of new regionalism can be regarded as the result of "a new regional political architecture" (Leslie & Wild, 2018, p. 20) that culminated with the establishment of the Framework for Pacific Regionalism in 2013 in which Pacific SIDS seek to collectively adhere to the SDGs and cooperate in responding to the impacts of climate change in the region. As a consequence, the emergence of the new regionalism means that there is a "growing interest in new paradigms suitable for small island developing states such as 'green growth', the blue (ocean) economy, and the traditional (village-centered) economy" (Maclellan, 2015, p. 5).

The new regionalism and the repositioning of SIDS as large ocean states is a significant change in the discourse surrounding SIDS, particularly with respect to climate change, fisheries policy and political sovereignty. The climate crisis lies at the heart of this change. For example, Prime Minister Bainimarama (2015) of Fiji stated, concerning Australia and New Zealand, traditionally dominant countries in Pacific relations, "As we see it, Australia and New Zealand have been put to the test on climate change and been found wanting … They simply do not represent our interests as we face this critical matter of survival". Yet, the rhetoric of large ocean states and greater demands for political sovereignty clearly represent a change in political positioning and, arguably, a rethinking of the resource make up of SIDS, the economic tensions surrounding sustainable development, and tourism within it, have not fundamentally changed (Hall, 2019). If anything, and as indicated in the SAMOA Pathway above, sustainable tourism is meant to contribute to economic growth without further exacerbating the environmental vulnerabilities of island states. As previously noted, the language of green growth is adopted in the SAMOA Pathway to stress the importance of improved efficiencies for sustainability, but no mention is made of the emissions resulting from international tourist travel to SIDS or of the stresses that tourism places on local resources such as biodiversity and water (Hall, 2015).

Balancing tourism and the effects of climate change presents a significant sustainable development policy challenge. Hall and Amore (2021) suggest that this appears to be managed under the notion that while tourism is a major source of emissions because of the aviation component, the overall low emissions of SIDS by international measures provide justification for further expansion. Nevertheless, the tensions created by the relationship between tourism and the climate crisis will clearly need to be managed in the long-term given the potential severity of the economic and environmental impacts on SIDS. As Maclellan (2015, p. 7), noted, "regional institutions will find it increasingly difficult to paper over contested visions for the future".

Conclusions

This chapter has noted the way in which the framing of sustainable development has shifted with respect to SIDS. Climate change has long been central to sustainable development thinking and policy, but the notion of crisis has also led to a change in the political positioning of SIDS. This arguably reflects Dornan, Morgan, Newton Cain and Tarte's (2018, p. 421) observation that "Climate-related challenges and responses will ... likely shape development policymaking in the coming decades". However, in the case of SIDS, the policy response also needs to consider the significance of the large ocean context and the coastal interface, i.e. the blue economy, and the role that this plays not only with tourism but with sustainable development overall in the Global South (Keen et al., 2018; Mallin, 2018).

Significantly, the promotion of a "sinking nation paradigm" has previously constrained the political vision of SIDS (Mallin, 2018), although this has now shifted. Instead, "adaptation" has permeated the economic development agenda, in which budgetary requirements and economic growth are prioritised. But this policy paradigm, in turn, has facilitated the emergence of highly environmentally problematic activities and has not served to resolve the basic conundrum of sustainable development with respect to the relationship between economic growth and the effects of global change. Tourism is inseparable from this debate (Hall, 2019). However, resolution in the foreseeable future seems unlikely given that the short-term economic vulnerability of SIDS is given priority over a longer-term environmental and climate disaster. Therefore, in the case of SIDS, political change has seen a fundamental reimagining of states and large ocean states with an associated shift in system thinking as to what constitutes an island state system, but this significant shift has not yet been accompanied by a change in economic thinking.

References

Alliance of Small Island States (AOSIS) (2009). *Declaration on Climate Change 2009*. Retrieved from http://www.aosis.info/documents/AOSISSummitDeclarationSept21FINAL.pdf

Bainimarama, V. (2015). *Building an Inclusive and Independent Institution for Pacific Islanders. Speech at Meeting on Agreement to Institutionalize PIDF*, 6 May 2015, Suva. Government of Fiji.

Bebbington, A. (1997). Social capital and rural intensification: Local organizations and Islands of sustainability in the rural Andes. *Geographical Journal*, *163*, 189–197.

Benhamou, Z. A., & Cassin, L. (2020). The impact of remittances on savings, capital and economic growth in small emerging countries. *Economic Modelling*. doi:10.1016/j.econmod.2020.02.019

Burnett, K., Endress, L., Ravago, M. L., Roumasset, J., & Wada, C. A. (2014). Islands of sustainability in time and space. *International Journal of Sustainable Society*, *6*(1–2), 9–27.

Chui, T. F. M., & Terry, J. P. (2013). Influence of sea-level rise on freshwater lenses of different atoll Island sizes and lens resilience to storm-induced salinization. *Journal of Hydrology*, *502*, 18–26.

Connell, J. (2018). Islands: Balancing development and sustainability. *Environmental Conservation*, *45*(2), 111–124.

Deschenes, P. J., & Chertow, M. (2004). An Island approach to industrial ecology: Towards sustainability in the Island context. *Journal of Environmental Planning and Management*, *47*, 201–217.

Dornan, M., Morgan, W., Newton Cain, T., & Tarte, S. (2018). What's in a term? "Green growth" and the "blue-green economy" in the Pacific Islands. *Asia & the Pacific Policy Studies*, *5*(3), 408–425.

Grydehøj, A., & Kelman, I. (2020). Reflections on conspicuous sustainability: Creating small Island dependent states (SIDS) through ostentatious development assistance (ODA)? *Geoforum*, *116*, 90–97.

Hall, C. M. (2010). Island destinations: A natural laboratory for tourism. *Asia Pacific Journal of Tourism Research*, *15*, 245–249.

Hall, C. M. (2015). Islands of sustainability or analogues of the challenge of sustainable development? In M. Redclift & D. Springett (Eds.), *Routledge International Handbook of Sustainable Development* (pp. 55–73). London: Routledge.

Hall, C. M. (2019). Constructing sustainable tourism development: The 2030 agenda and the managerial ecology of sustainable tourism. *Journal of Sustainable Tourism*, *27*(7), 1044–1060.

Hall, C. M., & Amore, A. (2021). Tourism and the new regional governance of Pacific large ocean states: From governance and development to metagovernance and resilience. In M. Stephenson (Ed.), *Routledge Handbook of Tourism and Small Island States in the Pacific*. London: Routledge.

Intergovernmental Panel on Climate Change (IPCC) (2013). Summary for policymakers. In Stocker, T. D., Qin, D., Plattner, G.-K., Tignor, M., Allen, S. K., Boschung, J., Nauels, A., Xia, Y., Bex, V., & Midgley, P. M. (Eds.), *Climate Change 2013: The Physical Science Basis*. Cambridge: Cambridge University Press, pp. 3–32.

Intergovernmental Panel on Climate Change (IPCC) (2014). Small Islands. In Stocker, T. D., Qin, D., Plattner, G.-K., Tignor, M., Allen, S. K., Boschung, J., Nauels, A., Xia, Y., Bex, V., & Midgley, P. M. (Eds.), *Climate Change 2014: Impacts, Adaptation, and Vulnerability. Working Group II Fifth Assessment Report*. Cambridge: Cambridge University Press, pp. 1613–1644.

Keen, M. R., Schwarz, A. M., & Wini-Simeon, L. (2018). Towards defining the blue economy: Practical lessons from Pacific Ocean governance. *Marine Policy*, *88*, 333–341.

Kerr, S. A. (2005). What is small Island sustainable development about? *Ocean & Coastal Management*, *48*, 503–524.

Leslie, H., & Wild, K. (2018). Post-hegemonic regionalism in Oceania: Examining the development potential of the new framework for Pacific regionalism. *The Pacific Review*, *31*(1), 20–37.

Maclellan, N. (2015). Transforming the regional architecture: New players and challenges for the Pacific Islands. *Asia-Pacific Issues, Analysis from the East-West Center. No. 118*, August.

Mallin, M. A. F. (2018). From sea-level rise to seabed grabbing: The political economy of climate change in Kiribati. *Marine Policy*, *97*, 244–252.

McGillivray, M., Naude, W., & Santos-Paulino, A. U. (2010). Vulnerability, trade, financial flows and state failure in small Island developing states. *Journal of Development Studies*, *46*, 815–827.

Moore, A. (2010). Climate changing small Islands: Considering social science and the production of Island vulnerability and opportunity. *Environment and Society: Advances in Research*, *1*(1), 116–131.

Nunn, P. D. (2013). The end of the Pacific? Effects of sea level rise on Pacific Island livelihoods. *Singapore Journal of Tropical Geography*, *34*, 143–171.

Pacific Island Forum Secretariat (PIFS) (2014). *The Framework for Pacific Regionalism*. Suva: PIFS.

Pelling, M., & Uitto, J. I. (2001). Small Island developing states: Natural disaster vulnerability and global change. *Global Environmental Change Part B: Environmental Hazards*, *3*(2), 49–62.

Péti, M. (2012). A territorial understanding of sustainability in public development. *Environmental Impact Assessment Review*, *32*(1), 61–73.

Pungetti, G. (2012). Islands, culture, landscape and seascape. *Journal of Marine and Island Cultures*, *1*(2), 51–54.

Tarte, S. (2014). Regionalism and changing regional order in the Pacific Islands. *Asia and the Pacific Policy Studies*, *1*(2), 312–324.

United Nations (1994). *Programme of Action for the Sustainable Development* of *Small Island Developing States (Barbados Programme of Action)*, A/CONF.167/9 – Report of the Global Conference on the Sustainable Development of SIDS. UN.

United Nations (2005). *Report of the International Meeting to Review the Implementation of the Programme of Action for the Sustainable Development of Small Island Developing States*, Port Louis, Mauritius, 1014 January 2005 (Mauritius Declaration). United Nations publication, Sales No. E.05.II.A.4 and Corrigendum, chap. I, Resolution 1, Annex II. UN.

United Nations (2014). *SIDS Accelerated Modalities of Action (SAMOA) Pathway*. UN General Assembly A/RES/69/15. UN.

United Nations Conference on Trade and Development (UNCTAD) (2012). *UNCTAD Handbook of Statistics 2012*. UNCTAD.

United Nations Department of Economic and Social Affairs (UN DESA) (2010). *Trends in Sustainable Development: Small Island Developing States*. UN DESA.

United Nations Development Programme (UNDP) (2013). *Human Development Report 2013. The Rise of the South: Human Progress in a Diverse World*. UNDP.

United Nations Development Programme (UNDP) (2019). *Human Development Report 2019. Beyond Income, Beyond Averages, Beyond Today: Inequalities in Human Development in the 21st Century*. UNDP.

UN Office of the High Representative for the Least Developed Countries, Landlocked Developing Countries and Small Island Developing States (UN-OHRLLS) (2015). *Small Island Developing States in Numbers Climate Change Edition 2015*. UN-OHRLLS.

UN Office of the High Representative for the Least Developed Countries, Landlocked Developing Countries and Small Island Developing States (UN-OHRLLS) (2018). *State of the Least Developed Countries 2018: Follow Up of the Implementation of The Istanbul Programme of Action for the Least Developed Countries*. UN.

UN Office of the High Representative for the Least Developed Countries, Landlocked Developing Countries and Small Island Developing States (UN-OHRLLS) (2020). Retrieved from https://www.un.org/ohrlls/content/list-sids

Wallner, H. P., & Narodoslawsky, M. (1996). Evolution of regional socio-economic systems toward "Islands of sustainability". *Journal of Environmental Systems*, *24*, 221–240.

Wallner, H. P., Narodoslawsky, M., & Moser, F. (1996). Islands of sustainability: A bottom-up approach towards sustainable development. *Environment and Planning. Part A*, *28*, 1763–1778.

7 The generative power of nurturing new connections

Indigenous tour operators learning across, learning deeply

Marnie Graham, Sandie Suchet-Pearson, Uncle Lexodious Dadd and Dirk Pienaar

Introduction

This chapter focuses on the generative power of nurturing new connections between Indigenous peoples and across Indigenous places within the Indigenous tourism sector. Indigenous peoples and their diverse experiences and relationships are largely absent within Indigenous tourism research and in tourism literature more broadly (Nielsen & Wilson, 2012). At the same time, tourism research more generally remains highly Eurocentric (Wijesinghe et al., 2019), and tourism knowledge production is persistently colonial (Chambers & Buzinde, 2015). These telling absences are indicative of deep-colonising legacies, whereby Indigenous peoples' presences, knowledges and agencies continue to be marginalised, alienated and denigrated in the academy and beyond (Rose, 1996a). In this chapter, we counter these realities to centre our experiences and relationships as Indigenous tourism operators to describe the powerful learning and sharing that emerged when we – an Indigenous tour operator from Darug Country in western Sydney, Australia, and a #Khomani San tourism officer from northern South Africa – came together on #Khomani San lands to share our places of care, our culture and our experiences of the tourism industry. As Indigenous leaders, custodians and tour operators, we write together here with non-Indigenous human geography researchers, Sandie Suchet-Pearson and Marnie Graham, in order to prioritise our Indigenous voices, aspirations and methodologies in tourism research.

To challenge the deep-colonising legacies of the tourism industry and tourism research, we make visible and celebrate the generative potential of interpersonal relationships and cultural exchange. Specifically, we explore what it means for Indigenous tour operators from opposite sides of the world – Australia and South Africa – to learn deeply from each other, and across these diverse contexts. We outline some important lessons for intercultural sharing and learning across diverse Indigenous tourism contexts, and advocate for tourism literature, research and practice to prioritise Indigenous operators and their voices, perspectives and priorities. We do this by first discussing the decolonising methodological framing for the research and then describing the way we worked together to enable it by utilising Indigenous methodologies. We then review the literature on Indigenous tourism and cross-cultural learning, before sharing the richness of the new connections our exchange nurtured. Finally, we conclude the chapter by highlighting some key recommendations for tourism research and practice from our particular generative exchange.

Working together

In January 2020, co-author Uncle Lex travelled from Darug Country in Sydney, Australia, to #Khomani San lands in the Kalahari Desert of northwestern South Africa to participate in a cross-cultural learning and sharing exchange with co-author Dirk Pienaar. Uncle Lex is an Indigenous Darug custodian from western Sydney who runs the Indigenous tourism business Burangilli Aboriginal Cultural Education Services, while Dirk is an Indigenous #Khomani San tour guide and Tourism Officer working around and within the Kgalagadi Transfrontier Park and surrounds in northern South Africa. Dirk and Uncle Lex were joined on this trip by non-Indigenous human geographers and co-authors Marnie Graham and Sandie Suchet-Pearson and their respective families.

For one week we stayed together at Erin Game Ranch near Askham and at Twee Rivieren within the Kgalagadi Transfrontier Park, South Africa. Erin Game Ranch is one of six farms returned to the #Khomani San community in 1999 as part of a successful land claim. The community also negotiated joint management arrangements and the return of a section of Kalahari Gemsbok National Park, now renamed Kgalagadi Transfrontier Park (#Khomani San, 2020). During our stay, Dirk and his #Khomani San colleagues led us on #Khomani San cultural education activities and walks around Erin Game Ranch, visits to local San communities in Askham and surrounds and drives around the Kgalagadi Transfrontier Park. This was an intensive and delightful week of learning and exchange, involving in-depth conversations, intimate sharing opportunities and time spent "deeply hanging out" in the phenomenally, stirringly beautiful Kalahari (see Figure 7.1).

Our approach to learning, sharing and co-authoring is guided by Indigenous research methodologies that prioritise Indigenous knowledges, experiences, worldviews and rights; all of which have been long denied by the academy (Chilisa, 2012; Kovach, 2011; Smith, 1999). Both theoretically and practically Indigenous geographies and methodologies aim to decolonise academic institutions and research practices by prioritising listening, learning and collaborating, with the goal of meeting Indigenous aspirations (Chilisa, 2012; Kovach, 2011; Smith, 1999; Tuck & Yang, 2012). In doing so, they support Indigenous communities in communicating and enacting their own futures (Howitt et al., 2009). Our research process therefore prioritises our – Uncle Lex and Dirk's – aspirations to meet each other, learn from and share our experiences as Indigenous tour operators. Our direct quotes are used throughout the paper to demonstrate the importance of connection, learning across different contexts, listening deeply and collaborating in our respective worldviews (see also Darug Ngurra et al., 2020a, 2020b).

Figure 7.1 Dirk Pienaar (second from left) sharing some of his knowledge with Uncle Lex (left), Marnie (middle), Sandie (taking the picture) and their families on Erin Game Ranch. Source: Authors.

Our approach builds on the work of other Indigenous–non-Indigenous collaborative teams working to decolonise academic practices (Austin et al., 2019; Burarrwanga et al., 2013; Darug Ngurra et al., 2019; Darug Ngurra et al., 2020a, 2020b, ; Woodward et al., 2020). This involves intercultural sharing processes that enable learning across diverse knowledges, backgrounds and experiences (Darug Ngurra et al., 2020b). Together, we chose to communicate with the core principles of kindness, humility, respect and leaving ego aside, so as to open the possibilities for learning and doing something new (Darug Ngurra et al., 2020b). Key to this methodology is respecting Indigenous leaders and cultural experts as research leaders, partners and co-producers of knowledges and outputs (Simpson, 2017; Smith, 1999; Tuck & Yang, 2014). Indeed, we write together as Indigenous tour operators and custodians (Uncle Lex and Dirk) and non-Indigenous researchers (Sandie and Marnie) in order to prioritise Indigenous tour operators and their experiences and knowledges within academic analyses of Indigenous tourism. These research processes and co-authorship align with Carr, Ruhanen and Whitford's (2016) assertion that Indigenous tour operators should be major contributors to and commentators on Indigenous tourism management, as well as with Wijesinghe et al. (2019), who identify and challenge the Eurocentric bias that continues to permeate much tourism research.

In attending to Indigenous geographies and the imperative to decolonise conventional academic practices our research together on ǂKhomani San lands predominantly utilised an Indigenous "yarning" methodology (Bessarab & Ng'andu, 2010; Mooney et al., 2018; Walker et al., 2014). To "yarn" is a "conversational process that involves the telling and sharing of stories and information" (Walker et al., 2014, p. 1217). Yarning involves deep respect, following protocol, ensuring cultural safety and enables deep sharing (Bessarab & Ng'andu, 2010; Mooney et al., 2018; Walker et al., 2014). We yarned together on the special homelands of the #Khomani San, where an important part of this process is being attentive to place – to the ways we are nurtured by these lands and the ways we sought to nurture them in return (Bawaka Country, Wright et al., 2015). This attentiveness means reading the signs and meanings in the landscape from our own Indigenous perspectives, not from those Western perspectives that seek to separate humans from nature and thinking from feeling. In doing so, we reflect on what matters to us as Indigenous tour operators.

To describe our deep learnings from this intercultural sharing, Uncle Lex's and Dirk's reflections are recorded through the co-authorship of this paper, informed by our notes from our yarning together in the Kgalagadi, one-on-one interviews and personal journal entries undertaken during and after the trip, and our deeply reflexive conversations in-person, via instant messaging and email about our experiences and the important learnings that emerged. Our analysis focuses on the generative power of new connections between people and place and the significance for Indigenous tour operators of learning across diverse contexts. That is, we focus on how this cultural exchange process inspired new connections between Uncle Lex and Dirk, their respective places of care and their cultures and communities, as well as enabled new reflections on their touristic offerings and their personal and community-based roles in the tourism industry. Through this generative potential, we conclude that these cross-cultural exchanges provide an important support process for Indigenous tour operators that has not been explored in the tourism literature.

Indigenous tourism and cross-cultural learning

Indigenous tourism initiatives and policies throughout the world have been the object of significant academic focus over the last three decades (see Butler & Hinch, 1996; Bunten & Graburn, 2018; Buultjens & Fuller, 2007; Carr et al., 2019; Ryan & Aicken, 2005; Whitford et al., 2017; Zeppel, 2006). Despite such intensive focus, Indigenous peoples and their voices remain limited in tourism research (Carr et al., 2016; Nielsen & Wilson, 2012; Stewart, 2008). Despite the work of the Bawaka Collective in emphasising the importance of self-determination in Indigenous tourism (Wright et al., 2009), and interrogating the transformative learning potential for visitors who undertake Indigenous tours (Bawaka Country, Wright et al., 2017; Lloyd et al., 2015), most of the research and literature on Indigenous tourism remains directed and written by non-Indigenous people. Indeed, in their extensive review of Indigenous tourism research, Nielsen and Wilson (2012, p. 1) state:

> although Indigenous tourism has been a topic of academic interest for over three decades, it is still predominantly driven by the needs and priorities of

> non-Indigenous people. Ultimately, it appears that Indigenous voice and presence as scholars and researchers within the Indigenous tourism literature remain elusive.

Alongside a lack of Indigenous presence, voices and priorities in tourism research, there is furthermore a paucity of culturally appropriate support for Indigenous tourism operators (Nielsen, 2010). In the Australian context, WAITOC (2018) identified the following:

> For a long time, Aboriginal tourism operators were not acknowledged as a segment of the tourism industry – these operators often struggled on alone without a support network. Information for visitors and for the travel industry about Aboriginal cultural experiences were fragmented and difficult to access.
>
> (WAITOC, 2018, p. 6).

Within the African context, tourism researchers find that access to formal tourism knowledge and networks are also often key barriers for Indigenous tour operators (Carlisle et al., 2013). While for Indigenous San communities specifically, Koot (2020) notes that state- and industry-led paternalism towards San still fundamentally underscores processes of tourism development on San lands. The Australian and South African tourism contexts are thus both suggestive of a lack of culturally appropriate support for Indigenous tourism operators.

In this chapter, we not only prioritise Indigenous presences, voices and research priorities, but also consider how opportunities to learn with and from each other present an important practice for nurturing culturally appropriate support for Indigenous tour operators. Certainly, this approach draws on the recognition that Indigenous tourism is widely acknowledged as offering opportunities for cultural exchange between tourists and operators (Hashimoto, 2015). However, this important learning opportunity has not been extended to Indigenous operators themselves. Black et al. (2019) argue tour guiding scholarship would benefit from researchers from different continents conducting collaborative cross-country and comparative studies, and we expand this approach further, centring the ways that Indigenous tour operators as custodians and knowledge-holders from cross-cultural contexts can learn deeply from each other. This centring is a key finding of this chapter.

Moreover, we situate our research within what Nielsen and Wilson (2012, p. 3) refer to as "Indigenous-driven research", whereby "Indigenous people driv[e] tourism research that facilitates their own needs and wants", focusing on "Indigenous self-determination through tourism; capacity-building through tourism and economic development; [and] Indigenist research methodologies". Indeed, this research process developed from the enthusiasm and determination of Uncle Lex to meet Dirk, see his places of care and learn and share together through cultural exchange, and Dirk's enthusiastic reception of this idea and the considerable planning effort he undertook for it to eventuate.

This cultural exchange trip was the result of personal connections and the relationships of deep trust that are at the heart of Indigenous research methodologies (Darug Ngurra et al., 2020a, 2020b). As research collaborators, friends and co-authors, we sustain strong connections to each other. Since 2016, Uncle Lex has led "culture camps" at Yarramundi Regional Park on Darug Country in Western Sydney in collaboration with Marnie and Sandie,[1] whereby visitors are invited to learn about and share in Darug culture (Darug Ngurra et al., 2019). Uncle Lex is also an Adjunct Fellow at Macquarie University, in recognition of his research and teaching contributions to the Department. Dirk and Sandie have known each other for seven years since Sandie spent time with Dirk in 2013 and 2017 learning about the #Khomani San's efforts in the tourism space – which built on research Sandie did in Southern Africa on Indigenous rights and wildlife management for her PhD in 1999.

Sandie and Marnie are non-Indigenous human geographers whose work focuses on Indigenous peoples' rights, knowledges and identities, including a particular analytical focus on Indigenous tourism (Bawaka Country et al., 2017). They have collaborated together in a supervisor–student relationship, respectively, for over ten years. Marnie has built strong connections with South Africa since 2009, through her PhD completed in 2015 and her postdoctoral research on Indigenous tourism and the role it plays in processes of reconciliation in both Australia and South Africa. Given our long and significant connections with each other, this learning and sharing exchange opportunity on #Khomani San lands in South Africa was a poignant experience for us all, and here we share our deep learnings with you. These learnings inspired a range of critically important connections and we share three of them now: (1) new connections to place and Country; (2) cross-cultural connections through sharing cultures and histories; and (3) generative connections to each other.

Inspiring new connections to place and Country

We share here stories of how this cultural exchange inspired new connections between Uncle Lex and the Kgalagadi and between Dirk and Darug Country. Indigenous tourism experiences involve Indigenous tour operators working hard to communicate to tourists a sense of connection to, and care for, place (Walker & Moscardo, 2016). For Uncle Lex and Dirk, however, these connections between Kgalagadi and Darug places are generated through their respective perceptions of the similarities (and differences) between #Khomani San and Darug places, worldviews and relationships.

When Uncle Lex, Marnie and Sandie first meet up with Dirk on Erin Game Ranch, we look out silently over the beautiful red sands and iridescent pink skies while Dirk explains with deep reverence: "I really love this place. This place, the red sand, is in my blood, I can feel it. Even when I leave here, I want to be back. It is always part of me". While many Western frameworks value nature as a utilitarian resource, fundamentally separate from humans (Adams & Mulligan, 2003), growing up in the #Khomani San community instilled a different understanding of land

and landscape in Dirk. He sees people as a natural part of the landscape, and always tries to search for and define the deeper connection to his immediate surroundings:

> "We as San people always feel that everything is part of its environment, especially in a natural context. It is thus very common that we feel more close, safe and part of a natural environment with which we associate often. For me, that place is this beautiful part of the Kalahari I call home".

These intimate relations between Dirk and the Kalahari are meaningful and familiar to Uncle Lex. He immediately transfers the Australian Aboriginal concept of "Country" to #Khomani San lands, in recognition of the familiar way that Dirk and his colleagues talk about and relate to their places of care. In Australian Aboriginal English, "Country" refers to the different territories of the more than 300 Indigenous Australian language groups, and incorporates complex relationships and ideas about Country as a sentient being, of which humans are an integral part:

> Country in Aboriginal English is not only a common noun but also a proper noun. People talk about country in the same way that they would talk about a person: they speak to country, sing to country, visit country, worry about country, feel sorry for country, and long for country. People say that country knows, hears, smells, takes notice, takes care, is sorry or happy ... [C]ountry is a living entity with a yesterday, today and tomorrow, with a consciousness, and a will toward life. Because of this richness, country is home, and peace; nourishment for body, mind, and spirit; heart's ease.
>
> (Rose, 1996b, p. 7)

In sharing these relationships with each other, Dirk recognises a familiarity with this sentient, living and inter-related Australian Aboriginal sense of Country; he relates:

> "Similar to Country, we Bushman see land as a bigger portion of the earth space divided into different territories. Remember: everything, including the land and humans are seen as one environment in the Bushman context. As nomadic descendants, our way of living was mostly influenced by the movement and characteristics of animals and the interrelationships of other organisms, and this relates with the concept of a defined territory. For San, however, territory was a temporary arrangement which was either defended or shared with the intruder".

For Uncle Lex, these similarities in San and Darug ways of knowing and talking about land/Country are also an embodied experience of calm familiarity. From before he even touched the ground in the Kalahari, Uncle Lex was inspired by the deep familiarities between Darug and #Khomani San places. This included a feeling of connection to all the living beings that inhabit this place:

> "Coming in to Upington was just amazing, because I saw these trees from the aeroplane, and I thought: "I know those trees". Then when I spoke to Dirk about the trees, I realised they are actually in a family of our trees in Australia. So I felt very comfortable, and at home, and at peace, and it wasn't as foreign as I thought it was going to be [see Figures 7.2 and 7.3]. The soil here looks similar. I heard birds the other night at Erin Game Ranch that were the same as our birds. I'm not saying anything's the same, but it's familiar. Lots of things are familiar."

However, the differences Uncle Lex experienced on ‡Khomani San lands were also awe-inspiring and deeply embodied. He was delighted by so much of the Kgalagadi's animal life and the connections he felt there (see Figure 7.4):

> "I've also seen things that are just blowing me away, that I haven't seen before. Like today on the game drive seeing lions, ground squirrels, so much animal life compared to Australia, and in such arid country. I've got about 30 photos of the yellow mongoose hanging around me, leaning up to me like this, and looking into my eyes. It was so special."

These reflections demonstrate how experiences of the unknown can nonetheless offer deep feelings of connection (Bawaka Country et al., 2017). Dirk reflects on Uncle Lex's emotional, embodied connections with the Kalahari's animal life:

Figure 7.2 Uncle Lex checking out the trees at Erin Game Ranch for which he felt such familiarity. Source: Authors.

Figure 7.3 #Khomani San tour guide Xopan and Uncle Lex discussing the Kalahari trees that Uncle Lex felt so connected to. Source: Authors.

Figure 7.4 Uncle Lex made a particular new friend in a yellow mongoose. This connection was a highlight of his trip. Source: Authors.

> "A special moment for me was when Uncle Lex saw his first lion. The awe and absolute respect that showed on his face was beautiful and with all the other lions we encountered his expression remain the same. With these encounters I could feel the undertone of emotional similarities from both San and Darug cultures and their relationships with the natural environment. This makes our connections stronger to each other's cultural and natural environments."

And while Dirk has not visited Australia – yet – he understands Uncle Lex's experience of feeling an immediate connection to the Kalahari:

> "I have similar experiences with different places and for different reasons. I particularly felt this connection when I presented and participated in the Festival of Images and Land Claims at the University of Fraser Valley in Canada 2014. When our group visited the Stalo Nation we went on a tour with a community elder. The elder took us to a creek and explained the ritual for salmon fishing. What connected me at the time was the elements of his interpretation to us. It was not only the vivid passion that captured me, but also how he made the environment come alive in my imagination. Not only was his interpretation of the natural environment similar to San, but he gave identity to the rocks, waves, trees and I felt the communication with the water resonated with me."

Dirk's experience in Canada and Uncle Lex's experience in the Kalahari reiterate our assertion of the generative possibilities emergent when Indigenous tour operators are able to share their places and tourism initiatives with each other. These possibilities include enabling diverse Indigenous places to learn from and with each other through the sharing of custodians' knowledges.

Inspiring cross-cultural connections

For Uncle Lex, this sense of familiarity and connection with the Kalahari also extended to #Khomani San culture and the shared experiences of colonial dispossession and devastation, despite the very different contexts and processes through which colonisation was enacted:

> "So many cultural things are so similar if not exactly the same, including experiences of colonisation, loss, the hardships. We've brought up a lot of history, talking about the ways we do things, the way we connect with Country, and it's very similar, if not the same. Then there are those important connections that, from First Nations people, even though we're worlds apart in a modern sense, we're doing similar things, and trying to care for Country on opposite sides of the world. So, I feel like I've been very fortunate to come over here, and make this connection. It's been very, very important. Even connections with the animals, the plants. (Figure 7.5 shows Uncle Lex, Dirk and #Khomani San tour guide Xopan sharing and learning together at Erin Game Ranch".)

Figure 7.5 Uncle Lex, #Khomani San tour guide Xopan and Dirk connect in the shade of a shepherd tree, Erin Game Ranch. Source: Authors.

For Dirk, there is a reciprocal feeling of connection to Indigenous Australians, both through meeting Uncle Lex and through his San colleagues' connections to Australia:

> "My first ever invitation to travel abroad was an opportunity to visit Australia for a conservation conference. I could not go unfortunately because I was studying full time. Two of my colleagues went to Australia for the World Indigenous Network Conference in 2013. They had a very rich experience with the Aboriginals around the Kakadu National Park. They both made lifelong friendships and connections which they still reflect on regularly. The most beautiful part for me is that both my colleagues felt that the San and Aboriginals from Australia are the same nation from two different countries. They both never used the word similar when they talk of the Aboriginal culture but always the words "the same"."

These connections to Australian Aboriginal culture for Dirk are also linked to the historical connections between the two continents of Australia and Africa:

> "Something that is also important for me is the continental connection which we shared with Australia before the break of Gondwanaland. This historical connection is of great conservation importance for me to experience personally."

Uncle Lex similarly notes:

> "We even spoke on a deeper level about how Australia was connected to Gondwana, and I know my Old People [ancestors] believe that everything comes from Australia."

For Uncle Lex, this new connection to ǂKhomani San Country and culture enabled him to gain an important new perspective on and connection to his own Darug Country as well as his Indigenous tour offerings at Yarramundi:

> "I sent a message from South Africa back to my wife in Australia and said 'I'm going to be coming home with new eyes'. I'm going to have far more appreciation - for our plants, our animals, and Country. I already had appreciation for it, but I'm going to have a different appreciation, and a deeper appreciation. I know what plants can heal us, what we use those plants for, et cetera. But I don't know all their names; I don't know all the names of our weeds. I learned so much from Dirk about what we could do. Imagine if we had those beautiful tents set up like at Erin Game Ranch!! [See Figure 7.6.] They were done so simply and they just fit in with the environment so beautifully, just imagine if we had that at Yarramundi, Marnie!! And the game hides they had! [See Figure 7.7.] I definitely want to build one here at Yarramundi, up in the bush. It was so cool, just watching animals do their own thing without interrupting them. We've gotta do it! I'm just inspired, and I want to do more."

Figure 7.6 Erin Game Ranch, safari tents. Source: Authors.

Figure 7.7 Looking out from the Hide at Erin Game Ranch towards the waterhole.

In addition to new connections between Countries and cultures, we now share how the exchange inspired new connections to each other.

Inspiring new connections to each other

Dirk is deeply involved in tourism and conservation in the Kalahari. He works hard to liaise with various organisations and government departments including the South African national parks authority SANParks, the National Department of Tourism (NDT), the Department of Economic Development and Tourism (DEDAT), the Northern Cape Tourism Authority (NCTA) and the Khomani San Communal Property Association (CPA) as the local stakeholder organisation. Through his work, in 2017, Dirk also represented the community at the 41st Session of the World Heritage Site Meeting in Krakow, Poland, to ensure the successful inscription of the #Khomani San Cultural Landscape as a UNESCO World Heritage Site (see Figure 7.8). Dirk and his colleagues work tirelessly and have a huge impact on the practices and outcomes of #Khomani San tourism and conservation in the Kalahari.

For Uncle Lex, this new experience of travel, sensing the familiarities and differences between home and the Kgalagadi, and learning about Dirk's work enabled a feeling of deep connection between him and Dirk and his #Khomani

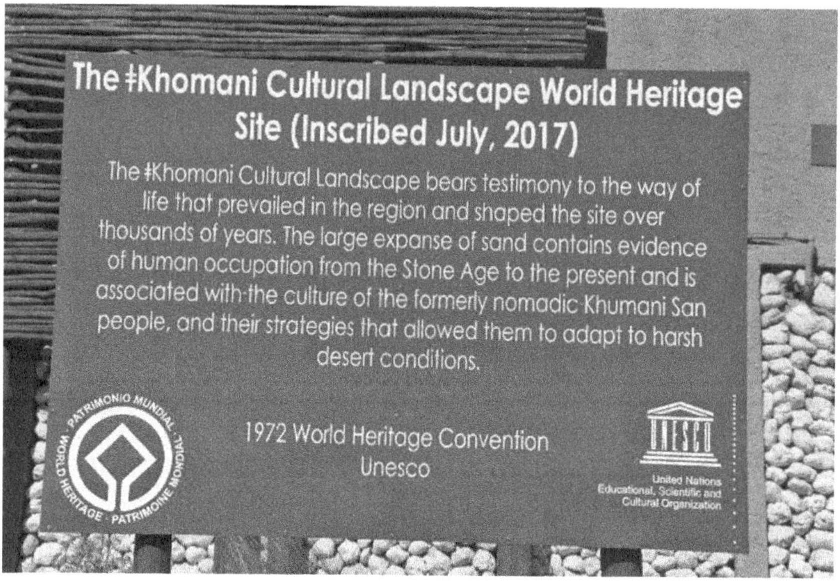

Figure 7.8 In 2017, #Khomani San efforts for recognition paid off with World Heritage listing of the #Khomani Cultural Landscape. Source: Authors.

San colleagues Xopan and Honeyboy. Indeed, the interpersonal relationships of connection and trust inspired feelings of admiration, reciprocity and solidarity. Uncle Lex elaborates:

> "Speaking with Dirk has been very important, to see what he's doing with conservation, and how he's rallying a lot of his people and so forth to look after Country, and to do the right thing. He is very inspiring. He is not just committed to tourism for his own self-indulgence, it's a commitment to his whole mob, his whole people, and to the Country, which is very inspiring. And I've been fortunate enough to spend, what? Only four or five days with him? But we've formed a really good relationship and we've got a really strong connection now that we're going to keep. We've made plans that he's going to help me with doing stuff on our own Country, but also I'll be able to help with stuff that he's doing here on his Country."

In the spirit of sharing and reciprocity in which Darug relationships are embedded, Uncle Lex, Sandie and Marnie have started planning and developing possibilities to bring Dirk and his #Khomani San colleagues to Australia to experience Uncle Lex's Country and his cultural camps, as well as those of other Indigenous tour operators within our Australian research networks. Uncle Lex, Sandie and Marnie want to extend to their #Khomani San colleagues the same generosity they were shown in South Africa, but also believe such an exchange will offer new insights of learning and sharing for the benefit not only of Darug and ǂKhomani San tour operators but will also support the Indigenous tourism industry more broadly by offering further suggestions on the generative importance of cross-cultural mentoring and learning. While we all wait for this opportunity to materialise, we have been utilising WhatsApp communication to stay in touch – discussing and co-authoring this paper as well as supporting each other through the COVID-19 crisis.

Learning and reflecting on the tourism industry

Uncle Lex had not really been "a tourist" before his trip to ǂKhomani San lands. Besides visiting family in New Zealand, where his wife grew up, he had not visited any other countries nor engaged in many touristic experiences, either in Australia or elsewhere. For Uncle Lex, this opportunity to travel to South Africa, meet Dirk and his colleagues and experience ǂKhomani San touristic offerings was a unique experience in "becoming a tourist". Becoming a tourist on ǂKhomani San lands offered distinctive lessons about the tourism industry, including deep reflections on the impacts of capitalist systems and lateral violence on Indigenous communities, as well as the importance of participatory tourism in supporting the intergenerational transfer of knowledge.

Capitalist imperatives and lateral violence

One of the starkest lessons for Uncle Lex was learning about the lived realities of ǂKhomani San and the limited benefits they receive from the tourism sector:

> "Reflecting on how much we "get" in Australia, and how much our dollar is worth, to see how poorly some ǂKhomani San people live, and yet they're happy – I see we just take so many things for granted in Australia. And we have so much. And we think we need so much. Even on our culture camps. I think even on our camps we live better than some people live in their everyday life in the Kalahari."

For Uncle Lex, this realisation led to reflections on the ways that Indigenous tourism is embedded within capitalist economic systems (Cave & Dredge, 2020; Higgins-Desbiolles et al., 2019), and that this reality does not always inspire solidarity within Indigenous communities:

> "Within all our Indigenous communities, whether it's in South Africa or in Australia, we fight. Dirk calls it "in-fighting" and in Australia we call it "lateral violence" (Atkinson, 2002). We have much conflict within our communities because we've come from a traditional point of view where we don't need money and competition, into trying to compete in a white man's world that is based on competition. And so people are trying to outdo each other, and there's a lot of competition, et cetera. So, within that capitalist framework,

> we're still finding our feet, and finding our way. And there are people in my community who whinge, and moan: "I want this, and I want that!" No, I'm going to go home and work even more on reducing lateral violence through sharing more of our traditional culture with my people and with non-Indigenous people; and even just taking a look at my material objects and going "I don't need that. Going to get rid of that"."

Certainly, academic attention is paid towards #Khomani San identities, identity politics and intra- and inter-community "in-fighting" in South Africa (Koot, 2016, 2020; Robins, 2008). This includes analyses of the multi-scalar politics of claiming Indigeneity in South Africa, entangled in questions of land rights, cultural revival and (in)authenticity, and the contentious possibilities for economic development in the impoverished Kalahari, of which tourism remains the primary economic focus (Koot, 2016, 2020; Robins, 2008). These analyses highlight the internal politics between San and other rural communities who share kinship ties. In Australia, the politics of claiming Indigeneity are also mired in debilitating "lateral violence"; this persistent feature of colonialism sees Indigenous people and communities pitted against each other, engendering deeply fractious and damaging behaviours and attitudes (Atkinson, 2002). In both contexts, Australia and South Africa, these are troubling relations. Yet, as argued in this chapter, Indigenous tourism – while certainly not a panacea for the many problems facing Indigenous communities and the deeply paternalistic context of tourism development (Koot, 2020) – is being harnessed by Dirk and Uncle Lex to enable opportunities for prioritising coming together to share culture and knowledge.

Participatory tourism

In response to #Khomani San community needs and relations, in the last three years, Dirk has worked deliberately to reframe San tourism towards "participatory tourism". From an academic perspective, participatory tourism has been discussed for over three decades and focuses on the ways that local communities and "stakeholders" are involved in processes of tourism development and planning (see Dragouni et al., 2018, for a comprehensive review). Dirk, however, embraces participatory tourism in order to create awareness that tourism as a participatory activity involves a willingness to share and learn from one another – tourists, guides and local communities alike – and especially for youth, and prioritises appropriate cultural knowledge transfer through this sharing. Dirk says:

> "A major concern of all Indigenous people in the world is the oral history element and how it is transferred to the younger generations in today's living context of traditional lifestyles. This is where tourism can continue to play a very critical role. Tourism enables me as an operator in the Kalahari to continue to transfer traditional knowledge and history to my children and to Indigenous and non-Indigenous youth through the various tourism activities. I have started to use the concept of participatory tourism because tourism

for me creates a perfect platform to enhance knowledge transfer for all participants. City parents share bits and pieces of their methods of doing things in the past to their children because the local custodians guide them in that direction. This is valuable knowledge transfer and it is worth mentioning in academic works."

Uncle Lex likewise runs cultural camps at Yarramundi specifically for sharing traditional culture with Indigenous Sydneysiders, including adults and children, many of whom have not had opportunities to receive such cultural learning. Inspired by his trip to the Kalahari, he has also extended this cultural sharing to running a weekly radio programme *Budyari Baya – Good Talk* – whereby he shares Darug culture, language and stories with Indigenous and non-Indigenous listeners. Thus, both Uncle Lex and Dirk are seeking to utilise tourism as a positive force within their respective communities and learning and sharing from each other has been a generative force in inspiring further collaboration and commitment.

Conclusion and recommendations

The highly Eurocentric nature of tourism research (Wijesinghe et al., 2019), the persistent absence of Indigenous voices and imperatives (Nielsen & Wilson, 2012) and the continuation of persistently colonial tourism knowledge production (Chambers & Buzinde, 2015) reflect the continued marginalisation, alienation and denigration of Indigenous knowledges and perspectives in the academy and beyond (Rose, 1996a). Certainly, we see these silences reinforce colonising processes of marginalisation and alienation because they do not speak to the heart of Indigenous tourism and what we try to do in our own tours and businesses. We, of course, want to earn money and have sustainable jobs to support ourselves, our families and our broader communities. But we also do our important work so that we can teach the world about our fundamentally important cultures, our histories, our stories, our languages, our connections to Country – which have stood the test of time, including, above all odds, the ravages of colonisation and apartheid. We continue to work in the tourism space to show our kids how proud we are of these things and hope to inspire them to continue to regenerate our culture and our places of care in this so-called "post-colonial" world (Darug Ngurra et al., 2020b) (see Figure 7.9). We do it because we know that some important connections have been disrupted, and we want to show people how to find and heal those connections again. And we do it so that "participatory tourism" can help us change how our communities, both near and far, Indigenous and non-Indigenous, can re-conceptualise tourism as a space for intimate and deep sharing and learning.

Little is written about these priorities in the academic literature, and when Indigenous tourism is of analytical focus, our Indigenous presence and voices go largely unseen and unheard (Carr et al., 2016; Nielsen & Wilson, 2012; Stewart, 2008). What we communicate here is the importance of these priorities, the deep learning and inspiring new connections we have encountered by meeting across

Figure 7.9 Sandie's, Marnie's and Dirk's kids learning together in the red sands of Erin Game Ranch. Source: Authors.

time, geography and space, intimately sharing our places of care, our cultures and our histories, and together finding new inspirations and new connections to each other and to Country near and far.

This chapter emphasised the importance of Indigenous tour operators learning and sharing together and recommends tourism research, practice and policy increase their support of opportunities for Indigenous people engaged in tourism throughout the world to connect in new ways. Underlying this support needs to be a decolonising imperative in which Indigenous people – researchers and tour operators – actively and directly shape research, practice and policy so that their perspectives, needs and knowledges are prioritised (Carr et al., 2019) and so that tourism shifts away from its deep-colonising legacies (Chambers & Buzinde, 2015).

The inspirations and connections shared here are generative of new ideas and approaches and provide cross-cultural learning that cannot be gained elsewhere. Being on-Country, forming deep relationships of care and connection, and deeply hanging out can enable Indigenous tour operators to gain new insights not only on our colleagues' places of care, but also on our own places, our own peoples, our own priorities and aspirations, and our own touristic activities. These are invaluable learning opportunities, and we cannot underestimate their impact on our lives. Indeed, as Uncle Lex summarises:

"That's the beauty about Xopan, Honeyboy and Dirk, that they were open to that sharing. And that's why every night I sat up with Dirk till the early hours of the morning, and we've been talking culture, and philosophies, and the way our country is and the challenges we have with governments, and legislations, and all those things, and he's taught me many, many things."

Note

1 Along with other central collaborators including Dr Paul Hodge (University of Newcastle, Australia) and Paul Glass (Blue Mountains Aboriginal Culture and Resource Centre).

References

Adams, W. M., & Mulligan, M. (2003). *Decolonizing Nature: Strategies for Conservation in a Post-Colonial Era*. London: Earthscan.

Atkinson, J. (2002). *Trauma Trails, Recreating Song Lines: The Transgenerational Effects of Trauma in Indigenous Australia*. Melbourne: Spinifex Press.

Austin, B. J., Robinson, C. J., Mathews, D., Oades, D., Wiggin, A., Dobbs, R. J., Lincoln, G., & Garnett, S. T. (2019). An indigenous-led approach for regional knowledge partnerships in the Kimberley Region of Australia. *Human Ecology*, *47*, 577–588. doi:10.1007/s10745-019-00085-9

Bawaka Country, Wright, S., Suchet-Pearson, S., Lloyd, K., Burarrwanga, L., Ganambarr, R., Ganambarr-Stubbs, M., Ganambarr, B., & Maymuru, D. (2015). Working with and learning from country: Decentring human authority. *Cultural Geographies*, *22*(2), 269–283. doi:10.1177/1474474014539248

Bawaka Country, Wright, S., Lloyd, K., Suchet-Pearson, S., Burarrwanga, L., Ganambarr, R., Ganambarr, M., Ganambarr, B., Maymuru, D., & Tofa, M. (2017). Meaningful tourist transformations with country at Bawaka, North East Arnhem Land, Northern Australia. *Tourist Studies*, *17*(4), 443–467. doi:10.1177/1468797616682134

Bessarab, D., & Ng'andu, B. (2010). Yarning about yarning as a legitimate method in Indigenous research. *International Journal of Critical Indigenous Studies*, *3*(1), 37–50. doi:10.5204/ijcis.v3i1.57

Black, R., Weiler, B., & Chen, H. (2019). Exploring theoretical engagement in empirical tour guiding research and scholarship 1980–2016: A critical review. *Scandinavian Journal of Hospitality and Tourism*, *19*(1), 95–113. doi:10.1080/15022250.2018.1493396

Bunten, A. C., & Graburn, N. (Eds.) (2018). *Indigenous Tourism Movements*. Toronto: University of Toronto Press.

Burarrwanga, L., Ganambarr, R., Ganambarr-Stubbs, M., Ganambarr, B., Maymuru, D., Wright, S., Suchet-Pearson, S., & Lloyd, K. (2013). *Welcome to My Country*. Allen & Unwin, Crows Nest, NSW.

Butler, R., & Hinch, T. (Eds.) (1996). *Tourism and Indigenous Peoples*. London: International Thomson Business Press.

Buultjens, J., & Fuller, D. (Eds.) (2007). *Striving for Sustainability: Case Studies in Indigenous Tourism*. Lismore: Southern Cross University Press.

Carlisle, S., Kunc, M., Jones, E., & Tiffin, S. (2013). Supporting innovation for tourism development through multi-stakeholder approaches: Experiences from Africa. *Tourism Management*, *35*, 59–69. doi:10.1016/j.tourman.2012.05.010

Carr, A., Ruhanen, L., & Whitford, M. (2016). Indigenous peoples and tourism: The challenges and opportunities for sustainable tourism. *Journal of Sustainable Tourism, 24*(8), 1067–1079. doi:10.1080/09669582.2016.1206112
Carr, A., Ruhanen, L., Whitford, M., & Lane, B. (Eds.) (2019). *Sustainable Tourism and Indigenous Peoples*. London: Routledge.
Cave, J., & Dredge, D. (2020). Regenerative tourism needs diverse economic practices. *Tourism Geographies, 22*(3), 503–513. doi:10.1080/14616688.2020.1768434
Chambers, D., & Buzinde, C. (2015). Tourism and decolonisation: Locating research and self. *Annals of Tourism Research, 51*(1), 1–16. doi:10.1016/j.annals.2014.12.002
Chilisa, B. (2012). *Indigenous Research Methodologies*. Thousand Oaks, CA: SAGE.
Darug Ngurra, Dadd, L., Glass, P., Scott, R., Graham, M., Judge, S., Hodge, P., & Suchet-Pearson, S. (2019). *Yanama budyari gumada*: Reframing the urban to care as Darug Country in Western Sydney. *Australian Geographer, 50*(3), 279–293. doi:10.1080/00049182.2019.1601150
Darug Ngurra, Dadd, L., Glass, P., Norman-Dadd, C., Hodge, P., Suchet-Pearson, S., Graham, M., Judge, S., Scott, R., & Lemire, J. (2020a). Yanama Budyari Gumada, walk with Good Spirit as *Method*: Co-creating local environmental stewards on/with/as Darug Ngurra. In A. Campbell, M. Duffy, & B. Edmondson (Eds.), *Located Research: Regional Places, Transitions, Challenges*. Singapore: Palgrave, 15–37.
Darug Ngurra, Dadd, L., Norman-Dadd, C., Glass, P., Hodge, P., Suchet-Pearson, S., Graham, M., Scott, R., Lemire, J., & Narwal, H. (2020b). Case study: Walking with Good Spirit at Yarramundi. In E. Woodward, R. Hill, & R. Archer (Eds.), *Our Knowledge Our Ways in Caring for Country: Guidelines*. Darwin: Northern Australia Environmental Resources Hub, National Environmental Science Programme. NAILSMA and CSIO, pp. 55–56.
Dragouni, M., Fouseki, K., & Georgantzis, N. (2018). Community participation in heritage tourism planning: Is it too much to ask? *Journal of Sustainable Tourism, 26*(5), doi:10.1080/09669582.2017.1404606
Hashimoto, A. (2015). Tourism and social development. In R. Sharpley & D. J. Telfer (Eds.), *Tourism and Development: Concepts and Issues* (pp. 205–236). Bristol: Channel View Publications.
Higgins-Desbiolles, F., Carnicelli, S., Krolikowski, C., Wijesinghe, G., & Boluk, K. (2019). Degrowing tourism: Rethinking tourism. *Journal of Sustainable Tourism, 27*(12), 1926–1944. doi:10.1080/09669582.2019.1601732
Howitt, R., Muller, S., & Suchet-Pearson, S. (2009). Indigenous geographies. In N. Thrift & R. Kitchen (Eds.), *International Encyclopedia of Human Geography*. Amsterdam: Elsevier.
#Khomani San (2020). *#Khomani San – Erin Game Ranch*. Retrieved from http://www.khomanisan.com/erin-game-ranch/
Koot, S. P. (2016). Contradictions of capitalism in the South African Kalahari: Indigenous Bushmen, their brand and baasskap in tourism. *Journal of Sustainable Tourism, 24*(8–9). doi:10.1080/09669582.2016.1158825
Koot, S. P. (2020). Articulations of inferiority: From pre-colonial to post-colonial paternalism in tourism and development among the indigenous Bushmen of Southern Africa. *History & Anthropology*. doi:10.1080/02757206.2020.1830387
Kovach, M. E. (2011). *Indigenous Methodologies*. Toronto: University of Toronto Press.
Lloyd, K., Suchet-Pearson, S., Wright, S., Tofa, M., Rowland, C., Burarrwanga, L., Ganambarr, R., Ganambarr, M., Ganambarr, B., & Maymuru, D. (2015). Transforming tourists and "culturalising commerce": Indigenous tourism at Bawaka in Northern

Australia. *International Indigenous Policy Journal*, 6(4), 1–21. doi:10.18584/iipj.2015.6.4.6

Mooney, J., Riley, L., & Blacklock, F. (2018). Yarning up: Stories of challenges and success. *Australian Journal of Education*, *62*(3), 266–275. doi:10.1177/0004944118803403

Nielsen, N. R. (2010). *Strengths, Support and Self-Determination: Indigenous Tourism Planning and the Biamie Dreaming Cooperative*. PhD Thesis. Lismore, Southern Cross University.

Nielsen, N. R., & Wilson, H. (2012). From invisible to indigenous-driven: A critical typology of research in indigenous tourism. *Journal of Hospitality and Tourism Management*, *19*(1), 67–75. doi:10.1017/jht.2012.6

Robins, S. L. (2008). *From Revolution to Rights in South Africa: Social Movements, NGOs and Popular Politics After Apartheid*. Pietermaritzburg: University of Kwa-Zulu Natal Press.

Rose, D. B. (1996a). Land rights and deep colonising: The erasure of women. *Aboriginal Law Bulletin*, *3*(85), 6–13.

Rose, D. B. (1996b). *Nourishing Terrains: Australian Aboriginal Views of Landscape and Wilderness*. Canberra. Canberra: Australian Heritage Commission.

Ryan, C., & Aicken, M. (2005). *Indigenous Tourism: The Commodification and Management of Culture*. Oxford: Elsevier.

Simpson, L. B. (2017). *As We Have Always Done: Indigenous Freedom through Radical Resistance*. Minneapolis: University of Minnesota Press.

Smith, L T. (1999). *Decolonizing Methodologies*. London: Zed Books.

Stewart, E. (2008). Book review: Tourism and Indigenous peoples: Issues and implications, Richard Butler and Tom Hinch (Eds.). *Tourism Management*, *29*(6), 1238–1246. doi:10.1016/j.tourman.2008.02.015

Tuck, E., & Yang, K. W. (2012). Decolonization is not a metaphor. *Decolonization: Indigeneity, Education & Society*, *1*(1), 1–40.

Tuck, E., & Yang, K. W. (2014). Unbecoming claims: Pedagogies of refusal in qualitative research. *Qualitative Inquiry*, *20*(6), 811–818. doi:10.1177/1077800414530265

WAITOC [Western Australia Indigenous Tourism Operator's Council] (2018). *WAITOC Strategic Plan 2019–2025*. Perth: WAITOC. Retrieved from https://www.waitoc.com/about-us/corporate/our-documents

Walker, K., & Moscardo, G. (2016). Moving beyond sense of place to care of place: The role of indigenous values and interpretation in promoting transformative change in tourists' place images and personal values. *Journal of Sustainable Tourism*, *24*(8–9), 1243–1261. doi:10.1080/09669582.2016.1177064

Walker, M., Fredericks, B., Mills, K., & Andersen, D. (2014). "Yarning" as a method for community-based health research with indigenous women: The indigenous women's wellness research program. *Health Care for Women International*, *35*, 1216–1226.

Whitford, M., Ruhanen, L., & Carr, A. (Eds.) (2017). *Indigenous Tourism: Cases from Australia and New Zealand*. Oxford: Goodfellow Publishers.

Wijesinghe, S. N., Mura, P., & Culala, H. J. (2019). Eurocentrism, capitalism and tourism knowledge. *Tourism Management*, *70*, 178–187. doi:10.1016/j.tourman.2018.07.016

Woodward, E., Hill, R., & Archer, R. (Eds.) (2020). *Our Knowledge Our Ways in Caring for Country: Guidelines*. Northern Australia Environmental Resources Hub, National Environmental Science Programme. NAILSMA and CSIRO, Darwin. Retrieved from https://www.csiro.au/en/Research/LWF/Areas/Pathways/Sustainable-Indigenous/Our-Knowledge-Our-Way/OKOW-resources

Wright, S., Suchet-Pearson, S., Lloyd, K., Burarrwanga, L., & Burarrwanga, D. (2009). "That means the fish are fat": Sharing experiences of animals through indigenous-owned tourism. *Current Issues in Tourism*, *12*(5), 505–527. doi:10.1080/13683500903042907

Zeppel, H. D. (2006). *Indigenous Ecotourism: Sustainable Development and Management*. Wallingford: CABI.

8 Roots tourism and the Year of Return campaign in Ghana

Moving belonging beyond the history of slavery

Matthias Gebauer and Marie Umscheid

Introduction

Diasporic consciousness, the condition of people living in one place while longing for home in another (Gilroy, 1994), can without a doubt be ascribed to those people of African descent whose ancestors were forcibly moved across the Atlantic Ocean as slaves on the Middle Passage. At the same time, reaching out for Africa – the strongly mythologised place of return called home – is an elusive undertaking as the specific places of origin of each person are in most cases almost impossible to pinpoint. It is exactly in this context where diaspora tourism with its specific form of roots tourism for the African diaspora offers the experience of reconnecting with one's African indigeneity at specific tourist destinations, especially in West Africa. Since the early 1990s, roots tourism from the Americas to African countries has evolved as an important field of economic revenue within which aspects of tourism experience, diasporic longing and desires of migration "back to Africa" intersect (Araujo, 2018; Handley, 2006). Countries such as Ghana focus on various tourism-oriented campaigns in order to address people of African descent, especially in the United States, thus steering the longing for roots and belonging towards the West African coast by offering a narrative of home and reconnection to an imaginary global African family. The overall success of these tourism-related economic development politics is very much linked to the fact that the roots narrative is made accessible through touristic consumption.

This chapter looks into Ghana's 2019 Year of Return campaign by analysing the official programme and presentation of the year-long tourism programme in official documents, speeches, newspaper reports and social media. We thereby differentiate between (a) the strategies of roots tourism and development and (b) the narratives of African diasporic consciousness and home that are told by the campaign. We argue that roots tourism is facilitated by Ghana within a mode that can be described as a post-modern eventification of the history of slavery that is built around a concept of making the imagined motherland Africa a tourist experience within which the peer group becomes an actor of the tourism performance. The overall goal is economic development through world-class tourism while at the same time stabilising the roots, heritage and kinship narrative of Ghanaian diaspora tourism. Beyond that, the Year of Return campaign reflects a crucial change within the strategies and narratives of commodification as the diasporic longing for roots and pan-Africanism shifts from its monolithic connection to the history of slavery to a more positive, event-oriented and less politicised tourist performance.

Roots tourism in Ghana: Common narratives, conflicting imaginations and the commodification of Africanness

Ghana has a long history of mobilising the African diaspora for roots tourism journeys.

In the early 20th century, such developments were linked to the first attempts of inducing *Back-to-Africa* dynamics. With Ghana's independence in 1957 and the presidency of Nkrumah, pan-Africanism and transatlantic travels in terms of people as well as politics and ideas induced a second wave of mobility that was not limited to tourism but also led to a migration of people of African descent to Ghana (see, among others, Benton & Shabazz, 2009; Dillette, 2020). Since the mid-1990s, roots tourism shifted away from these politicised beginnings towards a more place-based tourism experience which turned the slave forts or so-called slave castles along the Ghanaian coastline into an attraction for roots tourists. The historical contextualisation of such sites and their overall role within the transatlantic slave trade can be seen as an overrepresentation, but the touristic place-making of these points of departure as ports of call and doors of return to the mythologised motherland is of central importance for the economic success of Ghanaian roots tourism (Reed, 2015). The first analysis of tourism at the forts by Bruner (1996) mainly elaborates on the divergent conceptions of how their history should be represented. Accordingly, most roots tourists are solely interested in the connection to the transatlantic slave trade and their ancestors while Ghanaians and historians would like to see the whole of the fort's diverse history portrayed for tourism. Therefore, a struggle over the "correct" representation of history unfolds on these sites and leads to tensions and conflicts within the relation between touristic narrative and tourist recipient. Holsey (2008) similarly argues that the collective memory of the slave trade in Ghana is all too often left aside which conflicts with the historical reappraisal many roots tourists are seeking. In an earlier paper named "Transatlantic Dreaming" (Holsey, 2004), the author also discusses the issue of roots tourism – or kinship tourism as she terms it – from a different angle. She mostly focuses on the construction of Ghana as a homeland for and by African Americans and concludes that both Ghanaians and African Americans each have an idealised picture of the other group's way of life across the Atlantic. Schramm (2004) combines the topic of roots tourism with the motif of a pilgrimage and examines whether African American tourists who were visiting the first celebration of Emancipation Day in Ghana should be regarded as pilgrims. She explains that while these tourists are not on a pilgrimage in the sense of a spiritual journey to their homeland, they can be perceived as pilgrims en route to a stronger diasporic identity. In another work, Schramm (2010) concentrates on pan-Africanism and its role in the debate about African identity and a shared history of all Africans – on the continent and in the diaspora. Reed (2013, 2015) likewise discusses the portrayal of the history of the slave trade for tourism, as well as the motif of a pilgrimage and the debate surrounding the mythologised belief of a common African identity. She furthermore looks into the interests of the Ghanaian state in relation to roots tourism. Nti (2014) contrasts this with a more critical view. He mentions the known conflict about the representation of the history of the slave castles as well as their commodification, along with other examples such as the bestowal of traditional titles to tourists of African descent. He concludes that these performances are in place to generate development and that no true relationships emerge between Ghanaians and roots tourists (see also Delpino, 2011).

Roots tourism to Ghana very much fulfils the definition of diaspora tourism by making use of the historical background and offering the slave trade forts as destinations "primarily produced, consumed and experienced by diasporic communities" (Coles & Timothy, 2004, p. 1). As a variation of ethnic tourism – one within which the tourist becomes part of the performance – the longing for African roots by the tourists unfolds as a search for something exotic and peripheral (see Smith, 1978) but intrinsically homely and central at the same time and thus produces conflicts between the *diasporic expectations* and the *motherland presentations*. King (1994) describes this form of ethnic tourism as travels for the purpose of ethnic reunion with the motivation of exploring one's family history. This form of tourism therefore not only serves the purpose of providing a consumable realm of products, narratives and experiences but also has a much deeper meaning for the tourists pursuing it. The homeland, therefore, presents a key image for the creation of diaspora tourism. As Cohen (2008) points out, these homelands do not have to be specific countries since they rather represent a collectively shared idea or a myth about the home of their ancestors. It is thus an emotional connection to a country or a region that is important and makes a genealogic one almost insignificant. A trip to the homeland consequently bears great meaning for those in the diaspora who are looking for a (real or imagined) connection. The travels of people of African descent to their imagined homeland or the imagined homeland of their ancestors thus become laden with expectations from both the tourists and the organisations at the destination. While those living at the destination may regard the visitors first and foremost as tourists, the people from the African diaspora visiting Ghana may perceive this category as a contradiction to their endeavour, especially if the kinship voyage takes on the form of a pilgrimage or spiritual journey (Schramm, 2004). In the case of African Americans visiting the African homeland, the concept of healing a divided identity through exploring the history of their ancestors seems specifically relevant (Schramm, 2010).

An important aspect of diaspora tourism as a form of heritage tourism that is often left unmentioned is the overall aim of the country that is offering kinship destinations to commodify the narrative of roots, home and reconnection for economic development on the national level (Fredholm, 2016). Heritage may have a mythologised experience to it, but often (also) serves as an economic resource (see Silberman, 2013). While the above-mentioned studies on Ghana's roots tourism industry mostly derive from ethnographic perspectives, others analyse the roots tourism phenomenon to Ghana form the standpoint of tourism studies and tourism management. Mensah (2015), for example, pinpoints the particular demands and expectations of roots tourists at the slave castles. By using quantitative statistical methods, his study aims at developing recommendations for further advancement of the product of roots tourism. Furthermore, other authors also examine the developmental prospects of recent roots tourism activities and other more long-term efforts. Ankomah et al. (2012) highlight the positive economic effects of actively involving the African diaspora in the economic and political development of Ghana. They discuss various measures taken by the government as well as NGOs to consolidate, strengthen and nourish this relationship (see also Alex-Assensoh, 2010). Kleist (2013) also illustrates the methods and practices of engaging the diaspora for development. She repeatedly refers to pan-Africanism and reveals how it is used in politics by the Ghanaian government to especially mobilise African Americans.

Before the Year of Return: Plans and policies targeting the roots market

In addition to the overall development of roots tourism in Ghana, the 2019 Year of Return has to be put in context with African diaspora related policies in order to gain a proper understanding of the path-dependency of marketing strategies. As far back as 1996, African Americans were identified by Ghana as a market for diaspora or heritage tourism in a "National Tourism Development Plan for Ghana" for the period between 1996 and 2010 (Ministry of Tourism et al., 1996, p. 137). As a result, "The Strategic Action Plan – Tourism 2007" was drawn up in 2003, which put a clear focus on diaspora tourism and above all pursued the goal of positioning Ghana as "home" for all African diasporas (Kleist, 2013, p. 290). It is also interesting to note that the Ministry of Tourism was renamed the "Ministry of Tourism and Diaspora Relations" in 2006 and that during the time of this name change the strategy of the Ministry also changed towards a stronger focus on African diasporas (Reed, 2015). The extension of the name illustrates the connection between the country's tourism aspirations and diasporas. However, after a change of the ruling party in December 2008, the Ministry was renamed "Ministry of Tourism" again. The reason for this was that the extension of the name represented a purely profit-oriented use of pan-African ideas for the aims of the previous government (Delpino, 2011).

Beyond these early strategies for tourism mobilisation, additional measures were implemented aiming at a more long-term commitment towards people of African descent. These include, for example, the Right of Abode which was enacted in 1999 in the course of a new immigration ordinance (Kleist, 2013; Yankholmes & Timothy, 2017). This right to residence allows persons of African descent to live in Ghana for an unlimited period of time. However, it is linked to very strict regulations and follows a lengthy evaluation process during which, among other things, it is determined whether the applicant provides possible positive contributions to Ghanaian society (Delpino, 2011). Apart from officially granting Ghanaian citizenship, a similar practice has been established in civil society to encourage African diasporas to make long-term investments and "return" to Ghana. Thus, many traditional chiefs grant land or traditional titles to people form the diaspora (Ankomah et al., 2012). The underlying motive is the hope that wealthy African Americans will invest in local communities once they are holders of a traditional title of that community (Nti, 2014).

The historical development of diaspora tourism and African diaspora related policies by the Ghanaian state preceding the Year of Return campaign reveals a continuous combination of pan-African experiences, economic purposes and historical embeddedness. Nevertheless, it is the latter aspect of historicity that seems to evolve more and more into a *cul-de-sac* in terms of tourism development: the more the history of slavery became the major focus of the kinship narrative and the marketing of Ghanaian destinations, the more conflicts on historical representations and the commodification of African heritage came about. Moreover, marketing the shared history of suffering and the cruelty of slavery as a pull factor for people of African descent renders the roots tourism to Ghana almost as a form of dark tourism (Yankholmes & McKercher, 2015). The following insights into the most recent 2019 Year of Return campaign highlight how the Ghanaian state is trying to move the country into the position of a world-class tourist destination by redefining its already established roots tourism industry. At the same time, these strategies of reframing the kinship destination on the diaspora marketplace only seem to work by drawing on the common narratives – and associated conflicts – of slavery as the common history of suffering and the promise of a pan-African reunion.

The narratives of the Year of Return campaign

In 2019, 400 years after the first slave ship with African slaves arrived at Jamestown, –Ghana organised a year-long tourism campaign under the label Year of Return. Primarily organised by the Ghanaian Ministry of Tourism, Arts and Culture, the campaign revolved around the commemoration of the beginning of the commercial transatlantic slave trade to the United States. The format can be described as a cultural, historicised tourism experience of African indigeneity with the aim of marketing Ghana as the prime destination for those searching for their long-lost African homeland. The campaign addresses all globally dispersed people of African descent but very specifically aims at those living in North America. The introductory citation to this section originates from the Year of Return opening ceremony which took place in Washington DC in September 2018:

> We know of the extraordinary achievements and contributions they [Africans in the diaspora] made to the lives of the Americans, and it is important that this symbolic year – 400 years later – we commemorate their existence and their sacrifices.
>
> (President Akufo-Addo cited in Tetteh [2018])

The historical reference made is a symbolic representation of the first grand narrative of the campaign that is the common history of suffering and of remembering and overcoming these times. This narrative, however, is not conveyed from a victim's perspective, but rather moved into a positive and enabling sphere by focusing on the expression of the strength of Africans to free themselves from this situation. In this context, "resilience" is one of the most frequently used words by the Ghanaian president (Afeku, 2019, p. 5; Agyeman, 2019, p. 6; Akufo-Addo, 2019, p. 3). The Year of Return campaign thus not only tries to commemorate the beginning of the slave trade in 1619 but rather celebrates the overcoming of the many adversities that had to be endured by all people of African descent, in the past as well as today. Presenting Ghana as not only the gateway of return but also as the place to leave this history behind was already found in the preceding campaign, the 2007 Joseph Project (Schramm, 2008). Hence, the Year of Return is presented and justified by setting it in the light of an already existing reality of pan-African mobility affecting millions of Africans living in the diaspora and longing for home (Akufo-Addo, 2019). By drawing on the spiritual connection expressed by many visitors to the slave castles, Ghana positions itself as a "beacon for people of African descent living on the continent and in the Diaspora" (Akufo-Addo, 2019, p. 3). Against this background, a concentrated effort of communicating the aim of the campaign towards the United States and other areas, such as Jamaica and the Caribbean, where descendants of the former slaves live, can be found throughout the whole programme of the Year of Return. The call made is one of reaching out for long-lost brothers and sisters and for them to establish a (re)connection with Africa *through* Ghana. However, the reference to a common fate is surprising to a certain extent, since most ancestors of todays Ghanaians did not share the experiences of the enslaved ancestors of the African Americans, though the emotional aspect is nevertheless convincing. But it is without a doubt this emotional scale that creates the previously discussed conflicts on the representation of history. It is therefore all the more remarkable that the campaign continues to be based on the narrative of a common history, even though history is already in dispute on various sides of the spectrum of roots tourism. The reference to overcoming the adversities and miseries of all people of African descent is, therefore, an important aspect, as it can be related to Africans in general and their overcoming of colonisation, as well as to the still prevailing economic disadvantages of African states. Nevertheless, the campaign clearly focuses on peer groups in the United States, and while the narration may build on an all-encompassing Africanness with Ghana effectively *becoming* Africa for the diaspora, the overall aim is to increase tourism in Ghana.

Another narrative of the Year of Return campaign that aims at creating a connection between Ghana and people of African descent is that of an "African family" and its subsequent reunion. This aspect is strongly put into perspective with the pan-African history and the post-colonial formation of the Ghanaian state (Akufo-Addo, 2019; Agyeman, 2019). Such intersections also seem to justify the positioning of Ghana not only as the continuous leader of the pan-African movement but in its gateway role. The Year of Return campaign thus "presents a unique opportunity for Ghana to cement its Pan African leadership" (Afeku, 2019, p. 5). The term "African family" (Agyeman, 2019, p. 6; Ahu, 2019, p. 7; Akufo-Addo, 2019, p. 3) can be frequently found in the organisers' communications and advertising materials in this context of pan-Africanism and a common African identity of all Africans and African diasporas. The global scale of the imagined African community is often contextualised by terms such as "brothers and sisters" (Afeku, 2019, p. 5; Agyeman, 2019, p. 6; Ahu, 2019, p. 7; Akufo-Addo, 2019, p. 3) or "kith and kin" (Ababio, 2019, p. 7). While the transaction of tourism is an economic one, the connection is narrated as an act of familiarity and reunion. And even though locating one's own heritage to a specific place on the African continent continues to be a more than uncertain undertaking – an aspect which recently led to the emergence of an industry promising verifiable results through DNA testing – the promise of finding the imagined homeland by visiting Ghana is left undisputed (see Holsey, 2004). Two major Ghanaian events among many others are closely linked to this narrative and have been organised since the 1990s: PANAFEST, the Pan-African Historical Theatre Festival, and Emancipation Day. In the Year of Return, these events became central occasions for celebration. The Year of Return continues not only with the narrative of a pan-African reunion but enacts it through long-established festivities that entail a clear peer group focus on the North American continent. Within diaspora or roots tourism endorsed by the Ghanaian state, mythologised performances of homecoming are intersecting with the politics of migration and mobility, as well as touristic commodification.

However, such an address of African Americans as part of the "African family" and the positioning of Ghana as a home for the African diaspora entails a potential for conflict, not least of all over the various representations of the history of the slave trade. Furthermore, visitors' expectations of rejoining the "African family" may conflict with Ghanaian expectations of a strengthened economic development of the country – instead of forging long-term family-like relationships (Nti, 2014). Upholding the narrative of a pan-African reunion becomes an essential aspect of creating a common ground for the shared stage of the performed kinship experience. Such examples of conflict show that the expectations of the Year of Return campaign can only be fulfilled to a limited extent. In this context, Nti (2014, p. 114) already points out that the "Diaspora Project" gives the impression of a strategy to promote development rather than delivering a genuine commitment to a pan-African culture. However, this contradiction of discourse and practice is not only criticised by African Americans, but also by Ghanaian intellectuals who would like to see a different way of dealing with the issue of slave trade history in their country (Schramm, 2004).

The strategies of the Year of Return campaign

In addition to the discussed narratives, the public appearance and communication of the Year of Return campaign allow for the identification of marketing strategies. In this regard, both the manifold online advertisement activities together with the focus on getting African Americans involved are of significance. In addition, the country's political measures to call upon the diaspora in order to play a major role, not only as consumers or visitors but as performers became a crucial part of the campaign.

An outstanding aspect within the marketing of the Year of Return was and continues to be, the strong online presence and communication through social media such as Facebook and Instagram. Both official profiles were extremely active throughout 2019 and continuously promoted the upcoming events of the Year of Return, as well as the campaign in general, sometimes with several posts per day. Building on the follower-logics of social media channels, the significantly higher numbers of fans and followers of celebrities travelling from the United States to Ghana during the Year of Return and subsequently reporting about their experiences on their social media channels became an important incubator for the online campaign. Among high-profile visitors were actors and musicians such as Boris Kodjoe, Steve Harvey, Cardi B and Ludacris (Mensah, 2020). Through their positive publicity as well as interviews with American media concerning their experiences, they created awareness of Ghana and thus promoted the country as a prime tourist and cultural destination beyond the campaign (Greenwood et al., 2020). The CEO of the Tourism Authority, Akwasi Agyeman, was quoted in a newspaper article with the following statement: "It has been like a miracle if you look at even the media aspect of this; having Aljazeera, BBC, CNN, Ebony Magazine, Essence and all of them here, [...] we couldn't have paid for the media" (Mensah, 2020). For the organisers and the Ghanaian state, the celebrity factor within the web of communicating the Year of Return globally became a decisive element for the overall success of the campaign (Akufo-Addo, 2020).

In contrast to the narrative of reaching out to the global African diaspora, the promotional online materials such as the website of the Year of Return reflect a strong focus towards African Americans (Dzidzornu et al., 2019). The main reason for the campaign's focus on African Americans and the relative neglect of, for example, Jamaican and other Caribbean-African diasporas was linked to the peer group's disposable income (Reed, 2015). In addition, the factor of a common language plays an important role in marketing Ghana as a kinship tourist destination, creating a high degree of accessibility (Holsey, 2004). This renders Ghana a more attractive destination for African Americans wanting to travel abroad and the United States a sought-after diaspora market for the Year of Return campaign. In addition, the growing popularity of DNA testing for (supposedly) verifying the geographical heritage of the descendants of enslaved people and the resulting enthusiasm for the search of one's roots adds up to the regional focus. The special involvement with the United States can not only be found in the marketing approaches but within the institutional set-up of the overall organisation of the Year of Return itself. The Maryland-based Adinkra group became a major partner within the marketing of the Year of Return, offering pre-organised tour packages with destination Ghana specifically to people of African descent. It is worth noting that their "Birthright Journeys" packages include a DNA test with the company African Ancestry. Other organisations involved in the transatlantic promotional partnership were lobby groups such as the National Association for the

Advancement of Colored People (NAACP) and the Congressional Black Caucus, which were specifically approached in order to reach out to African Americans (GhanaWeb, 2019). The figures on the development of international tourism to Ghana so far point to a successful implementation of US-focused strategies with a 26% increase of Americans entering the country from January to September 2019 in comparison with the previous year (Greenwood et al., 2020).

Ghana has long been pursuing the goal of attracting African diasporas and their investments to the country. Through online and social media strategies targeting the African diaspora in general and the United States in particular, the Year of Return became the most well-known, and consequently, in all probability, the most successful measure of the Ghanaian government so far. Proclaiming a "Decade of African Renaissance" through the already initiated follow-up campaign "Beyond the Return" (Beyond the Return, 2020) underlines Ghana's intentions to further invest in the field of roots tourism.

Conclusion: Roots tourism to Ghana as a new stage for global cosmopolitans

In his classic account "Who is a Tourist?", Cohen (1974, pp. 542–543) identifies a special form of tourist, which he calls "old-country visitors". According to his definition, which specifically includes African Americans travelling to Africa, most "old-country visitors" that did not emigrate from the homeland themselves, primarily engage in touristic activities and consume touristic products during their travels. Roots tourism to Ghana as it evolved since the 1990s brought about such a market of old-country visitors with the places and traces of the common heritage of slavery as a history of suffering being the consumptive aspect in the touristic endeavour. In addition, a highly politicised narrative of pan-Africanism makes it difficult to offer heritage tourism to people of African descent without referring to the grand narratives of overcoming the history of slavery and reuniting the African family – the previously presented conflicts included. The 2019 Year of Return campaign, if judged by the promotional strategies, can in a way be seen as a shift from a politicised narrative of pan-African reunion to offering a culturalised experience of rather un-political pan-Africanism as a way of life and leisure. This focus on enjoying the present moment while at the same time operating for the future economic development of Ghana is strongly reflected in the follow-up campaign "Beyond the Return". It comprises seven pillars that address future visitors from the African diaspora: experience and investment in Ghana, find pathways to it, celebrate the West African country, brand it, give back to it and further improve tourism and heritage assets (Beyond the Return, 2020). This agenda not only reflects the extension of a so-far successful strategy of making the African diaspora work for development but moves heritage tourism into a new, post-modern setting where the tourist is a recipient of and actor within the spectacle. Thereby, Ghana continues to perfect its role as a tourist destination with a significance beyond the national realm, while at the same time offering an idea of the African homeland that renders the rest of the continent almost obsolete. By eventifying the history of slavery, Ghana becomes a representation of roots and indigeneity in a world of more cosmopolitan, diasporic Africanness.

References

Ababio, A. A. (2019). Message by Akwasi Awuah Ababio, the Director of Diasporan Affairs, Office of the President, Ghana. *Voyages Afriq Special Edition*, January-December, 7.

Afeku, C. A. (2019). Message by the Minister for Tourism, Arts and Culture, Hon. Mrs. Catherine Abelema Afeku [MP]. *Voyages Afriq Special Edition*, January-December, 5.

Agyeman, A. (2019). Message from Akwasi Agyeman, Chief Executive Officer of the Ghana Tourism Authority. *Voyages Afriq Special Edition*, January-December, 6.

Ahu, B. A. (2019). Message by Mrs. Bella Ayayee Ahu, President of Ghana Tourism Federation (GHATOF). *Voyages Afriq Special Edition*, January-December, 7.

Akufo-Addo, N. A. D. (2019). Foreword. *The Year of Return Brochure*. Retrieved from https://www.yearofreturn.com/wp-content/uploads/2019/06/YOR-Brochure.pdf.zip

Akufo-Addo, N. A. D. (2020). President Akufo-Addo's Message on the State of the Nation. *The Presidency Republic of Ghana*. Retrieved from http://www.presidency.gov.gh/index.php/briefing-room/speeches/1511-message-on-the-state of-the-nation-by-the-president-of-the-republic-nana-addo-dankwa-akufo-addo-on-the-occasion-of-the-state-opening-of-parliament-on-thursday-20th-february-2020

Alex-Assensoh, Y. M. (2010). African Americans, African immigrants and homeland-diaspora development in Africa. *African Diaspora*, *3*(2), 207–234. doi:10.1163/187254610X526922

Ankomah, P., Larson, T., Roberson, V., & Rotich, J. (2012). A creative approach to development. *Journal of Black Studies*, *43*(4), 385–404. doi:10.1177/0021934711425488

Araujo, A. L. (2018). Tourism and heritage sites of the Atlantic slave trade and slavery. In D. Dean (Ed.), *A Companion to Public History* (pp. 277–288). Chichester: Wiley Blackwell.

Benton, A., & Shabazz, K. Z. (2009). "Find their level": African American roots tourism in Sierra Leone and Ghana. *Cahiers d'études Africaines*, *XLIX*(1–2), 477–511. doi:10.4000/etudesafricaines.18791

Beyond the Return (2020). About "Beyond the return". *Visit Ghana*. Retrieved from https://visitghana.com/beyond-the-return/

Bruner, E. M. (1996). Tourism in Ghana: The representation of slavery and the return of the black diaspora. *American Anthropologist*, *98*(2), 290–304.

Cohen, E. (1974). Who is a tourist?: A conceptual clarification. *The Sociological Review*, *22*(4), 527–555.

Cohen, R. (2008). *Global Diasporas: An Introduction*. New York: Routledge.

Coles, J., & Timothy, D. J. (2004). "My field is the world": Conceptualizing diasporas, travel and tourism. In J. Coles & D. J. Timothy (Eds.), *Tourism, Diasporas, and Space* (pp. 1–29). London: Routledge.

Delpino, G. (2011). Building up belonging: Diasporic "homecomers", the Ghanaian government and traditional rulers: A case of return. *African Diaspora*, *4*(2), 163–184. doi:10.1163/187254611X606409

Dillette, A. (2020). Roots tourism: A second wave of double consciousness for African Americans. *Journal of Sustainable Tourism*. doi:10.1080/09669582.2020.1727913

Dzidzornu, A., Moore, C., Forson, P. K., & Abeiku Aggrey, G. (2019). Ghana proclaims year of return. *Voyages Afriq Special Edition,* Jan-Dec, 8–9.
Fredholm, S. (2016). Assets in the age of tourism: The development of heritage planning in Ghanaian policy. *Journal of Contemporary African Studies, 34*(4), 498–518. doi:10.1080/02589001.2017.1285011
GhanaWeb(2019).WespentaboutGH₵6mon"YearofReturn"campaign–TourismAuthority. *GhanaWeb.* Retrieved from https://www.ghanaweb.com/GhanaHomePage/business/We-spent-about-GH-6m-on-Year-of Return-campaign-Tourism-Authority-826231
Gilroy, P. (1994). Diaspora. *Paragraph, 17*(3), 207–212.
Greenwood Davis, H., & Williams, S. (2020). How Ghana became the hottest destination for African-American travelers. *National Geographic.* Retrieved from https://www.nationalgeographic.com/travel/destinations/africa/black-americans-are-going-to west-africa-in-search-of-roots/
Handley, F. J. L. (2006). Back to Africa: Issues of hosting "roots" tourism in West Africa. In J. B. Haviser & K. C. MacDonald (Eds.), *African Re-Genesis. Confronting Social Issues in the Diaspora* (pp. 20–31). London: Routledge.
Holsey, B. (2004). Transatlantic dreaming: Slavery, tourism, and diasporic encounters. In F. Markowitz & A. Stefansson (Eds.), *Homecomings. Unsettling Paths of Return* (pp. 153–168). Maryland: Lexington Books.
Holsey, B. (2008). *Routes of Remembrance: Refashioning the Slave Trade in Ghana.* Chicago: University of Chicago Press.
King, B. (1994). What is ethnic tourism?: An Australian perspective. *Tourism Management, 15*(3), 173–176. doi:10.1016/0261-5177(94)90101-5
Kleist, N. (2013). Flexible politics of belonging: Diaspora mobilisation in Ghana. *African Studies, 72*(2), 285–306. doi:10.1080/00020184.2013.812883
Mensah, I. (2015). The roots tourism experience of diaspora Africans: A focus on the Cape Coast and Elmina Castles. *Journal of Heritage Tourism, 10*(3), 213–232. doi:10.1080/1743873X.2014.990974
Mensah, N. A. S. (2020). Nana Adwoa Sarkodie writes: The year of return, social media as a tool to promote Ghana's tourism. *Citi Newsroom.* Retrieved from https://citinewsroom.com/2020/01/nana-adwoa-sarkodie-writes-year-of-return-social-media as-a-tool-to-promote-ghanas-tourism/
Ministry of Tourism, United Nations Development Programme, & World Tourism Organization (1996). *National Tourism Development Plan for Ghana (1996-2010).*
Nti, K. (2014). "Uniting the African family": Monuments, tourism, development, and neoliberalism in Ghana – historic African diaspora relations. *Journal of Third World Studies, 31*(2), 113–125.
Reed, A. (2013). Diaspora tourism: The heritage of slavery in Ghana. In A. Quayson & G. Daswani (Eds.), *A Companion to Diaspora and Transnationalism* (pp. 524–538). Oxford: Wiley Blackwell.
Reed, A. (2015). *Pilgrimage Tourism of Diaspora Africans to Ghana.* New York: Routledge.
Schramm, K. (2004). Coming home to the motherland: Pilgrimage tourism in Ghana. In S. Coleman & J. Eade (Eds.), *Reframing Pilgrimage: Cultures in Motion* (pp. 135–151). London: Routledge.
Schramm, K. (2008). Das Joseph-Project. *Historische Anthropologie, 16*(2), 227–246. doi:10.7788/ha.2008.16.2.227
Schramm, K. (2010). *African Homecoming: Pan-African Ideology and Contested Heritage.* London: Routledge.

Silberman, N. A. (2013). Discourses of development: Narratives of cultural heritage as an economic resource. In R. Staiff, R. Bushell, & S. Watson (Eds.), *Heritage and Tourism: Place, Encounter, Engagement* (pp. 213–225). Abingdon: Routledge.

Smith, V. L. (1978). *Hosts and Guests: The Anthropology of Tourism*. Philadelphia: University of Pennsylvania Press.

Tetteh, B. (2018). 2019: Year of return for African Diaspora. *Africa Renewal,* December 2018–March 2019.

Yankholmes, A., & McKercher, B. (2015). Rethinking slavery heritage tourism. *Journal of Heritage Tourism, 10*(3), 233–247. doi:10.1080/1743873X.2014.988159

Yankholmes, A., & Timothy, D. J. (2017). Social distance between local residents and African-American expatriates in the context of Ghana's slavery-based heritage tourism. *International Journal of Tourism Research, 19*, 486–495. doi:10.1002/jtr.2121

9 "The culture of the souq is lost"

Valuing social capital and a legacy within tourism development

Manuela Gutberlet

Introduction

In light of the worldwide COVID-19 pandemic and the consequent economic slowdown, tourism researchers have called for tourism in the service of the people of destinations (Higgins-Desbiolles, 2020). It is time to redefine and reorient tourism operations and planning, representing the interests of locals (Higgins-Desbiolles, 2020). Communities should be valued as complex systems that require continuous interactions and common solutions (Musavengane & Kloppers, 2020), while locals should be allowed to determine their development through broad community involvement and bottom-up approaches (Telfer, 2003). The current COVID-19 pandemic has made clear that local safety nets, social caring and social solidarity are important assets of communities (Higgins-Desbiolles, 2020). Strong social capital can promote long-term community resilience and the ability of the community to overcome disturbances while bridging governance, financial and professional skills (Musavengane & Kloppers, 2020). In addition, nostalgia as a tool for community tourism can enhance bonding with the place, creating stronger local identities (Christou et al., 2018). An understanding of the respect and emotional connections locals have with tourist sites is integral to meaningful, successful local community involvement and planning (Zang & Smith, 2019). Often under-represented within communities are specific ethnic groups or "diaspora". They are groups of people living abroad, outside their home countries, being connected with each other either literally or figuratively by spatial concentration, culture, religion, ethnicity or national identity (Timothy, 2011).

Researchers have highlighted the importance of social capital, social identity and social networks in community-based tourism development, providing important assets for business exchange, economic development and positive community attitudes, while, on the other hand, challenging power issues, marginalisation and social exclusion (Colic-Peisker & Robertson, 2015; Hall et al., 2008; Musavengane & Kloppers, 2020; Palmer et al., 2013; Park et al., 2015; Soulard et al., 2018; Telfer, 2003; Waldinger, 1995). If not managed well, social capital can create unsustainable development (Park et al., 2015, p. 128). Conversely, social changes in urban settings are ubiquitous and constant. They are part of the lived experience of locals, featuring in everyday narratives of changed local life

(Colic-Peisker & Robertson, 2015). The control of tourism businesses by certain ethnic groups may lead to the marginalisation or exclusion of others (Waldinger, 1995). Long-term qualitative research on social capital and legacies within community-based tourism is scarce globally and non-existent in the Arab world. My study follows previous research (Gutberlet, 2016a, 2017, 2019).

The main purpose of my research is to deepen the local knowledge asset about how well-established Omanis and the diaspora communities from India and Pakistan perceive their social capital within a fast developing, emerging heritage destination by identifying narratives and life stories. It also responds to research by Causevic and Neal (2019) that calls for illuminating some of the recent history of Souq Muttrah, which seems to be hidden or veiled. The "local community" as such consists of multi-ethnic, well-established shopkeepers and owners, residents as well as decision-makers. Through my long-term, emic approach, my study is valuable for tourism researchers, practitioners, policymakers and urban planners in other developed and small destinations who wish to include the local, multi-ethnic community and their social capital. In the following sections, I will outline the main concepts, methods and results, followed by my concluding thoughts.

Concepts

Social capital

The economic term "capital" suggests that communities can protect, sell or consume the natural, physical and social assets of their environments in order to promote tourism development (Soulard et al., 2018, p. 191). Within capitalism and urban living, social capital and community cohesion have seen an increasing research interest.

High social mobility, ethnically diverse populations, socio-economic fragmentation and the pursuit of individual success dominating local community considerations have been some of the main reasons behind decreasing social capital (Colic-Peisker & Robertson, 2015). Compared with individualistic cultures such as in Europe, the United States or Australia where people value social memberships, norms and values less, collectivist societies like those in Asia put more emphasis on group memberships and social relations (Park et al., 2015). Characteristics such as trust, social norms and interpersonal networks are important to enhance a positive economic, social, cultural and political development within collectivist societies (Putnam, 1993). These non-materialistic components of an exchange relationship such as tourism's contribution to community well-being have a significant influence on individual support for tourism development (Park et al., 2015, p. 129). However, the business culture of the destination needs to be considered, limiting the involvement of a wide range of stakeholders (Soulard et al., 2018, p. 198). Hence, the social identity of community members influences their active involvement and advocacy for the destination, conceptualised in the sharing of knowledge, word-of-mouth promotion of the destination or social media (Palmer et al., 2013). As such, social capital can be seen as an important tool to promote sustainable, inclusive tourism and to analyse the linkages between communities and other stakeholders (Soulard et al., 2018). It offers opportunities to evaluate perspectives of how groups are included within the community and whether they have the confidence to reach out to other groups and government entities (Soulard et al., 2018, p. 198).

Bonding social capital enhances community ties through trust, reciprocity and direct involvement or cooperation. It provides a tool to evaluate power struggles and a lack of trust within communities. It reinforces relationships between government institutions and an informal web of networks, relationships and norms, thus creating a valuable link with tourism businesses. For example, an informal community tourism support group meets regularly to share information about government initiatives, enhancing bonding social capital (Soulard et al., 2018, p. 189). A lack of consultation and communication decreases bonding social capital, creating feelings of exclusion and being left out of the development process (Soulard et al., 2018, p. 197).

Bridging social capital includes social ties that link community networks, organisations and governmental representation through trust, reciprocity and cooperation, for example, between regions, industries and government entities. Possible negative effects of social capital are that social networks and bonding create a demand for conformity, where community members are expected to behave according to well-established norms (Portes, 1998). Reciprocity is a sense of mutual and fair exchange, for example, through the representation of cultural and ethnic minorities such as Aboriginal culture (Soulard et al., 2018). As a case in point, bridging social capital increased when different tourism stakeholders like hotels, attractions and destination management companies cooperated to make tourism more politically visible (Soulard et al., 2018, p. 197).

Bonding and bridging social capital are important for valuing heritage sites and conservation, for example, through skills training (Musavengane & Kloppers, 2020). Social memories and experiences can be generated, conceptualised and reconceptualised through shared heritage tourism and heritage tourism experiences can contribute to the process of forming a local identity. The symbols and meanings attached to physical manifestations of heritage resources are important for shaping the essence of identity and history. In the next part, I will explain the methods used.

Methods

I have applied ethnography in order to understand the social world of people within their environments (Colic-Peisker & Robertson, 2015; Snape & Spencer, 2003). As part of my fieldwork, I applied participatory observation, in-depth interviews, photography and informal conversations. I also participated in the local flow of life in the souq, especially during the tourist timings in winter. I started my initial fieldwork in Souq Muttrah between 2012 and 2014 (Gutberlet, 2016a, 2016b, 2017) and conducted follow-up research between 2017 and 2019. Such research increases the credibility, trustworthiness and robustness of the concepts initially used (Baxter, 2010, p. 90). I revisited the same shops continuously, especially during the cruise seasons. I observed the scene along the main streets, took photographs and conducted informal conversations and formal follow-up interviews in both English and Arabic with the remaining, well-established vendors along the main streets of the souq. The semi-structured interviews were conducted individually, with business owners and with both well-established and new shop vendors of various ethnicities. My interview questions asked about the history of the shop, the family history, the product range, the type of customers, the views of people about tourism development and the intercultural interactions with tourists. The aim was to enable the interviewees to provide insights into their family history, the legacy of their shops, their experiences with customers and their everyday lives, as well as other memories of Souq Muttrah. For the interviews with the male vendor community, I used snowball sampling, in particular, with the diaspora communities. Each interview took between 10 and 60 minutes. The interviewees were given different names to preserve their anonymity. Moreover, no recording was made, so that they were more open to speak freely and develop their lines of thought. The entire interaction was documented in writing and coded manually shortly afterwards. In addition, I interviewed local residents, a heritage expert and local decision-makers. I have been seen as a female "local" on the one side and as an "insider" who knows Arabic and has been living and working in Oman, including many visits to the souq as part of my work and leisure time, participating in the "local flow of life" for the past 16 years. Throughout the interviews, I applied a social constructivist approach, which allowed me to understand how the local community constructs their life experiences, narratives and life stories. In the beginning, I conducted 45 interviews in 45 shops with shop vendors and owners of different age groups and nationalities, all of them male. The shops covered in the interviews sold local perfumes like frankincense, handicrafts, spices, kummas (traditional small hats for Omani men), pashmina scarfs, fabrics and foodstuffs. All interviews were conducted anonymously and not recorded. I also interviewed local customers (n=8) and Omani residents (n=7). Most of the shopkeepers interviewed were from the Banyan diaspora community, originating from Pakistan and India, but who have not yet been granted Omani nationality. They were chosen by the snowball sampling method. Being introduced to them helped me as a female researcher to get trust and thus access to the well-established community. Through a "critical pedagogical approach", I tried to generate knowledge from an endemic, community angle, while giving an educational impulse to the community in Muscat, including decision-makers, through cultivating values and identifying existing social identities (Wilson & Hollinshead, 2015). To organise and analyse the dataset, I applied cross-sectional codes and retrieve methods, using codes and labels (Spencer et al., 2003).

Research setting

Until the outbreak of the COVID-19 pandemic and the worldwide lockdown, the Middle East and Asia were seen as the world's fastest emerging tourism markets.

The Middle East recorded US$73 billion in tourism receipts and the region witnessed high investment in tourism infrastructure (UNWTO, 2019). With 3.5 million international visitors in 2019, compared with 1.4 million in 2010, the Sultanate has been focusing on a growth-oriented tourism strategy. The National Tourism Strategy 2040 (NTS) was launched in 2016. Its target is to increase the number of intentional tourist arrivals to 11.7 million by 2040 and to raise the sector's share of GDP to more than 6% (Oxford Business Group, 2019, p. 212). The mission of the Oman Tourism Strategy is to strengthen local pride and identity while developing tourism SMEs and cultivating Omani culture, heritage and traditions. The vision is "to diversify the Omani economy and to create jobs by offering the world enriching tourism experiences with Omani personalities" (Innovative Tourism Advisors, 2016, p. 10).

Located on the Arabian Peninsula, the Sultanate of Oman has been a seafaring country. Trade has therefore long been a major part of the economy, and people are accustomed to interacting with different cultures. Until the 1970s, Souq Muttrah was the central trading place in Oman, linking traders and craftsmen from outside Oman with traders, craftsmen and agriculture in the interior of the country. Until that time, the Sultanate was lacking in basic infrastructure, having only a few kilometres of paved roads between Muttrah and Old Muscat, in addition to a small airport. Traders arrived by boat or by camel in Souq Muttrah. Under the rule of His Late Majesty Sultan Qaboos bin Said Al Said, the country developed into the modern state that it is today (Oxford Business Group, 2019). Souq Muttrah has been mentioned as a meeting place for a multi-ethnic community, originating from India and Pakistan. Males flock to the souq: Arab and Gulf traders, merchants and labourers from India and Baluchistan, crews and passengers from ships seeking anchorage and supplies in Muttrah as well as Omani villagers, Bedouin tribesmen from the interior and urban residents alike come to the souq to buy, sell and barter (Richardson & Dörr, 2003, p. 214). Muscat harbour in the district of Muttrah was colonised by foreign powers, principally the Portuguese, Persian and Ottoman empires (Causevic & Neil, 2019, p. 505). Along with the city of Sur, both locations were key hubs for the slave trade, a sensitive taboo topic in Oman, where social divisions and inequalities are not discussed publicly (Causevic & Neil, 2019, p. 512).

At the end of 2019, the total population in Oman was 4.675 million, comprising 57.4% Omanis and 42.6% expatriates. The governorate of Muscat has the largest number of expatriates. Compared with the number of Omanis, Muttrah is the district where the largest number of expatriates live within a densely populated area, with the majority coming from India, Pakistan and Bangladesh. The Sultanate's current development strategy, Oman Vision 2020, focuses on the diversification of the economy through privatisation, including increased support for small and medium-sized companies, and "Omanisation", which means prioritising Omanis and gradually replacing expatriates with Omani nationals in the job market. Due to a large number of young Omanis entering the job market each year, there has been a strong focus on the "Omanisation" policy, mainly in managerial positions and less in the retail sector or other low-paid jobs. For example, new vendors from

India and Bangladesh are selling in Souq Muttrah. In the next part, I will analyse my research results.

Results

Bonding social capital: The young generation is not interested

Until the 1970s and 1980s, the beginning of the economic renaissance and the reign of Sultan Qaboos bin Said Al Said, Souq Muttrah was the main business centre of Oman. The souq served as a warehouse, a trading place for merchandise, a meeting place and a place for the exchange of information, creating strong social capital and social cohesion. It can be compared with a kind of "heterotopia", a place of dark corners, mystery, a labyrinth with changing meanings and poignant memories (Urry, 2000, p. 228).

An elder member of the Al Lawati community in Souq Muttrah sees the hardship of the past "through rose-tinted glasses". The community seemed closer to each other: Families and their neighbours worked side by side in the souq, which formed a "loci of informal community development" (Colic-Peisker & Robertson, 2015):

> This is a sad story: The young generation. In the past all of them worked in the souq. Our fathers and grand-fathers worked in the souq. It was a tradition. They had their shops in the souq. In the mornings they went to the souq, before their prayers. The old generation were working for example as carpenters. The new generation wants to work for the government. At that time everybody wanted to have his own business.

Nowadays, the majority of the Al Lawati community members have left the souq. Their large business cooperations are dispersed in Muscat, owning monopolies, instead of bridging their knowledge with the souq, where they are unwilling to continue the legacies of their families.

> The young who are in the capital are not keen to sit for hours working in the souq. I think the government programmes to support SMEs need a direction from everybody, a push to encourage them… They see the Bengalis making good money.

An Omani said this and stressed the importance of setting up a small business and bridging social capital with the government, referring to the business practices of new Asian vendors. An elder Omani halwa maker remarked that salaries and living standards are low for those who work in the souq.

> Lulu hypermarket pays higher salaries, maybe 300 RO. Here in Muttrah the salaries are only 100–150 RO. In Muttrah five foreigners live in one room, maybe sometimes 15 or more live in one apartment. They don't stay at home, they work the whole day.

To increase the attractiveness of the place and to push for more Omanis in the workforce, he suggested the creation of laws and financial incentives to raise the working and living conditions for Omanis in Souq Muttrah. The souq has witnessed continuous physical and social changes throughout time. People travelled for days to the souq for trade and shopping, creating a legacy. The Omani halwa maker remembers:

> Since the 70s the souq has changed several times. In the past the souq was roofed with palm branches. There was no Corniche road. People came with a camel, a donkey or a horse and it sometimes took them two weeks to travel from Nizwa to Muscat. In the souq, they sold dried dates, cooking oil, dried fish and even mishkak [meat-sticks].

The well-established community retain fond memories of their simple lifestyle and trade. With an increase in modernity and tourism development, changes in the architecture occurred. With the increase in tourist numbers, the well-established vendor community became nostalgic about the past, their attachment to the place and social bonding with the other members of their ethnic, "kin-based community network".

> "In the past, the souq was closed at sunset. Everybody slept at 8 pm. Each area had large doors that were closed at 6 pm. You could then only pass through small doors with a secret password. There were four doors in Muttrah", remembers an Omani shop owner who is in his 50s.

The "Banyan" community and the Omani Balushi community created a sense of mutual bonding social capital (Soulard et al., 2018), similar to African Americans in the United States. In the following, I will outline the life stories and narratives of the diaspora community.

Being part of a legacy: "My father started the shop in 1922"

Sanjay (not his real name), who is around 70 years old, is a thin small person without a beard. I had already met him when I worked as a tourist guide. He relocated his small shop from the entrance to the inside of the souq some years ago (Figure 9.1). There he works along with his son (Gutberlet, 2016a). I interviewed him in 2012 in his old shop close to Corniche road.

> "My father started the shop in 1922. I took over about 40 years ago. Since then everything has changed. In the 1950s and 1960s there was no Corniche road, the seawater came up to here", he said pointing to the outside of his shop. He continued: "There were no cars, everybody arrived on a camel or on a donkey. Since around 10 years ago, tourists have been coming here".

He described the changing use of the space, from a traditional trading place, where hardship and a slow life prevailed, developing into a commodified place where

Figure 9.1 A shop of the well-established trader community in Souq Muttrah. Source: Gutberlet (2013).

many visitors came once and stayed for a short time, reflecting on the fast development and modernisation of the surrounding environment, its infrastructure and ethnic diversity. An Omani tourist guide said about the souq:

> In the past the souq was multi-cultural. Indian traders, ancestors of the Khimji and Bahwan family brought food and electrical items to Oman,

However, in official marketing brochures, the souq is only promoted as a traditional, Omani destination (Gutberlet, 2016a), demonstrating a lack of representation of ethnic minorities and increased ethnic fragmentation. Due to an influx of newly arrived expatriate vendors, the bonding social capital through reciprocity decreased (Colic-Peisker & Robertson, 2015). Sanjay said, in the past, only Omani and Banyan men worked in the souq. Omani customers trusted their quality and only bought from certain shops, confirming a bonding and bridging social capital between the customer and the shop vendor.

> I have many Omani customers. All my customers come from the interior of Oman. Tourists buy other items. Tourists buy pashmina scarfs and caps.

Sanjay said this with a smile, stressing the differentiation between his shop and the rest of the souq. During the interview, an elder Omani woman enters Sanjay's shop. She sits down on a small chair along with her son. She tries some Arabic oud perfume (a local perfume) on her wrist. They decide to buy "oud" and "bokhoor" that are used to perfume clothes and homes. They leave and carry many plastic bags with them. Another Omani customer arrives to pay for his purchase. Sanjay calculates the amount on a piece of paper. He does his accounting manually, inside a book, placed on a cupboard. There is no technology involved in the process. Another Omani buys a package of seashells and packages of bokhoor. Here the authenticity of the place goes deep beyond the commercial purpose and the surface appearance. It is an emotional exchange and bonding social capital between the customer and the business owner. In 2013, Sanjay's perfume shop closed down. It was renovated, adjusted to mass tourism and turned into a pashmina shop. New vendors from India are selling with little emotional exchange and bonding involved (Gutberlet, 2016a).

A legacy vanishes: "The culture of the souq is lost"

Akshay (name changed) and his family have a small grocery and herbal shop, located in the heart of the souq, selling herbs, dried limes, nutmegs, dates, tea, spices, dry ginger, henna leaves, rose water and herbal medicine, as well as local herbal beauty products from India and rock salt, soaps and rose petals from the Omani Jebel al Akhdar mountains. In summer, he sells fresh dates from their own farm. The items are displayed naturally, in open wooden boxes, as in the past. The shop also has no entrance door or window. It does not have an air conditioner, only a fan, preserving its "old character". There are only a few remaining simple shops like that. He said:

> Our shop is still the same as it was in the seventies, the furniture, the doors. The only change is the electric signboard.

Akshay works alongside his father. His family shop was established around 250 years ago. Originally, they are a Sindhi family from Pakistan, a "diaspora community" (Timothy, 2011). Akshay grew up in Oman and the souq. In June 2013, he told me about his family legacy:

> We are a marginal community in Oman. We are early traders, around 79 families. Around 12–15 families still work in Muttrah. The first traders came to Oman about 600 years ago. We established our business in Sohar, then we moved to Musannah and then to Muttrah. We still live here in Muttrah, around 10 minutes from the shop. Our community was the main reason why the Portuguese traders gave nuts the name "Muscat nuts".

Akshay talked about his attachment to the souq and the community, which has shaped the legacy of the trade in the souq. Akshay was in his late 30s when I interviewed him for the first time. While he was sitting in his open shop, he often acted as a storyteller for tourists, transforming the small, elevated shop into a theatre stage. He then quickly attracted an interested crowd of listeners. For Akshay, the transformation of the souq occurred at the expense of social values and traditional community ties, which were more prevalent in the past.

> "The culture of the souq is lost", he said, adding: "Tourism is totally money-based. You can no longer see the culture of the souq".

By culture, he means the local traditions, community trust, networks and cooperation within the legacy of his shop. Through "bonding social capital", his family was connected with the community network that secured their survival. The vendors became more profit-oriented and less social. Joint everyday activities created a bond or cohesion between the community members. In 2014, he said:

> You don't see people playing together or sitting together for a coffee. It is modernisation. I believe we need to move on. But I wish we find a balance between modernisation and our culture.

Akshay referred to the modernisation of the souq, linked with a neoliberal, capitalism-based development lacking the slower encounters between people that were a characteristic of the place. Similar to the other Omani shops, Akshay's shop vanished from Souq Muttrah during my field research. It closed at the end of 2013 when his father became sick. I then interviewed him again in 2014.

> "I went through hell", said Akshay explaining that he had to throw all his merchandise into the sea and the boxes were dumped into the garbage. "It was an old trading store. It had to be dirty to give an impact", he said, mourning the loss of the past, his attachment to the souq and his fond memories.

Akshay wanted to leave the past behind and to start something new. He had invested Omani Rials 10,000 (around Euro 22,000) and was waiting for a reply from a bank to invest further in the site.

> "Nobody wants to invest. We have to move on" he said, adding, "For the shopkeepers tomorrow is not stable, so they take everything today. They are more aggressive than two to three years ago. In the past, we had no problem of sales. Today, there are too many restrictions on them".

He said this referring to the economic slowdown, a lack of social stability and cohesion within the souq. To my surprise, Akshay installed an ATM money machine for some time in his inherited family shop, adjusting to the capitalist development in the souq. Then, in 2016, the shop turned into a flashy pashmina shop with windows and new vendors from Bangladesh, leaving behind the legacy. In 2016, the shop transformed into a tourist shop and Akshay was not seen again. He left the souq and opened up another business. This indicates a loss of "bonding and bridging social capital" through a lack of cooperation and trust. It developed along with a shift from the traditional long-term attachment to the souq towards "liquid modernity" and fragmentation of individual lives, an infinite series of short-term projects and episodes (Bauman, 2000). The old customers were replaced with new customers from the cruise liners within a neoliberal economy, enhancing the selling of hybrid, global products and the loss of local knowledge. During an interview in March 2013, a Western expert from the Ministry of Heritage highlighted the long-standing predominance of Banyan artisans and their knowledge.

> In the past a high percentage of silver and goldsmiths were from the subcontinent. That is nothing new. Even Sayyida Salma was talking about life in Zanzibar in the 1800s. All her goldsmiths, who she called "Banyans", were Indians. Significantly, she did not mention any traditional Omani silver or goldsmith in her memoir. Evidently, I think this is a long-standing tradition.

According to the expert, their handicraft expertise should be managed through international networks:

> I don't think Oman has a chance to compete for cheap souvenirs. But it has a very good chance to compete for very fine and traditional products, but they have to be absolutely wonderful and absolutely authentic.

Due to modernisation and the increase in mass tourism, well-established shopowners like Saleh, Sanjay and Akshay have left their shops, diminishing the local social capital.

Being part of the flow of everyday life

Similar to Sanjay and Akshay, Abdul (name changed) from India has been working in the souq for a long time. To reach Abdul's shop, I have to climb up a few stairs. Inside the tiny shop, there is a smell of washing powder. It seems time has come to a standstill; the air conditioner is running fast. Abdul's shop has been in existence for around 40–50 years.

> "I arrived with the Akbar Line in Muscat", he recalled his arrival by ship. "We had to anchor the ship far away and were brought in with a small boat. At that time, Muttrah harbour did not exist" he said, adding that his brother had worked as an engineer for an Austrian company in Muscat. That was the main reason why he came to Oman. Previously, he had been assisting his father in a shop in Mumbai.

One of his long-term customers, a female Omani dressed in a black abaya explained one day:

> "In this shop you can buy everything, this is very nice. I always buy my stuff here" she said. "When you go to Lulu hypermarket, you buy more than what you want to buy. Here, I buy exactly what I need and it is not expensive".

This reflected her close connection and bonding social capital with the shop. She orders a list of fresh vegetables that are not available in the shop. Abdul explains that he will buy them and deliver all groceries to her home, an individualised customer service with added value.

In February 2019, I noticed that more well-established shop-owners had shifted their business to the backs of their shops, while others were selling in person, during certain times, online or via phone. In the front of their shops newly arrived Asian vendors sold imported, "global" souvenirs (Figure 9.2).

The face-to-face interaction became secondary and handled by expatriates who were aggressively approaching customers, giving Souq Muttrah a rather "transitional and disruptive identity". Some well-established vendors told me they felt left out from the official tourism planning, also they were not included within a governmental Muttrah Tourism Committee, formed a few years ago. To counteract the disappearance of traditional shops, the Omani Handicraft Authority has set-up a small, "mobile Omani handicraft kiosk" close to the main entrance of the souq, thus mitigating any conflict with new vendors. However, the small kiosk also vanished in 2018.

During a public event on urban development in Souq Muttrah held at a public museum in Muttrah in March 2019, different stakeholders from the local community were invited by the author to create a platform for discussion. None of the well-established Omani and expatriate vendors and owners attended, compared to a number of government representatives. This may translate into a lack

Figure 9.2 Flashy tourist shops at the entrance of Souq Muttrah. Source: Gutberlet (2019).

of trust and reciprocity regarding establishing an equal platform for a constructive exchange, bridging and bonding social capital between different stakeholders. It seems tourism development in Souq Muttrah has divided the community, their products, their generations, their legacy, their bonding social capital and their ethnic backgrounds. The souq developed into a staged, commodified place where new vendors are at frontstage, while the older generation of well-established Omani vendors and those from diaspora communities and their legacy have taken a back seat.

Conclusions

My research aims at contributing knowledge to the analysis of valuing the bonding and bridging social capital (Soulard et al., 2018), based on narratives and life stories of a well-established trader community within community-based tourism in an emerging destination. It also fills a research gap towards more just, diverse and inclusive tourism development. Social capital can be seen as an important tool to promote sustainable tourism and to analyse the cohesion of communities. According to the community members interviewed, everyday life in the souq created a bond or cohesion between families, which has decreased over time. A lack of consultation and communication decreased the bonding social capital, creating feelings of exclusion, that is, of being left out of the development process. In the past there was more attachment to the place and other members of the ethnic, "kin-based community network", creating a sense of mutual social capital (Soulard et al., 2018). There has been a loss of bonding social capital through an increased ethnic fragmentation, a lack of trust and networks enhanced by a shift from long-term, local customers with a long-term attachment to the souq to short-term visitors who are disconnected from the place and its products. Due to increased tourism, the product range for the local community has decreased. The face-to-face interaction has become secondary, handled by new Asian expatriates who are aggressively approaching customers, giving Souq Muttrah a character of a "transitional neighbourhood" (Colic-Peisker & Robertson, 2015, p. 88) with a socially "disruptive identity" (Table 9.1).

The socio-cultural changes are in line with "liquid modernity", a collapse of long-term thinking, where planning and acting have resulted in a fragmentation of individual lives, a series of short-term projects and episodes that seem to be infinite (Bauman, 2000, p. 3). It appears extremely difficult to disentangle and disrupt the power structures that have developed in Souq Muttrah. Consequently, it seems easier to construct a new place with new vendors from Asia and little social capital. Communities are complex; it is therefore important to further study the relationship between social capital, race, identity and community-based tourism. My study used data collected from Omani and expatriate stakeholders in Souq Muttrah. As such, my findings have limited applicability to other societies, but they can act as a reference to other local markets in the region as well as to other markets in emerging destinations that have turned into commodified tourist sites. The different stakeholders involved in the planning and management of Souq

Table 9.1 Framework: Dimensions of bonding and bridging social capital in Souq Muttrah

High bonding and bridging social capital	*Low bonding and bridging social capital*
Handicraft: locally made	Ready-made, global
Shops: kin-based, owned and operated by a family, handed over from one generation (Omani and non-Omani) to the next one – promoting continuity	Sold or operated by new expatriates – lack of continuity
Government intervention: little, social mix	High, e.g. Omanisation
Lifestyle in the souq: simple, sharing the flow of everyday lives, working, playing cards	Modernisation, little sharing of everyday lives, anonymity, sales-oriented behaviour
Networks: kin-based, tight, local mutual bonding between vendors and vendors and with customers, based on trust, reciprocity and collaboration	Loose global networks, social mobility Capitalist-driven development, low work conditions and cheap labour Little trust, cooperation and involvement, isolation
Customers: predominantly local, emotional exchange and people-focused (the storyteller)	International tourists, superficial, money-focused and online exchange
Community resilience	Little community resilience
"Loci of informal community development"/cohesion	"Transitional neighbourhood"/disruptive identity

Source: Gutberlet (2020).

Muttrah should protect, preserve and promote the soul and spirit of the place. The souq is the ancient heart of a fast-growing city and an inspiration for future and current generations. It should be marketed as a multi-ethnic, heritage and trading space with its social capital and legacy. It constitutes a valuable place to unite and present the rich, diverse and little-discussed trading history within the "living space" of the souq. Innovative ideas and bonding social capital including trust, the involvement of the local community and cooperation are necessary to unite and present the heritage and its multi-ethnic community. Strong social capital can promote community resilience and the ability of the community to resist disruption. My research is especially relevant for tourism development within political and social planning after the COVID-19 pandemic. Between the end of March and mid-June 2020, the souq area was completely locked down as one of the hotspots of the pandemic. The shops had no income and many expatriates returned to their home countries. A valorisation of the social capital within tourism development faces a more than uncertain future.

References

Bauman, Z. (2000). *Liquid Modernity*. Cambridge: Polity Press.

Baxter, J. (2010). Case studies in qualitative research. In I. Hays (Ed.), *Qualitative Methods in Human Geography*. Oxford: Oxford University Press.

Causevic, S., & Neal, M. (2019). The exotic veil: Managing tourist perceptions of national history and statehood in Oman. *Tourism Management, 71*, 504–517.

Christou, P., Farmaki, A., & Evangelou, G. (2018). Nurturing nostalgia?: A response from rural tourism stakeholders. *Tourism Management, 69*, 42–69.

Colic-Peisker, V., & Robertson, S. (2015). Social change and community cohesion: An ethnographic study of two Melbourne suburbs. *Ethnic and Racial Studies, 38*(1), 75–91.

Gutberlet, M. (2016a). Socio-cultural impacts of large-scale cruise tourism in Souq Muttrah. *Fennia, 194*(1), 46–63.

Gutberlet, M. (2016b). Cruise tourist dress behaviors and local-guest reactions in a Muslim country. *Tourism Culture & Communication, 16*, 15–32.

Gutberlet, M. (2017). Staging the oriental other: Imaginaries and performances of German-speaking cruise tourists. *Tourist Studies, 19*(1), 110–137.

Gutberlet, M. (2019). Geopolitical imaginaries and cultural ecosystem services (CES) in the desert. *Tourism Geographies*. doi:10.1080/14616688.2018.1545250

Gutberlet, M. (2020). "They just buy a karak and leave" – Overtourism in Souq Muttrah, The Sultanate of Oman. *Zeitschrift Fuer Tourismuswissenschaften*. doi:10.1515/tw-2020-0004

Hall, C., Prayag, G., & Amore, A. (2008). *Tourism and Resilience: Individual, Organisational and Destination Perspectives*. Bristol: Channel View Publications.

Higgins-Desbiolles, F. (2020). Socialising tourism for social and ecological justice after COVID-19. *Tourism Geographies, 22*(3), 610–623. doi:10.1080/14616688.2020.1757748

Innovative Tourism Advisors (2016). *Oman Tourism Strategy*. Executive Summary for the Ministry of Tourism of the Sultanate of Oman, Doc. 309, extended version, 8 July 2016.

Musavengane, R., & Kloppers, R. (2020). Social capital: An investment towards community resilience in the collaborative natural resources management of community-based tourism schemes, *Tourism Management Perspectives, 34*, 100654.

Oxford Business Group (2019). *The Report Oman*. London: Oxford Business Group.

Palmer, A., König-Lewis, N., & Jones, L. E. M. (2013). The effects of residents' social identity and involvement on their advocacy of incoming tourism. *Tourism Management, 38*, 142–151.

Park, D. B., Nunkoo, R., & Yoon, Y. S. (2015). Rural residents' attitudes to tourism and the moderating effects of social capital. *Tourism Geographies, 17*(1), 112–133.

Portes, A. (1998). Social capital: Its origins and applications in modern sociology. *Annual Review of Sociology, 24*, 1–24.

Putnam, R. (1993). *Making Democracy Work: Civic Traditions in Modern Italy*. Princeton: Princeton University Press.

Richardson, N., & Dörr, M. (2003). Crossroads of culture. In H. H. Seyyid Shihab bin Tariq Al Said (Ed.), *The Craft Heritage of Oman*. Dubai: Motivate Publishing.

Snape, D., & Spencer, L. (2003). The foundations of qualitative research. In J. Ritchie & J. Lewis (Eds.), *Qualitative Research Practice: A Guide for Social Science Students and Researchers* (pp. 1–23). London: Sage.

Soulard, J., Knollenberg, W., Boley, B. B., & Perdue, R. R. (2018). Social capital and destination strategic planning. *Tourism Management, 69*, 189–200.

Spencer, L., Ritchie, J., & O'Connor, W. (2003). Analysis: Practices, principles and processes. In J. Ritchie & J. Lewis (Eds.), *Qualitative Research Practice* (pp. 199–218). London: Sage.

Telfer, D. J. (2003). Development issues in destination communities. In S. Sing, D. Timothy, & R. K. Dowling (Eds.), *Tourism in Destination Communities* (pp. 155–180). Wallingford: CABI.

Timothy, D. (2011). *Cultural Heritage and Tourism: An Introduction*. Bristol: Channel View Publications.

Urry, J. (2000). *Consuming Places*. London: Routledge.

Waldinger, R. (1995). The "other side" of embeddedness: A case-study of the interplay of economy and ethnicity. *Ethnic and Racial Studies*, *18*(3), 555–580.

Wilson, E., & Hollinshead, K. (2015). Qualitative tourism research: Opportunities in the emergent soft sciences. *Annals of Tourism Research*, *54*, 30–47.

World Tourism Organization, UNWTO (2019). *Tourism in the MENA Region*. Madrid: UNWTO.

Zang, R., & Smith, L. (2019). Bonding and dissonance: Rethinking the interrelations among stakeholders in heritage tourism. *Tourism Management*, *74*, 212–223.

10 The evolution of township tourism in South Africa

Irma Booyens

Introduction

One of the most distinctive elements of tourism in South Africa relates to the phenomenon of what is known as township tourism. This chapter analyses the evolution of visitations to townships, hereby contributing to the literature on the history of "slumming" in the Global South (see Frenzel et al., 2015; Steinbrink, 2012). In this chapter, both township visitation and tourism are discussed. A reflection on early township visits enlightens us on how this has paved the way for contemporary forms of the phenomenon. The analysis draws on a review of secondary sources which includes academic literature, archival material (i.e. newspaper articles, reports and policy documents), contemporary news media and other online sources. This work is further informed by the author's field observations in townships for over a decade.

The development of township tourism is discussed in light of Butler's tourism area lifecycle (see Butler, 1980). The stages of the lifecycle involve the (1) "discovery"; (2) "local control" (or local involvement); 3) "institutionalisation" (also maturity or consolidation); and then (4) "stagnation" followed by either the (5) "rejuvenation" or "decline" of tourism destinations. Drawing on evolutionary approaches, the development of tourist areas can be understood as a complex, path-dependent process characterised by co-evolution (Brouder, 2014; Brouder & Ioannides, 2014; Ma & Hassink, 2014). Path dependency refers to the cumulative effects of past developments that create evolutionary patterns towards a critical mass for tourism area development (Ma & Hassink, 2014). Multiple paths typically emerge in a destination and can be complementary and/or contesting (Brouder, 2014). Paths are discernible within tourism contexts and also in relation to broader economic paths in an area (Brouder & Ioannides, 2014). This contribution maps evolutionary patterns or paths for the dominant drivers that spurred the development and major processes of change in township tourism.

Phases in township tourism development

Phases in township tourism development are outlined in Table 10.1. Tourism under apartheid can be understood as political tourism (see Frenzel, 2016, 2020).

Table 10.1 Phases in the development of township tourism

Phase	*Description*	*Period*	*Characteristics*
1	Under apartheid	1960s to early 1990s	• Propaganda visits by leisure and official/political visitors • Political awareness and solidarity vis-à-vis of the plight of townships and their residents • Limited leisure tourism evident in the late 1980s
2	(Re)emergence (early post-apartheid)	Post-1994	• Political solidarity pre and post the 1994 democratic election • High-profile visits to township residents, places symbolic in the struggle and persons in need • Emergence of leisure tourism, mostly foreign visitors • Charity a characteristic driver
3	Commercialisation	2000s	• Growth of township tours as a leisure activity • Growth of volunteer tourism • Government involvement in tourism development, focus on heritage and culture
4	Diversification	2010s to present	• Diversification of the township tourism product and destinations • Growth of experience-based, incl. creative tourism • Emergence of educational and research tourism • Continued "politically correct" visits by high-profile visitors

Source: Author.

The highly politicised gaze on racial segregation in South Africa led to the formation of two contradictory modes of political tourism, namely propaganda and solidarity tourism as discussed in Phase 1.

Phase 1: Township visits under apartheid

The formation of townships is the result of deliberate planning for racialised, spatial segregation by the colonial and apartheid regimes that designed townships as "Native" group areas (Beavon & Rogerson, 1990). Townships under apartheid design were dormitory "labour reserves", constructed an "appropriate" distance away from economic activity in towns or suburbs occupied by white residents, and provided with minimal infrastructure and services such as electricity, water, sanitation, tarred roads, social services and public transport (Beavon & Rogerson, 1990; Phillip, 2014).

Evidence for the first guided township tours can be traced back to the mid-1960s (Dondolo, 2002; Johannesburg City Council, 1964). The tours showcased "Bantu life in the city": viewing township areas and stopping at a "Bantu" art gallery while also having a cup of coffee at one of the city's landmarks, the Oppenheimer Tower (Rogerson & Rogerson, 2021). While these tours had aspects of leisure, they can be interpreted as propaganda tourism. That is, politically motivated tours to normalise townships, making them more acceptable in the eyes of privileged, mostly white onlookers.

The second manifestation of propaganda visits that emerged during the 1960s was official trips facilitated and "mediated" by apartheid authorities for foreigners, usually politicians and reporters. During the 1960s, authorities went to great pains to showcase certain townships as "model" settlements. One such "model" township was Orlando – established in 1923 and extended in 1954 to become the area known as Soweto. Another model township was Sharpeville, established in 1946 about a mile from Vereeniging and planned for 4,000 houses with schools, clinics, shops, sports grounds and other open spaces (*Airdrie and Coatbridge Advertiser*, 7 September 1946). One example of a visit where the propaganda motive was unmistaken occurred in Sharpeville in 1961. A British reporter from the *Daily Herald* was shown around Sharpeville by the Director of Non-European Affairs in Vereeniging, a certain Mr Ferreira, who at the time boasted that Sharpeville was the "best township in South Africa" (Baistow, 1961). Mr Ferreira's astonishingly callous remarks, as reported in the article, came one year after the Sharpeville massacre when police opened fire on Black residents who protested against pass laws, killing or wounding up to 250 people.

Foreign tourism to South Africa dropped dramatically in the 1970s, under 300,000 in 1973 and remained low throughout the 1980s (Dondolo, 2002). During this time, leisure visits to townships all but ceased. By and large, townships became "no-go" areas not only for foreigners but also for non-Black South Africans. Overall, the mobilities of South Africans were severely restricted during apartheid (Rogerson & Rogerson, 2020). While Africans had to carry passbooks and had curfews, whites on their part were restricted from entering townships (Kgagudi, 2019). Whites who worked in townships regularly, like medical doctors and missionaries, needed permits. Foreigners usually needed permission to enter townships and were often accompanied by officials (see Baistow, 1961; Kgagudi, 2019). Those who did visit townships typically were activists and journalists who sought to raise political awareness of the plight of Black residents with respect to poor living conditions and human rights abuses, in addition to the ongoing violence in the townships. There was push-back from apartheid officials and police who did of course not endorse such visits at a time of strict media and literary censorship by the state. At times, police checkpoints were set up to monitor and restrict the mobility of people entering and exiting townships.

In the context of growing international criticism of the apartheid regime, townships became symbolic places of oppression and resistance. Soweto, the largest township in South Africa, emerged centre stage in the anti-apartheid struggle (the struggle for short), receiving much international media and political attention.

Magubane and Lee (1979, p. 22) note that by the early 1970s, people were increasingly "peering" into Soweto:

> The naturally inquisitive ones such as the journalists had begun to dissect the place with their analytical investigations and exposés, and added to its aura with a hundred emotive labels.

In 1972, *The Star* reported: "Soweto is one of the most disgraceful pieces of 20th-century town planning to be found in any advanced nation today" (Magubane & Lee, 1979, p. 22).

Then came the events of the 16 June 1976 youth uprising, which shook the country and the world. About 15,000 schoolchildren and students took part in a protest march against the imposition of Afrikaans as the medium of instruction in township schools (Ndlovu, 2006). The police intercepted the marching youth when they neared Vilakazi Street and after the protestors resisted to disperse, the police opened fire killing up to 500 students and school children. One of the first shot, 13-year-old Hector Pieterson, became the iconic image of the uprising. A photograph taken by Sam Nzima of the dying Hector in the arms of a fellow student, with his sister Antoinette running alongside, was published around the world. In South Africa, the event became catalytic in the changing political landscape with the effects felt in several other townships during the months that followed. Thousands of young people joined the liberation movement via the ANC (African National Congress) and its military arm MK (Umkhonto weSizwe), and also the PAC (Pan-African Congress) (Ndlovu, 2006). Civil unrest and accompanied violence continued during international sanctions in the 1980s, culminating in a State of Emergency in 1985 when strong police and army presence collided with the insurgency in townships.

Dondolo (2002) notes that by the late 1980s, white South Africans along with international funders and thrill-seekers were visiting townships around Cape Town to see the "reality" of township conditions and to experience "danger". These tours seemingly occurred on a limited scale, arguably as a form of "dark" or "atrocity" tourism. However, these leisure motived tours appear to be in stark contrast to the work of human rights activists, social scientists and journalists who continued to frequent townships and who acted in solidarity with the liberation movement.

The year 1990 ushered in an era of new hope with the ban on the ANC lifted and Nelson Mandela released from prison. However, township violence flared up again in the early 1990s and escalated before the first democratic election. Violence spread from locations in KwaZulu-Natal to the townships around Johannesburg and Pretoria. Urban townships were marred with bloodshed resulting primarily, but not exclusively, from an apparent rivalry between ANC and IFP (Inkatha Freedom Party) factions.[1] The victims included ordinary township residents, senior members of political parties, human rights activists, religious leaders, peace negotiators and journalists (Human Rights Watch, 1993). During this time of extreme violence, photographers and journalists braved the townships to tell of the "township killings". Among them was a group of four white South Africa photojournalists dubbed the Bang-Bang Club.[2]

Phase 2: (Re)emergence

This phase outlines the emergence of townships as a destination on the post-1994 tourism landscape with townships becoming a "must-see" for high-profile and leisure visitors alike when visiting South Africa.

The first democratic election was held on 27 April 1994, after which Nelson Mandela took the helm of the coalition government. Economic sanctions were dropped, and South Africa dramatically re-entered the world stage. The "Mandela factor" arguably played a pivotal role in popularising townships (McClarence, 1999). Kgagudi (2019, p. 39) suggests that the Mandela "syndrome" or "factor" was a powerful motivator for foreign visitors to look into the "lifestyles of people living in a country recently liberated from a racist colonial rule".

The early post-1994 years saw the rise of township visits, initially by world leaders, foreign diplomats and politicians, humanitarians and celebrities. These foreign visitors toured the townships accompanied by ANC representatives to meet township residents and struggle stalwarts, visit symbolic places associated with the struggle and also residents in need. Indeed, itineraries often included visits to upliftment projects, schools, orphanages, clinics and hospitals along with some cultural display: food, crafts, singing and dancing. Queen Elizabeth visited townships in 1995 and again in 1999; Tony Blair also visited in 1999. The first American president to visit South Africa was Bill Clinton; he visited Soweto with Hillary Clinton in 1998 (see Table 10.2).

High-profile visits can also be understood as a form of political tourism. These visits are symbolic in nature, highly publicised and arguably the "politically correct" and/or "socially conscious" thing to do when visiting South Africa. The drivers behind such visits appear to be solidarity and charity. Persons who style themselves as humanitarians actively support township projects of various kinds, sometimes initiating these and/or donating substantially to causes. In addition to the "Mandela factor", the media attention given to high-profile visits no doubt further popularised and created a certain fascination with townships. However, the presence and role of such visitors in shaping township tourism have escaped scrutiny.

Leisure visits in the form of township tours gained popularity from the mid-1990s (Frenzel et al., 2015) alongside political tourism. In the late 1990s, advertisements and travel journalism on trips to Soweto and Cape Town's townships started to appear in the foreign newspapers, particularly in the travel supplements (McClarence, 1999; *Sunday Independent*, 27 February 2000; *The Times*, 3 January 1998; 25 September 1999). One company offered a "township and winelands" day trip from Cape Town (*The Times*, 3 January 1998), and another a "township and Robben Island tour" (*Sunday Independent*, 27 February 2000) which included a visit to the political prison where Nelson Mandela and other struggle stalwarts were incarcerated. It is argued that the emergence of leisure tourism post-1994

Table 10.2 Township visits by foreign dignitaries, politicians and celebrities (1990s to present)

Person/s	*Year*	*Township*	*Place/City*
Douglas Hurd	1991	Several townships	KwaZulu-Natal province
John Major	1994	Alexandra	Johannesburg
Queen Elizabeth II	1995; 1999	Khayelitsha; Langa; Alexandra; Soweto	Cape Town; Johannesburg
Bill and Hillary Clinton	1998	Soweto	Johannesburg
Tony Blair	1999	Alexandra	Johannesburg
Prince of Wales and Duchess of Cornwall	2001	Soweto	Johannesburg
Niall Mellon (philanthropist)	2001/2	Several townships	Cape Town
Beyoncé and Bono	2003	Soweto	Johannesburg
Oprah Winfrey	2005	Soweto	Johannesburg
Jamie Oliver	2005; 2009	Soweto; Gugulethu	Johannesburg; Cape Town
Kofi Annan	2006	Soweto	Johannesburg
Angela Merkel	2010	Khayelitsha	Cape Town
Michelle Obama	2011	Soweto	Johannesburg
Barack Obama	2013	Soweto	Johannesburg
Brian Chesky (Airbnb)	2017	Langa	Cape Town
Bill and Melinda Gates	2019	Khayelitsha	Cape Town
Duke and Duchess of Sussex	2019	Khayelitsha and Nyanga; Thembisa	Cape Town; Johannesburg

Source: Author – compiled from media reports.

can be understood as part of the "discovery" stage of the tourism area lifecycle. The discovery phase refers to new arrivals to a newly developed tourism destination marked by an increase in marketing efforts as seen in the late 1990s with respect to township tourism.

Phase 3: Commercialisation

From the 2000s, the commercialisation of township tourism as an urban leisure activity is increasingly evident. In this phase, the government enters the stage as an actor in the development of township tourism. The involvement of government (at national, provincial and local spheres) in township tourism development is in line with Butler's "local control or involvement" stage in the tourism area lifecycle.

In 2002, Rogerson (2002, p.177) averred that while townships have limited tourism infrastructure, they did have a strong attractiveness for international tourists who equate Soweto in particular as the "the spatial embodiment of the struggle for democracy". Early academic observers of township tourism describe township tours as a form of cultural tourism (Dondolo, 2002; Ramchander, 2004; Rogerson, 2002). Ramchander (2004, p. 8) maintains:

> Political violence may have made black township no-go areas for foreign tourists in the days of apartheid, but in 1976 and the political strife of the 1980s subsequently made townships such as Soweto world-famous, and it is not surprising that township tourism has increased significantly since the first democratic election in South Africa in 1994. Their legacy of violence and pain has made townships unlikely tourist destinations, yet busloads of visitors arrived every day to sample the renewed vitality of township life. Most leave with a very different impression from the one with which they arrived, having gained new insights following tours led by local entrepreneurs, and discovering that townships are not depraved areas of violent crime, but vibrant centres populated by friendly people with inspirational stories to tell.

Elements of culture are included in township tours through visits to *shebeens* (drinking) and jazz venues, artists, craft markets and dancing shows (Ramchander, 2004; Rogerson, 2004). Scheyvens (2002) perceives township tourism as a form of "justice tourism": tourists visiting sites of significance associated with the struggle and residents telling their stories to a receptive audience after decades of having their voices silenced by an oppressive and hostile government. This theme has reoccurred in a recent investigation in Cape Town which focused on the perceptions of residents: Muldoon (2020) observes that touristic encounters can create liminal opportunities for hosts and facilitate a level of "reconciliation" between Black hosts and white visitors. Nonetheless, criticism of township tourism abounds in academic circles and sometimes in the media. The infamous township coach tour, described by Ramchander (2004) as "safari-style drive-through tours", is widely regarded as voyeuristic, that is viewing poverty from a position of luxury.

During this commercialisation phase, charity continued to be a strong driver. Following on from the activities of high-profile visitors, townships tours undertaken by leisure visitors characteristically included visits to crèches, schools, clinics, hospitals and other upliftment projects. The charity trend is further manifested through the growth of volunteer tourism (Frenzel, 2016, 2020; Frenzel et al., 2015). At first, tourists donated to schools, families, community projects, clinics and artists (Ramchander, 2004), and later became involved in development initiatives as volunteers. A prominent example is the widely reported Niall Mellon Township Trust which repeatedly called on volunteers to build houses in Cape Town's townships as part of a "building blitz" (Keaney, 2005; O'Donoghue, 2008). It is evident from news articles that these volunteers often visited townships first, were moved by the need they observed and then decided to return with others in an attempt to do something about it. Sympathy sparked action by those with resources and a desire "to make a difference" are alluded to in the articles cited.

Succeeded by the growth of township tours, B&Bs (bed-and-breakfast establishments) broadened the township tourism offering (Rogerson, 2004). The hand of government is identifiable in the promotion of township B&Bs. Delegates to the 2002 World Summit on Sustainable Development in Johannesburg were encouraged to stay in townships both in Soweto and Alexandra (Rogerson, 2004). During the Summit, Soweto was regarded as a "showcase" for international visitors (Kgagudi, 2019). A "Get off the bus" campaign was accordingly launched by government agencies to encourage more people to overnight in townships (Mgibisa, 2007). However, the campaign achieved limited success and B&Bs continued to have low occupancies (Booyens, 2010; Mgibisa, 2007).

From 2000 onwards, authorities increasingly invested in government-led tourism development projects particularly in the townships around Johannesburg, Cape Town and Durban in a response to the growth of foreign visitor demand in these cities. Projects were predominantly focused on struggle heritage (see Table 10.3 for prominent examples). Not all developments have necessarily succeeded as tourism products – the Kliptown precinct development is an example

Table 10.3 Key cultural and struggle heritage products in townships

Type	*Year*	*Township*	*City*
Guga S'thebe Cultural Centre	Constructed in 1999	Langa	Cape Town
Hector Peterson Museum	Opened in 2002	Soweto	Johannesburg
Sharpeville Township and Memorial	Opened in 2002	Sharpeville	Vereeniging
The Red Location Museum	Opened in 2006, now closed	New Brighton	Port Elizabeth
Mandela House	First opened a heritage site in 1999. Restored, with new visitor centre opened to the public in 2009	Soweto	Johannesburg
Vilakazi Street Precinct	Upgrade of public and tourism infrastructure completed in 2010	Soweto	Johannesburg
Cato Manor Heritage Centre	Initially constructed in 2012, currently being extended	Cato Manor	Durban
Kliptown Precinct: • Freedom Square (Walter Sisulu Square of Dedication) • Freedom Charter Monument • Soweto Hotel and Conference Centre (four-star boutique hotel) • Struggle heritage museum	Site opened 2005, the completion of sites and structures followed	Soweto	Johannesburg
Mpumalanga Heritage Centre	Unveiled in 2017	Mpumalanga	Durban

Source: Author – compiled from various sources.

Figure 10.1 Freedom Square, Kliptown Precinct. Source: Author.

of this (Figure 10.1). Mention of the sites is included to show that politicians and local officials actively promote "struggle heritage", particularly by marking sites of significance and building museums or interpretation centres. Government officials routinely regard these initiatives as local economic development (LED) projects even in areas where there is no real tourism potential. It is evident that during the commercialisation phase, township tourism has become part of pro-poor, tourism-led LED strategies (Booyens, 2010; Rogerson, 2006). This approach draws on tourism policy which stresses that tourism development should be "responsible" and promote "culture" to showcase a destination's "real African character" with a focus on arts and crafts, heritage, historical and political aspects (DEAT, 1998; RSA, 1996).

Phase 4: Diversification

Township tourism has increasingly diversified since 2010 with townships offering a wider set of services, activities and experiences. This corresponds with the "institutionalisation" or "consolidation" stages in the tourism area lifecycle.

In 2010, the country experienced a boom in foreign arrivals because of the FIFA World Cup. In anticipation of increased foreign tourist flows, authorities embarked on physical upgrading projects in townships close to major cities (Kgagudi, 2019). Projects involved the expansion of social housing, transport infrastructure, sports facilities, tourism attractions and also beautifying township spaces. For instance, the upgrade of Mandela House was completed in 2009 and construction of the Vilakazi Street Precinct in Soweto was completed in 2010 (Kgagudi, 2019). The Orlando Stadium in Soweto which was rebuilt at a cost of R280 million (Gauteng Tourism Authority, 2020) is located in close proximity to the Vilakazi Precinct in Orlando West where tourist attractions and activities are clustered. In the run-up to the World Cup, the government also embarked on renewed efforts to promote township B&Bs and heritage routes (Briedenhann, 2011). While townships reportedly experienced an influx of visitors during the World Cup and despite all the initial optimism, the expected economic returns to locals proved disappointing (Briedenhann, 2011). Nonetheless, the World Cup coincided with the diversification phase of township tourism which acted as a catalyst for events-based tourism, sport and cultural, along with the expansion of other leisure and hospitality services increasingly in townships. The geographical spread of township tourism beyond the townships around Cape Town, Durban and Johannesburg to those around small towns and rural areas is also evident in the diversification phase (Frenzel, 2020; Rogerson, 2014).

The diversification of township tourism in Soweto, one of the first slum tourism destinations (Frenzel et al., 2015; Steinbrink, 2012), is most pronounced. The township boasts heritage sites, adventure activities like bungee jumping and paintball, restaurants, festivals and cultural events, concerts, food markets, craft breweries, specialist shops, marathons and sporting events (Booyens & Rogerson, 2019a, 2019b). Additionally, townships around Cape Town have cultural and craft centres such as the Guga S'thebe centre in Langa, Makukhanye Art Room and the Shack Theatre in Khayelitsha. Initiatives such as these are often run by city councils, or as NGOs, often with the support of donors with the aim to support arts and youth development in impoverished communities (Booyens et al., 2021). These initiatives draw foreign visitors, some of whom become involved as volunteer tourists. Other activities and services on offer in Cape Town's townships include Airbnb accommodation; arts and food experiences; artisan coffee shops and restaurants such as the Department of Coffee and Streetside Bakery in Khayelitsha; recreation activities like kayaking and cycling in townships; and festivals and concerts of various kinds. Indeed, a manifestation of township tourism diversification is the emergence of creative tourism in townships (Booyens & Rogerson, 2019a, 2019b). Prominent examples are craft making experiences, street-art tours, drumming sessions, visiting jazz or visual artists, etc. The media has picked up on the experience-based township tourism trend as a way to benefit locals (*Herald Live*, 13 June 2019; Mowlana, 2018; Niselow, 2019; Omarjee, 2019). At the same time, there is evidence of tourism precinct development which results in the spatial bundling of tourism activities and attractions, notable areas are the Vilakazi precinct in Soweto and the area around Guga S'thebe Arts and Cultural Centre in Langa (Booyens & Rogerson, 2019b).

Another growing trend associated with tourism diversification is education or research tourism in townships, associated with charity and volunteering. This is a way for foreign students to gain credits towards their courses while overseas over the summer holidays (Budworth, 2012). A recent Soweto study found that over a third of visitors interviewed were students, with several indicating that they visited for education or research purposes (Hoogendoorn et al., 2020). Further recent observations in Langa revealed that local female entrepreneurs who market themselves on Airbnb typically house students and/or volunteers (see Gebauer & Hofäcker, 2021). An example of a tourism NGO from Port Elizabeth that has broadened its development focus is the Calabash Trust. Calabash started offering townships tours in the late 1990s. They won the overall Responsible Tourism Awards[3] in the category "best for poverty reduction" in 2004 (Chesshyre, 2004). The trust has since diversified its portfolio to support community schools (skills, infrastructure and facilities) and facilitating story-telling workshops for parents and students through creative means (theatre/drama, art, music) while offering environmental education and facilitating permaculture design and implementation (www.calabashtours.ac.za). They offer volunteer opportunities to schools, universities and individuals and also for research.

Contributions

This chapter offers insights into the evolution of township tourism. The observation that multiple, co-evolving tourism paths can be identified (Figure 10.2) is the first contribution of this research. The development of paths (often complementary) is the natural outcome of evolutionary processes over time (Brouder, 2014; Brouder & Ioannides, 2014) and is not surprising in the light of tourism area development.

Three dominant drivers of township tourism are propaganda, solidarity and leisure. The propaganda and solidarity drivers are both underscored by political motives. While Frenzel (2016, 2020) refers to township tourism during apartheid as political tourism, this chapter demonstrated that political tourism is a path-dependent thread running through the evolution of township tourism from its beginnings in the 1960s until the present day with high-profile visitors

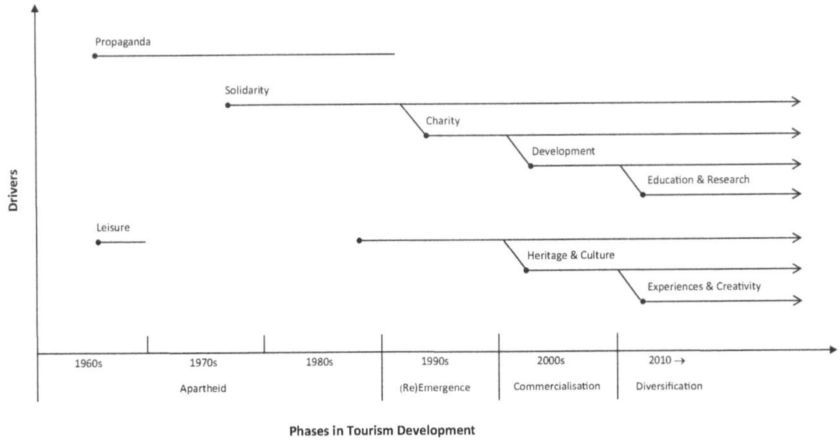

Figure 10.2 Township tourism development paths. Source: Author.

continuing to frequent townships. Politics are part and parcel of the place narrative of townships and township tourism as a phenomenon. The formation of townships was in itself a political act and in the case of Soweto, Kgagudi (2019, p. 2) points to its centrality in "contemporary political and social post-apartheid South Africa". The nature of political tourism, however, has shifted from propaganda in the early years to solidarity during apartheid and most acutely after democracy, and then to charity and development emphases in later years with the growth of volunteer tourism which has undergone further diversification to focus on education and research. Leisure tourism coincided with the propaganda visits in the 1960s but ceased in the 1970s and 1980s. While there is some evidence of leisure visits in the 1980s, the 1990s saw the re-emergence of township tours as an urban leisure activity. The commercialisation of township tourism gained traction from the 2000s with the growth of leisure tourism pushing political tourism into the background and while foregrounding heritage and culture heritage. Leisure tourism has diversified in recent years to include experience-based and creative tourism.

The second contribution of this chapter centres on the changing character of township tourism in the last decade: the township tourism product is diversifying and tourism precincts are expanding, which points to maturing township destinations. Volunteer and experience-based tourism (cultural, creative and adventure) has raised the level of participation in the consumption of tourism experiences. Visitors who take part in volunteer and experience-based tourism are not simply onlookers or gazers who consume poverty, they co-create experiences with hosts which entail more complex motives. Tourist consumption patterns are accordingly shifting in certain respects from poverty to leisure consumption (see Booyens & Rogerson, 2019a, 2019b; Jürgens & Donaldson, 2012). In addition, tourism precincts with heritage sites, up-market leisure facilities and services have developed in townships, along with a number of events, which are consumed by locals, South African day visitors and foreign visitors alike. It should, however, be recognised that township tourism diversification remains unequal. Tourism development has expanded in Soweto, and to a lesser degree in Alexandra, in the Johannesburg area, and also in some townships around Cape Town; but remain minimal in many other townships. The township tourism experience in Soweto, which has expanded dramatically, should, for instance, not be compared with the experience in a place like Imizamo Yethu (Hout Bay, Cape Town) where touring poverty remains the dominant feature (see Huysamen et al., 2020).

Critiques of township tourism highlight that township tours typically result in limited, if any, benefits for local communities. However, it is argued that tourism diversification affords more opportunities for local entrepreneurship and employment. While investigations underscore that local Black entrepreneurs face various barriers and benefit little (Hikido, 2018; Koens & Thomas, 2016), there are some examples of success. Booyens and Rogerson (2019b) draw attention to Black entrepreneurs who own successful restaurants and craft breweries in Soweto and also facilitate creative experiences in Soweto and Langa. In a recent investigation in Soweto, Kgagudi (2019) claims that there are growing opportunities for many small entrepreneurs and businesses. She interviewed several, active local Black tour guides. In another recent study in Soweto, several residents reported that they have benefitted from tourism through "the sale of food and drinks in restaurants or cafés, working in B&Bs, selling crafts and clothing to tourists and providing forms of tourism-related entertainment" (Hoogendoorn et al., 2020, p. 9). Emergent indications of greater local benefit and embeddedness along with government support for township tourism development, however, need to be explored in more depth in future investigations on township tourism.

As the last word, the caveat should be added that at the time of writing township tourism has all but collapsed because of the COVID-19 pandemic. Foreign tourism to townships has ceased and the mobility of domestic visitors severely restricted. The recovery outlook of township tourism remains uncertain. It is too soon to know whether township tourism enterprises will rebound when tourist demand returns and to what extent if and when it does. However, as argued in this chapter, since township tourism development is path-dependent, tourism will likely continue alongside similar patterns when it re-emerges (again).

Notes

1 Nelson Mandela among others suspected a "third force" linked to the apartheid government of stirring up unrest in townships to undermine the negotiations and processes steering towards a democratic transition.
2 Greg Marinovich, Kevin Carter, Ken Oosterbroek and João Silva. Among media prizes won by members of the group, Greg Marinovich won a Pulitzer Prize for his depictions of the Hostel War.
3 In association with *The Times*, *World Travel Market* and *Geographical Magazine*.

References

Airdrie and Coatbridge Advertiser, 7 September 1946.

Baistow, T. (1961). Sharpeville: I go back a year later. *Daily Herald*.

Beavon, K., & Rogerson, C. M. (1990). Temporary trading for temporary people: The making of hawking in Soweto. In D. Drakakis-Smith (Ed.), *Economic Growth and Urbanization in Developing Areas* (pp. 263–286). London: Routledge.

Booyens, I. (2010). Rethinking township tourism: Towards responsible tourism development in South African townships. *Development Southern Africa*, *27*(2), 273–287. doi:10.1080/03768351003740795

Booyens, I., & Rogerson, C. M. (2019a). Re-creating slum tourism: Perspectives from South Africa. *Urbani Izziv*, 30, 52–63.

Booyens, I., & Rogerson, C. M. (2019b). Creative slum tourism: South African township explorations. *Tourism Review*, *74*(2), 256–267. doi:10.1108/TR-12-2017-0200

Booyens, I., Mkuzo, N., & Morgan, M. (2021). The cultural centre of GugaS'thebe as a transformative creative space. In B. J. Hracs, R. Comunian, & L. England (Eds.), *Spaces and Working Practices for Creative Economies in Africa*. Abingdon: Routledge.

Briedenhann, J. (2011). Economic and tourism expectations of the 2010 FIFA World Cup: A resident perspective. *Journal of Sport & Tourism*, *16*(1), 5–32. doi:10.1080/14775085.2011.568085

Brouder, P. (2014). Evolutionary economic geography and tourism studies: Extant studies and future research directions. *Tourism Geographies*, *16*(4), 540–545. doi:10.1080/14616688.2014.947314

Brouder, P., & Ioannides, D. (2014). Urban tourism and evolutionary economic geography: Complexity and co-evolution in contested spaces. *Urban Forum*, *25*(4), 419–430. doi:10.1007/s12132-014-9239-z

Budworth, D. (2012). Why studying abroad is becoming a matter of course for Britons. *The Times*.

Butler, R. W. (1980). The concept of a tourist area cycle of evolution: Implications for management of resources. *The Canadian Geographer*, *24*(1), 5–12.

Chesshyre, T. (2004). Triumph from the townships. *The Times*.

DEAT (Department of Environmental Affairs and Tourism) (1998). *Tourism in GEAR: Tourism Development Strategy 1998–2000*. Pretoria: DEAT.

Dondolo, L. (2002). *The Construction of Public History and Tourism Destinations in Cape Town's Townships: A Study of Routes, Sites and Heritage*. Unpublished master's dissertation. University of the Western Cape.

Frenzel, F. (2016). *Slumming It: The Tourist Valorization of Urban Poverty*. London: Zed.

Frenzel, F. (2020). Touring poverty in townships, inner-city, and rural South Africa. In J. M. Rogerson & G. E. Visser (Eds.), *New Directions in South African Tourism Geographies* (pp. 167–181). Cham: Springer.

Frenzel, F., Koens, K., Steinbrink, M., & Rogerson, C. M. (2015). Slum tourism: State of the art. *Tourism Review International*, *18*(4), 237–252. doi:10.3727/154427215X14230549904017

Gauteng Tourism Authority (2020). *Orlando Stadium*. Retrieved from https://www.gauteng.net/attractions/orlando_stadium

Gebauer, M., & Hofäcker, J. (2021). Airbnb in South Africa's townships: From "no-go" to "stay" area. In C. M. Rogerson & J. M. Rogerson (Eds.), *Urban Tourism and the Global South – South African Perspectives*. Cham: Springer.

Herald Live (2019). *Real-Life Experiences Can Boost Township Tourism*. Retrieved from https://www.heraldlive.co.za/news/2019-06-13-real-life-experiences-can-boost-township-tourism/

Hikido, A. (2018). Entrepreneurship in South African township tourism: The impact of interracial social capital. *Ethnic and Racial Studies*, *41*(14), 2580–2598. doi:10.1080/01419870.2017.1392026

Hoogendoorn, G., Letsatsi, N., Malleka, T., & Booyens, I. (2020). Tourist and resident perspectives on "slum tourism": The case of the Vilakazi precinct, Soweto. *GeoJournal*, *85*, 1133–1149. doi:10.1007/s10708-019-10016-2

Human Rights Watch (1993). *Half-Hearted Reform: The Official Response to the Rising Tide of Violence*. New York: Human Rights Watch.

Huysamen, M., Barnett, J., & Fraser, D. S. (2020). Slums of hope: Sanitising silences within township tour reviews. *Geoforum*, *110*, 87–96. doi:10.1016/j.geoforum.2020.01.006

Johannesburg City Council (1964). *Happy Living*. Johannesburg: City Council Non-European Affairs Department.

Jürgens, U., & Donaldson, R. (2012). A review of literature on the transformation processes in South African townships. *Urban Forum*, *23*(2), 153–163. doi:10.1007/s12132-012-9149-x

Keaney, H. (2005). West Sligo men on house-building project for poor in South Africa. *The Sligo Campion*.

Kgagudi, S. M. (2019). *Soweto, The "storybook Place": Tourism and Feeling in a South African Township*. Unpublished doctoral dissertation. University of Pennsylvania.
Koens, K., & Thomas, R. (2016). "You know that's a rip off": Policies and practices surrounding micro-enterprises and poverty alleviation in South African township tourism. *Journal of Sustainable Tourism, 24*(12), 1641–165. doi:10.1080/09669582.2016.1145230
Ma, M., & Hassink, R. (2014). Path dependence and tourism area development: The case of Guilin, China. *Tourism Geographies, 16*(4), 580–597. doi:10.1080/14616688.2014.925966
Magubane, P., & Lee, M. (1979). *Soweto*. Cape Town: Don Nelson.
McClarence, S. (1999). What price is the Mandela Factor? *The Times*.
Mgibisa, M. (2007). Soweto B&B owners lose out. *City Press*.
Mowlana, S. (2018). Youth Summit 2018: Township tourism and the value of our roots. *Traveller24*. Retrieved from https://www.traveller24.com/FindYourEscape/SAHeritage/youth-summit-2018-township-tourism-and-the-value-of-our-roots-20180712
Muldoon, M. L. (2020). South African township residents describe the liminal potentialities of tourism. *Tourism Geographies, 22*(2), 338–353. doi:10.1080/14616688.2019.1648545
Ndlovu, S. (2006). *Mapping the Soweto Uprising Routes*. Emba Project Management on behalf to the Gauteng Department of Public Works, Johannesburg.
Niselow, T. (2019). Moving township tourism away from "poverty porn" to an economic goldmine. *Mail&Guardian*.
O'Donoghue, C. (2008). Mellon recruit for his "building blitz": Millionaire's charity seeking 2,000 volunteers for its latest South African project. *Irish Independent*.
Omarjee, L. (2019). How a fresh take on township tourism can boost SA's economy. *Fin24*. Retrieved from https://www.fin24.com/Entrepreneurs/5-questions-how-a-fresh-take-on-township-tourism-can-boost-sas-economy-20190331
Phillip, K. (2014). A history of townships in South Africa. In S. Mahajan (Ed.), *Economics of South African Townships: Special Focus on Diepsloot* (pp. 31–49). Washington, DC: World Bank.
Ramchander, P. (2004). *Towards the Responsible Management of the Social-Cultural Impact of Township Tourism*. Unpublished doctoral dissertation. University of Pretoria.
Rogerson, C. M. (2002). Urban tourism in the developing world: The case of Johannesburg. *Development Southern Africa, 19*(1), 169–190. doi:10.1080/03768350220123927
Rogerson, C. M. (2004). Urban tourism and small tourism enterprise development in Johannesburg: The case of township tourism. *GeoJournal, 60*(3), 249–257. doi:10.1023/B:GEJO.0000034732.58327.b6
Rogerson, C. M. (2006). Pro-poor local economic development in South Africa: The role of pro-poor tourism. *Local Environment, 11*(1), 37–60. doi:10.1080/13549830500396149.
Rogerson, C. M. (2014). Rethinking slum tourism: Tourism in South Africa's rural slumlands. *Bulletin of Geography. Socio-economic Series, 26*(26), 19–34. doi:10.2478/bog-2014-0042
Rogerson, C. M., & Rogerson, J. M. (2020). Racialized landscapes of tourism: From Jim Crow USA to apartheid South Africa. *Bulletin of Geography. Socio-economic Series, 48*(48), 7–21. doi:10.2478/bog-2020-0010
Rogerson, C. M., & Rogerson, J. M. (2021). Urban tourism under apartheid: The Johannesburg Chapter. In C. M Rogerson & J. M. Rogerson (Eds.), *Urban Tourism in the Global South: South African Perspectives*. Cham: Springer.

RSA (Republic of South Africa) (1996). *White Paper on the Development and Promotion of Tourism in South Africa*. Pretoria: Government Printer.

Scheyvens, R. (2002). *Tourism for Development: Empowering Communities*. Harlow: Prentice Hall.

Steinbrink, M. (2012). We did the slum!: Reflections on urban poverty tourism from a historical perspective. *Tourism Geographies, 14*(2), 1–22. doi:10.1080/14616688.2012.633216

Sunday Independent (London), 27 February 2000.

The Times (United Kingdom), 3 January 1998; 25 September 1999.

Part III

Tourism for change

11 Impacts of tourism on the livelihoods of women in Victoria Falls, Zimbabwe

Tariro Sibanda and Joseph M. Cheer

Introduction

Tourism has been recognised for its potential to bring about economic and social empowerment and enhance local livelihoods (UNWTO, 2020). However, the industry has also been criticised for being "exclusive" by not involving marginalised groups in the development and sharing of benefits from the tourism that occurs in their communities (Biddulph & Scheyvens, 2018). The poor are often marginalised, and women are the largest group of people that are systematically disadvantaged due to social and cultural norms (UNDP, 2019). Extreme inequality is becoming increasingly evident across the world today and is seen clearly in the concentration of wealth and decision-making power among a small elite few. To ensure that no one is left behind, it is essential to break the links between economic and social exclusion and decision-making power (United Nations, 2018). Scheyvens and Biddulph (2018) pointed out that tourism can only be considered inclusive if marginalised groups are involved in its production or consumption and share directly in the benefits that flow from it.

Zimbabwe offers tourists a wide variety of attractions, ranging from beautiful scenery, wildlife and national heritages (Muzapu & Sibanda, 2016). Zimbabwe has five UNESCO World Heritage Sites, of which Mosi-oa-Tunya/Victoria Falls and Great Zimbabwe National Monument are key. In 2019, tourism contributed 6% to Zimbabwe's US$21 billion economy (World Bank, 2020; WTTC, 2020). This is significant due to Zimbabwe's ongoing economic challenges and high rates of unemployment (African Development Bank, 2020). Tourism has emerged as the country's fastest-growing industry – it was recognised in the Medium-Term Plan as one of the key pillars for economic transformation (Zimbabwe Ministry of Tourism, 2015). The majority of Zimbabwe's tourism products are nature-based, which makes the industry vulnerable to environmental changes that may intensify as a result of climate change (Dube & Nhamo, 2020).

Zimbabwe is arguably and enduringly a patriarchal society, where there has been a high tendency for women to be marginalised (UN Women, 2015). A large percentage of households in Southern Africa are female-headed due to men migrating for work (O'Laughlin, 2009). In the case of Zimbabwe, 38% of the country's households are female-headed (UNICEF Zimbabwe, 2016). Consequently, this has led to women becoming key to maintaining rural livelihoods, with the primary responsibility for household affairs (Mutopo, 2011). It is generally acknowledged that when women are decision-makers, there is often an improvement in family livelihoods (World Bank, 2014). The number of women participating in the workforce is limited by poor access to education opportunities (United Nations, 2012). Although this is slowly improving, the current workforce reflects the historically low levels of educational attainment among women (UN Women, 2015). This exploratory study investigates the impacts of tourism on the livelihoods of women in Victoria Falls, particularly regarding the extent to which tourism contributes to their empowerment.

At a community level, the most significant impacts of tourism on the town's livelihoods were those pertaining to well-being and income. The findings show that the town is heavily dependent on tourism, leaving the local economy vulnerable to internal and external shocks. The extent to which tourism has impacted women's livelihoods was analysed by disaggregating the data to compare the results between men and women. Sara Longwe's (1995) empowerment framework was employed to access women's level of empowerment and results indicate that women had not reached the control level of empowerment, which is the highest level of empowerment according to Longwe's (1995) framework. It was found that women's access to employment and business opportunities was limited due to historical constraints such as their lack of access to formal education. This study acknowledges the links between enhanced livelihoods and more meaningful participation of women in tourism towards the overall empowerment and gender equality agenda. While Longwe's (1995) empowerment framework is employed, without more meaningful participation in tourism and the security of more sustainable livelihoods, progress towards empowerment and gender equality will languish.

Literature review

Livelihood strategies

Several approaches to facilitate tourism benefits among the marginalised have been used in the past. Among these is pro-poor tourism that seeks to "generate net benefits for the poor" (Ashley et al., 2001). Community-based natural resource management (CBRM) has been used in conjunction with ecotourism to involve local communities in order to maximise their benefits from natural resources. CBRM can be an effective strategy to transfer tourism benefits to local communities (Rogerson & Saarinen, 2018). However, if the local community is not included, CMBRM can further marginalise communities. This is a major concern, particularly for communities that have been displaced to make way for protected areas, thereby losing access to resources they traditionally used for their livelihoods (Matiku et al., 2020). Chiutsi and Saarinen (2017) found that locals living in close proximity to protected areas believed that tourism had the potential to greatly benefit their communities, yet the benefits they experienced were limited due to poor resource governance. Scheyvens and Biddulph (2018) called for inclusive tourism in order to transform the industry to include those that did not have opportunities to participate in the tourism development that was occurring in their communities. Scheyvens and Biddulph (2018, p. 4) define inclusive tourism as:

> Transformative tourism in which marginalized groups are engaged in ethical production or consumption of tourism and the sharing of its benefits.

Inclusive tourism goes beyond job creation and seeks to transform oppressive power relations (Biddulph & Scheyvens, 2018). Butler and Rogerson (2016) found that when an inclusive tourism development approach was implemented in Dullstroom, South Africa, the growth in the capacity for local women to participate meaningfully led to livelihoods enhancement and social empowerment. Tourism development can be inclusive and contribute towards alleviating poverty if a broad range of stakeholders participate in the creation of opportunities and sharing benefits (Rogerson & Rogerson, 2020).

Women's livelihoods

The sustainable livelihoods approach was born out of the Brundtland Commission Report of 1987 in which sustainable development was proposed (Shen et al., 2008). Chambers and Conway (1992, p. 6) stated that a livelihood can be considered sustainable:

> when it can cope with and recover from stress and shocks, maintain or enhance its capabilities and assets, and provide sustainable livelihood opportunities for the next generation; and which contributes net benefits to other livelihoods at the local and global levels and in the short and long term.

The Sustainable Livelihoods Framework developed by the UK's Department for International Development (DFID, 1999) clearly outlines that one of the fundamental outcomes that should result from development is the generation of sustainable livelihoods. These outcomes, which also informed this research, include increased income, heightened well-being, reduced vulnerability, improved food security and more sustainable use of natural resources (DFID, 1999). Importantly, the sustainable livelihoods approach does not take into account gender and power relations, and their impacts on men and women attaining viable livelihoods (Harcourt, 2016). To address these gaps, we argue that the sustainable livelihoods approach should be used in conjunction with a feminist analysis that does not solely focus on women but also considers social and material inequalities (Hansen & Buechler, 2015). Feminist political ecology, for example, views gender as a distinct social category and considers the distribution of resources vital and assesses power relations and how these underline decision-making capacity (Lutz-Ley & Buechler, 2020).

Participation of women in tourism

Tourism is considered to be experienced differently by men and women, be they producers or consumers (Pritchard et al., 2007). Tourism in Africa, for the most part, is a male-dominated activity, with men in the majority of executive and management positions, while, in contrast, there are limited opportunities for women to gain the experience needed for career progression (Carvalho et al., 2019; UNWTO, 2019). Women are usually found in entry-level and non-managerial jobs in the tourism industry and it is argued that they are constrained and their limited participation threatens the sustainability of the tourism industry more broadly (Alarcón & Cole, 2019). Moswete and Lacey (2015) found that women in neighbouring Botswana were empowered through their involvement in cultural tourism, although they were constrained due to a lack of capital and poor education. Similarly, Movono and Dahles (2017) reported that Indigenous Fijian women were empowered through micro-entrepreneurship, leading to beneficial socio-cultural changes within their respective communities. Conversely, others argue that tourism drives the exploitation of women through abusive and precarious employment arrangements (Pritchard, 2014). Most poignantly, Figueroa-Domecq et al. (2015) argue that although women have come to play a vital role within tourism, the sector suffers gender blindness, resulting in the diminishment of the potential for their meaningful engagement.

The characterisation of UN Women (2011, p. 11) regarding how the greater inclusion of women can have an impact on livelihood outcomes is a point of reference for gender empowerment more broadly and is defined as:

> Empowerment means that people – both women and men can take control over their lives: set their own agendas, gain skills (or have their own skills and knowledge recognized), increase self-confidence, solve problems, and develop self-reliance, it is both a process and an outcome.

This agrees with Ferguson (2011) who states that although tourism has the potential to contribute to development goals through the empowerment of women, the industry would require restructuring in order to realise that potential. Similarly, Sinclair (1997) argues that tourism reinforces gender inequalities due to the clear division of men and women's work. Tourism can be detrimental to women as it exacerbates inequalities, although in limited cases, female involvement in tourism has reduced male control over households (Ferguson & Alarcón, 2015; Ferguson, 2010), supporting assertions that tourism has the ability to shift existing power structures (Boley et al., 2017). Importantly, while tourism offers an opportunity for women to gain employment and social independence, this can also lead to household conflict (Duffy et al., 2015).

In their exposé of women in tourism in the Dominican Republic, Duffy et al. (2015), found that women were employed in jobs that mirrored their traditional domestic contexts as housekeepers or servers, suggesting the entrenchment of distinct gendered roles. Women also assumed triple roles or "third shifts" by working in paid employment, assuming care for their children and undertaking the management of the household (Duffy et al., 2015; León, 2007). In some cases, the reproductive roles of women intensify demands, causing them to defer or curtail professional aspirations (Morgan & Winkler, 2020). Cornwall (2016, p. 356) highlights the importance of building critical consciousness in women collectively; "the process of changing the way people see and experience their worlds to stimulate indignation about injustice and generate the impetus to act together". While the empowerment of women does not automatically result in a reduction of gender-based inequalities, building economic independence is central to gender equity (Grasmuck & Espinal, 2000; UN Women, 2020). Ferguson (2010) argues that the emergent new power dynamics occurring in households is related to the increasing empowerment of women.

Research context

Victoria Falls is located in Matabeleland North Province of Zimbabwe and has a population of 35,761 (Worldometers, 2020). The town is surrounded by the Victoria Falls National Park, a designated UNESCO World Heritage Site (UNESCO, 1989). Victoria Falls accounts for more than 62% of the 957,752 visitors to Zimbabwe's national parks and is clearly the country's most popular icon (ZTA, 2018). According to UNICEF Zimbabwe (2016), Matabeleland North is among the poorest provinces in Zimbabwe, while curiously, Victoria Falls has the lowest unemployment rates in the province. Unsurprisingly, the Victoria Falls economy has come to wholly rely on tourism and this has been the case since the early 1900s. Mutana et al. (2013) argue that at a provincial level, tourism has had a generally positive impact on poverty alleviation in rural Zimbabwe. On the flipside, Mudimba and Tichaawa (2017) highlight that the negative environmental and socio-cultural impacts that tourism has had on the community of Victoria Falls has been considerable. This study acknowledges the significance and history of Mosi-Oa-Tunya (Victoria Falls) and what it meant for the Khoisan and Tonga people of Zimbabwe before David Livingstone arrived in 1855 (Livingstone and Santos, 1940). More broadly, this study acknowledges the call by Chambers and Buzinde (2015) for more Indigenous voices to be heard.

Methodology

Sara Longwe's Women's Empowerment Framework

Sara Longwe's Women's Empowerment Framework was employed to identify the differences between the tourism impacts on men and women. Longwe (1995) developed her women's empowerment framework in response to the programmes that were being implemented in developing countries that seemingly failed to consider the needs of women. The five levels of the gender empowerment framework are (1) welfare, (2) access, (3) conscientisation, (4) mobilisation and (5) control. Welfare refers to the lowest level of intervention; it addresses the need for women to have equal access to food, income, shelter and medical care (March et al., 1999). Access is underlined by the principle of equal opportunity that requires the removal of all forms of discrimination, increasing women's access to resources. Conscientisation refers to the process by which women become aware of their disproportionate access to opportunities relative to men. It is the realisation of discriminatory practices that give priority to men, highlighting that gender roles are cultural and can be altered (March et al., 1999). Mobilisation ensures that where a women's community is affected, they are able to play a part in decision-making processes. At this juncture, women form groups to create space that allows them to tackle issues together. Control is the level that women who have taken action reach and where it allows them to participate in decision-making. Much like any methodological framework, Longwe's (1995) approach has distinct limitations – it is static and does not consider changes in the context in which it is applied and also sparsely acknowledges other forms of equality (ILO, 1996). Notwithstanding these shortcomings, the premise of the framework to distinguish between the impacts of tourism on men and women is pertinent to the study.

The research design was exploratory by nature and follows strategies used in similar studies regarding the impacts of sustainable tourism for development (Ashley et al., 2001; Mutana et al., 2013). Sampling was aligned with a rapid rural appraisal approach with two sample populations studied – the residents of Victoria Falls and key tourism sector informants in the area. The residents of Victoria Falls were sampled using stratified random sampling, with questionnaires administered to 104 respondents comprising 57 men and 47 women. Key informants were selected through judgemental sampling and comprised of 12 key tourism stakeholders and community leaders (Table 11.1). Data were collected over a period of two weeks where the main author was ensconced at Victoria Falls. Language constraints were overcome as the main author was fluent in Shona, and as a Zimbabwean national, was well versed with etic and emic views.

Overall, a mixed-method approach was employed, leveraging both quantitative and qualitative data alongside an intimate understanding of complex local-level social interactions. Quantitative data were collected through a questionnaire and the survey results were supported by the use of semi-structured key informant

Table 11.1 Description of key informants interviewed

Respondent	*Description of respondent*
R1	Business Development Manager
R2	Local business owner
R3	Private Pilot
R4	Tour Operator Human Resources Manager
R5	Zimbabwe Tourism Authority Official
R6	Zimbabwe Tourism Authority Board Member
R7	Zimbabwe Tourism Council Representative
R8	Hotel General Manager
R9	Hotel Assistant General Manager
R10	Local Tourism business owner
R11	Overland Tour Guide
R12	Public Relations Manager

interviews. The questionnaire utilised indicators of poverty as proxies for quality of life (Cahyat et al., 2007). Women's empowerment levels of the respondents were measured by integrating Sara Longwe's Women's Empowerment Framework (1995).

Quantitative data were analysed using the SPSS statistical package where frequencies and means were analysed, while the second level of analysis employed cross tabulations to investigate responses to each variable by men and women. Disaggregating the data allowed analysis of differences in impact between men and women. Conversely, qualitative data were analysed using NVivo software to categorise and code data to elicit key themes. The quantitative and qualitative data were synthesised to provide a clearer picture of the overall situation in Victoria Falls. One limitation to the approach taken was that respondents to the survey were accessed in public only – the survey was not authorised to occur within households according to research ethics stipulations.

Findings and discussion

Demographics

The surveyed respondents consisted of 54% men and 46% women. The majority of survey respondents (49%) were below the age of 34, while 27.5% were aged between 35 and 44 years. The remaining 23.5% of respondents were in the age range of 45–54 years. The proportion of married respondents was 56%, while 7% were divorced, 6% were widowed and 31% were unmarried. A larger proportion of men (71%) than women (36.2%) were married and there was a higher percentage of single women (38.3%) and only 25% of surveyed men were single. This may be due to married women in Zimbabwe being expected to stay home to look after their households, as seen in Duffy et al. (2015). The higher proportion of men that were married could be explained by the fact that men tend to migrate from other cities and rural areas in Zimbabwe to Victoria Falls where opportunities for employment or trade-related work are stronger. As far as educational attainment, men without formal schooling accounted for 5.8%, while for women this was around 6.4%. There was a higher proportion of women that had completed primary school, 10.6% compared with 6.8% of men. More than half of surveyed men (55.4%) had been educated up to secondary school compared with 40.4% of women, while a total of 38.2% of women had attended university, slightly higher than the 34% of men that had tertiary education.

Livelihoods

The proportion of respondents listing livestock farming as a key livelihood activity was 9.7%, while 5.8 % of respondents indicated that crop farming was one of their livelihood activities. Craft makers, whose work was sold to tourists, constituted 23.3% of respondents, while 15.5% cited vending as a source of income. The proportion of men involved in craft-making were 21.4%, while 25.5% of women made crafts. There was no significant gap between men and women that were formally employed, with 62.5% of men formally employed and 59.5% of women in formal employment. However, a key informant (R2) mentioned that it was easier for men to get jobs and that more men were employed in management positions. Victoria Falls experiences the impact of this historical imbalance as the older generation of women are poorly educated and, therefore, are unable to be employed in many of the positions held by men. These findings accord with the United Nations (2012) finding that the number of women found in Zimbabwe's workforce is hindered by the historical legacy of poor access to education with the resultant effect of a pronounced gender income disparity. Another informant (R10) remarked that women were sent traditionally by their husbands to sell curios to tourists, while men worked in formal employment (Figure 11.1).

The levels of monthly income indicate that 68% of respondents earned above US$300 per month, with 24.4% earning between US$201 and US$300 and 11.6% earning below US$200. There were no significant differences between men and women earning between US$301 and US$500 monthly. The proportion of men and women earning between US$301 and US$500 per month was 32.1% and 36.2%, respectively while 41% of men and 25% of women earned more than US$500. In general, and compared across the country, this demonstrated the potential of tourism to underline higher incomes (Table 11.2).

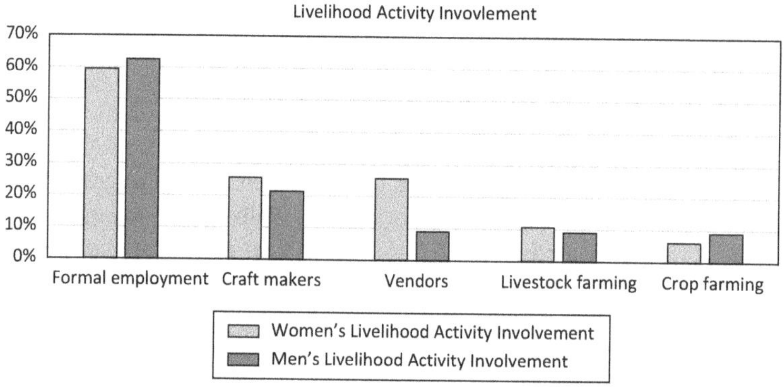

Figure 11.1 Livelihood activity by proportion of men and women. Source: Authors

Table 11.2 Monthly income of respondents

Monthly Income (USD)	*Percentage of Respondents*	*Men*	*Women*
$0–$100	1.9%	3.6%	0%
$101–$200	9.7%	5.4%	12.5%
$201–$300	20.4%	17.9%	23.4%
$301–$500	34%	32.1%	36.2%
$501+	34%	41%	25.5%

Quality of life

The proportion of respondents that cared for three or four dependents was 37.4%, while 16.5% provided for more than four dependents. The women who had no dependents comprised 17%, while 12.5% of the men had no dependents; 64.1% of respondents reported that they were able to afford school fees. There was no significant difference between the ability of men and women to afford their dependents' education. The proportion of respondents who were able to afford health care for themselves or their dependents was 68%. Most respondents (70%) said they did not own the homes they were living in and there was little difference between the proportion of men and women that owned homes at 31.9% and 28.6%, respectively. The majority of surveyed women (80.9%) said they decided alone on how their income was used; however, only 26% of married women reported making their own decisions regarding the use of income. Significantly fewer women indicated that tourism had increased their participation in decision-making. This suggests that women already had the decision-making power in households or their decision-making participation had not been affected by their involvement in tourism. A total of 37.5% of men and 10.6% of women said that there were other ways in which decisions were made on expenditure, including those that decided with their spouses and those that had families decide for them.

Impacts of tourism

Respondents that had been involved in tourism for three to four years represented 24.3%, while 20.4% had been involved in tourism for one to two years. More than 50% of men (55.4%) had been involved in tourism for more than five years, while 38.3% of women had been involved for the same period (see Figure 11.2). The higher percentage of men than women who had been involved in tourism for more

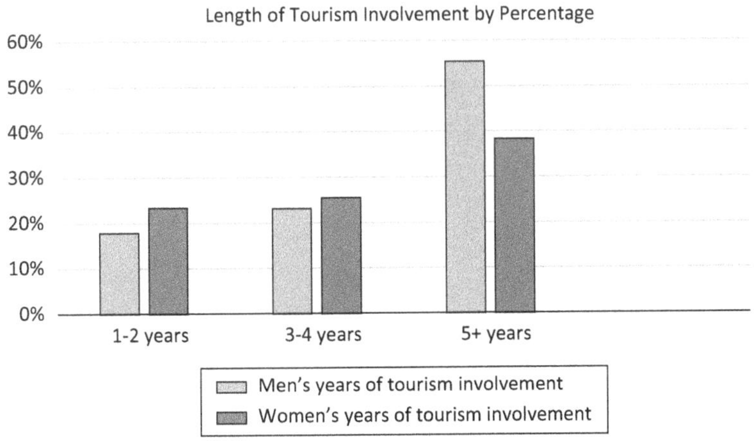

Figure 11.2 Length of men's and women's involvement in tourism. Source: Authors.

than five years could indicate that women have a shorter window to be involved in tourism due to family commitments. Alternatively, this could point to the changing dynamics of the tourism workforce in Victoria Falls over the last few years, with more women becoming involved in the sector as opportunities that women did not have in the past become more open.

A comparable increase in income as a result of being involved in tourism was reported by 43% of respondents, with 44% citing increased well-being as a result. Increased well-being was reported by 46.4% of men and 40.4% of women, while 34% of men and women said they now had less time for home duties (see Figure 11.3).

The proportion of respondents who cited fluctuating tourist numbers as the main challenge that they faced was 52.4%, while 22.3% felt that they did not have enough resources and a further 28% of the respondents noted that the English language was a challenge. Women indicating the English language was a constraint stood at 34% (Figure 11.4) and this probably relates to lower levels of

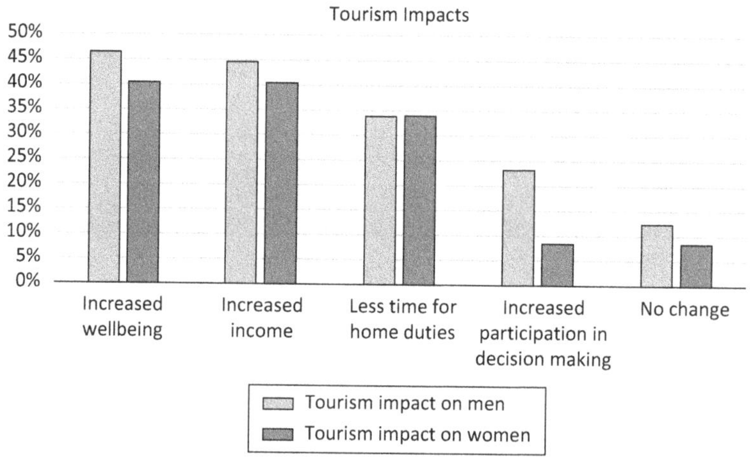

Figure 11.3 Impacts of tourism on men and women. Source: Authors.

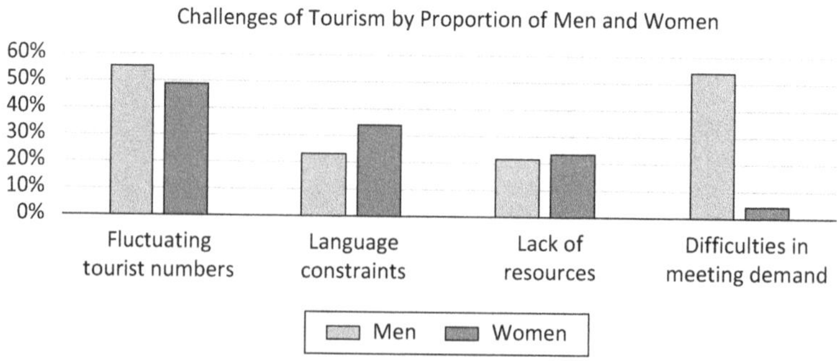

Figure 11.4 Challenges of tourism by proportion of men and women. Source: Authors.

education. Among the women who had not attended secondary school education, 12.8% reported that the English language was a constraint, compared with 3.5% who had attained secondary school education.

Women in tourism

The majority of respondents (58%) thought that tourism had a different effect on men and women in Victoria Falls, aligning with extant literature. Men who believed that tourism had the same effect on women were 44.6%, while 61.7% of women indicated that tourism had a different effect on women. The higher proportion of female respondents indicates that they have reached the conscientisation level of empowerment and are able to identify the inequalities and disparities in opportunities. Women are also aware of the burden of the triple gender roles they are playing. When asked to explain why they believed tourism had or did not have a different effect on men and women, the answers revealed entrenched local-level attitudes and biases regarding gender equality. Although respondents tended to believe that the benefits for men and women were more or less the same, they thought that women should not assume some roles. Female respondents remarked that unfair treatment included the fact that there were more jobs for men and that men tended to get more favourable conditions. A key informant (R4) remarked:

> Equal opportunity, depending on trade and gender, tends to default to the job suitable for each person.

Most of the key informants believed that women were found at the lower end of the tourism industry hierarchy, with low representation in upper management. Key informants also suggested that there were roles in the tourism industry that few women, if any, were found in, including rangers, drivers, hunters and guides. Many believe such roles would take women away from their families and that they would be unable to fulfil the reproductive labour expected of them. The sentiment that women's involvement in tourism would interfere with their family roles agrees with previous research that pointed to the gendered roles that occur in the tourism sector in developing countries (Duffy et al., 2015). Long periods away from home that interfere with family roles, personal safety and long hours were raised as the reasons. A key informant (R7) reflected on their experience:

> Having grown up as a girl child in Zimbabwe, as a woman, you were told that you have to be a housewife, so the task is to take that culture away from the men themselves.

A strong theme that emerged from surveys and interviews was that women could not work in roles where they may come into contact with wildlife due to safety concerns and cultural beliefs, which resulted in limited access to such roles.

Involvement of the community in tourism planning

The proportion of respondents who participated in association meetings concerned with tourism planning at Victoria Falls was 40% in total. Key informants tended to report that they were involved in tourism planning via industry associations through which they communicated to the Zimbabwe Tourism Board. Community involvement was characterised by consulting chiefs of nearby villages and the only way that residents could be involved in tourism planning was through their ward councillor, who conveyed their views to the town council. One respondent (R12) stated that not every local person could be involved in tourism planning because:

> One needs to be trained, one needs to be taught how to present community issues to a formal setting.

Unsurprisingly, a greater proportion of men participated in council and association meetings, with 21.4% of men involved in tourism planning through council meetings. More men (46%) were involved in tourism planning through association meetings, as opposed to 34% of women, indicating that women had low levels of participation in public spheres (Figure 11.5).

Level of women's empowerment

In order to determine the impact of tourism on livelihoods, and the overall empowerment of women in Victoria Falls through tourism, data were analysed and contextualised based on Sara Longwe's Framework (1995). In the case of Victoria Falls, data shows that women have reached the welfare level of empowerment as they are able to provide for their needs, as the majority of surveyed women were able to afford meals, school fees and medical care for their household. The

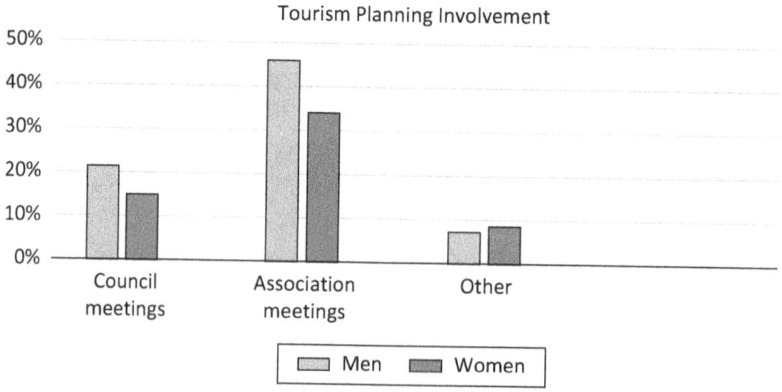

Figure 11.5 Tourism planning involvement by proportion of men and women. Source: Authors.

majority of surveyed women had reached the conscientisation level of empowerment as they were aware of inequalities they faced when compared with men, however, they had not mobilised to form groups to address the issues they face. Drawing from respondents' statements there are expected gender norms within the community that may be preventing women from moving to mobilisation. The challenges experienced by women in tourism were all rooted in social constructs and will require time to change (R1):

> Our culture still favours the men as the provider in our family. Sadly, because of where we are located, things tend to take a lot longer. This is a developing nation and, as such, we cannot make comparisons with developed nations when it comes to gender equality and the gender gap, but I have no doubt in the next few years we will see that gap becoming less and less.

The socialisation that has occurred at childhood limits the aspirations of women and influences the attitudes of men regarding what is allowable for women. In order for women in Victoria Falls to move to mobilisation, changing the mindsets of both local men and women is critical. The findings indicate that women have not reached the control level of empowerment. This may be due to the lack of opportunities for women in Victoria Falls due to their limited access to education and employment. An improvement in the income of women will increase self-esteem and control over assets (Grasmuck & Espinal, 2000; UN Women, 2020).

Conclusion

This study set out to understand the impacts of tourism on women's livelihoods and in turn, their overall empowerment. The findings indicate that women in Victoria Falls have achieved the welfare level of empowerment and that there were no significant gaps between men and women in terms of their ability to afford daily meals, medical services and their dependents' education. The women of Victoria Falls have also reached the access level of empowerment, although there are historical legacies tied to education (or lack of) that are still harming employment and business opportunities for women. Women in Victoria Falls are aware of the inequalities that exist, and they have reached the conscientisation level of empowerment according to Longwe's (1995) framework. They have begun to mobilise by forming a Women's Association and understand the need to speak with one voice but are yet to reach the mobilisation level of empowerment. These results do not indicate that women have reached the control level of empowerment. Although Sara Longwe's Women's Empowerment Framework (1995) has been criticised for viewing women's empowerment as hierarchical, the case of Victoria Falls implies the hierarchical nature of women's empowerment.

Disaggregating the data between men and women highlights the gender differences in the impact of tourism. There is a need for further research on the impact of marriage on women's involvement in tourism as indicated by the higher proportion of single and divorced women involved in tourism. The data that emerged indicated that men are placed higher than women in the tourism value chain. The proportion of women vendors was greater than that of men, confirming that there are gendered roles within tourism in Victoria Falls. A large proportion of women in Victoria Falls were placed in the informal sector of tourism, therefore, there is a risk that their levels of empowerment may be affected by reduced income and limited social protection that has since intensified due to the impact of COVID-19.

While the focus of the research was pointed at identifying the empowerment of women in Victoria Falls, the findings support the view that overall, the community of Victoria Falls has been positively impacted by tourism. Respondents indicated that the most significant livelihood impacts experienced as a result of tourism included increased well-being and additional income. The negative impacts that were identified were high costs of living, changes to the natural heritage being developed for tourism activities and dependency on tourism. As a result of tourism being the main industry in Victoria Falls, residents are shifting their livelihood activities to focus more on tourism. As more residents turn to tourism as a core livelihood strategy, it is intensifying competition. Additionally, there is limited community involvement in tourism planning – this means that control over governance is limited.

Finally, there is an urgent need for livelihood diversification in Victoria Falls to ensure that the town develops resilience to external and internal shocks that may impact tourism visitation. The importance of reducing dependency on tourism and building resilience has been amplified by the COVID-19 pandemic as tourism has all but come to a halt. As a nature-based attraction, Victoria Falls is susceptible to changes that may occur due to climate change effects, such as the recent prolonged drought. The tourism sector must help facilitate linkages to other sectors such as agriculture and local tourism businesses must go beyond corporate social responsibility and create shared value for the community. Participation in tourism planning will allow for greater opportunities for the community to benefit from tourism. The implications for tourism research are therefore many and varied but a priority must be placed on longitudinal investigations that measure and track the ongoing impacts of tourism on the livelihoods of women, and in the long run, their enduring empowerment.

References

African Development Bank (2020). *Zimbabwe Economic Outlook*. Retrieved from https://www.afdb.org/en/countries/southern-africa/zimbabwe/zimbabwe-economic-outlook

Alarcón, D. M., & Cole, S. (2019). No sustainability for tourism without gender equality. *Journal of Sustainable Tourism*, *27*(7), 903–919. doi:10.1080/09669582.2019.1588283

Ashley, C., Roe, D., & Goodwin, H. (2001). *Pro-Poor Tourism Strategies: Making Tourism Work for The Poor A Review of Experience*. London: Overseas Development Institute.

Biddulph, R., & Scheyvens, R. (2018). Introducing inclusive tourism. *Tourism Geographies*, *20*(4), 583–588. doi:10.1080/14616688.2018.1486880

Boley, B. B., Ayscue, E., Maruyama, N., & Woosnam, K. M. (2017). Gender and empowerment: Assessing discrepancies using the resident empowerment through tourism scale. *Journal of Sustainable Tourism*, *25*(1), 113–129. doi:10.1080/09669582.2016.1177065

Butler, G., & Rogerson, C. M. (2016). Inclusive local tourism development in South Africa: Evidence from Dullstroom. *Local Economy*, *31*(1–2), 264–281. doi:10.1177/0269094215623732

Cahyat, A., Gönner, C., & Haug, M. (2007). *Assessing Household Poverty and Wellbeing: A Manual with Examples from Kutai Barat, Indonesia*. Bogor: CIFOR.

Carvalho, I., Costa, C., Lykke, N., & Torres, A. (2019). Beyond the glass ceiling: Gendering tourism management. *Annals of Tourism Research*, *75*, 79–91. doi:10.1016/j.annals.2018.12.022

Chambers, D., & Buzinde, C. (2015). Tourism and decolonisation: Locating research and self. *Annals of Tourism Research*, *51*, 1–16. doi:10.1016/j.annals.2014.12.002

Chambers, R., & Conway, G. (1992). *Sustainable Rural Livelihoods: Practical Concepts for the 21st Century*. Brighton: Institute of Development Studies.

Chiutsi, S., & Saarinen, J. (2017). Local participation in transfrontier tourism: Case of Sengwe community in great Limpopo transfrontier conservation area, Zimbabwe. *Development Southern Africa*, *34*(3), 260–275. doi:10.1080/0376835X.2016.1259987

Cornwall, A. (2016). Women's empowerment: What works? *Journal of International Development*, *28*, 342–359. doi:10.1002/jid.3210

DFID (1999). *Sustainable Livelihoods Guidance Sheets*. Retrieved from https://www.livelihoodscentre.org/documents/114097690/114438878/Sustainable+livelihoods+guidance+sheets.pdf/594e5ea6-99a9-2a4e-f288-cbb4ae4bea8b?t=1569512091877

Dube, K., & Nhamo, G. (2020). Vulnerability of nature-based tourism to climate variability and change: Case of Kariba resort town, Zimbabwe. *Journal of Outdoor Recreation and Tourism*, *29*, 100281. doi:10.1016/j.jort.2020.100281

Duffy, L. N., Kline, C. S., Mowatt, R. A., & Chancellor, H. C. (2015). Women in tourism: Shifting gender ideology in the DR. *Annals of Tourism Research*, *52*, 72–86. doi:10.1016/j.annals.2015.02.017

Ferguson, L. (2010). Interrogating "gender" in development policy and practice: The World Bank, tourism and microenterprise in Honduras. *International Feminist Journal of Politics*, *12*(1), 3–24. doi:10.1080/14616740903429080

Ferguson, L. (2011). Promoting gender equality and empowering women?: Tourism and the third Millennium Development Goal. *Current Issues in Tourism*, *14*(3), 235–249. doi:10.1080/13683500.2011.555522

Ferguson, L., & Alarcon, D. M. (2015). Gender and sustainable tourism: Reflections on theory and practice. *Journal of Sustainable Tourism*, *23*(3), 401–416. doi:10.1080/09669582.2014.957208

Figueroa-Domecq, C., Pritchard, A., Segovia-Pérez, M., Morgan, N., & Villacé-Molinero, T. (2015). Tourism gender research: A critical accounting. *Annals of Tourism Research*, *52*, 87–103. doi:10.1016/j.annals.2015.02.001

Grasmuck, S., & Espinal, R. (2000). Market success or female autonomy?: Income, ideology, and empowerment among microentrepreneurs in the Dominican Republic. *Gender & Society*, *14*(2), 231–255. doi:10.1177%2F089124300014002002

Hanson, A. M., & Buechler, S. J. (2015). Introduction: Towards a feminist political ecology of women, global change, and vulnerable waterscapes. In S. Buechler & A-M. Hanson (Eds.), *A Political Ecology of Women, Water and Global Environmental Change* (pp. 2–16). London: Routledge.

Harcourt, W. (2016). Sustainable livelihoods. In A. Wong, M. Wickramasinghe, R. Hoogland & N. A. Naples (Eds.), *The Wiley Blackwell Encyclopedia of Gender and Sexuality Studies*. doi:10.1002/9781118663219.wbegss003

International Labour Organization (ILO) (1996). *Unit 1: A Conceptual Framework for Gender Analysis and Planning*. Retrieved from http://www.ilo.org/public/english/region/asro/mdtmanila/training/unit1/empowfw.htm

León, Y. M. (2007). The impact of tourism on rural livelihoods in the Dominican Republic's coastal areas. *The Journal of Development Studies, 43*(2), 340–359. doi:10.1080/00220380601125214

Livingstone, D., & Santos, C. (1940). *David Livingstone*. Friedrich.

Longwe, S. H. (1995). Gender awareness: The missing element in the Third World development program. In C. March & T. Wallace (Eds.), *Changing Perception: New Writings on Gender and Development* (pp. 149–157). Oxford: Oxfam.

Lutz-Ley, A. N., & Buechler, S. J. (2020). Mining and women in northwest Mexico: A feminist political ecology approach to impacts on rural livelihoods. *Human Geography, 13*(1), 74–84. doi:10.1177/1942778620910901

March, C., Smyth, I., & Mukhopadhyay, M. (1999). *A Guide to Gender-Analysis Frameworks*. Oxford: Oxfam.

Matiku, S. M., Zuwarimwe, J., & Tshipala, N. (2020). Sustainable tourism planning and management for sustainable livelihoods. *Development Southern Africa*, 1–15. doi:10.1080/0376835X.2020.1801386

Morgan, M. S., & Winkler, R. L. (2020). The third shift?: Gender and empowerment in a women's ecotourism cooperative. *Rural Sociology, 85*(1), 137–164. doi:10.1111/ruso.12275

Moswete, N., & Lacey, G. (2015). "Women cannot lead": Empowering women through cultural tourism in Botswana. *Journal of Sustainable Tourism, 23*(4), 600–617. doi:10.1080/09669582.2014.986488

Movono, A., & Dahles, H. (2017). Female empowerment and tourism: A focus on businesses in a Fijian village. *Asia Pacific Journal of Tourism Research, 22*(6), 681–692. doi:10.1080/10941665.2017.1308397

Mudimba, T., & Tichaawa, T. M. (2017). Voices of local communities regarding their involvement and roles in the tourism development process in Victoria Falls, Zimbabwe. *African Journal of Hospitality, Tourism and Leisure, 6*(4), 1–15.

Mutana, S., Chipfuva, T., & Muchenje, B. (2013). Is tourism in Zimbabwe developing with the poor in mind?: Assessing the pro-poor involvement of tourism operators located near rural areas in Zimbabwe. *Asian Social Science, 9*(5), 154–161. doi:10.5539/ass.v9n5p154

Mutopo, P. (2011). Women's struggles to access and control land and livelihoods after fast track land reform in Mwenezi District, Zimbabwe. *Journal of Peasant Studies, 38*(5), 1021–1046. doi:10.1080/03066150.2011.635787

Muzapu, R., & Sibanda, M. (2016). Tourism development strategies in Zimbabwe. *Management, 6*(3), 55–63.

O'Laughlin, B. (2009). Gender justice, land and the agrarian question in Southern Africa. In A. H. Akram-Lodhi, & C. Kay (Eds.), *Peasants and Globalization. Political Economy, Rural Transformation and the Agrarian Question* (pp. 190–213). London: Routledge.

Pritchard, A. (2014). Gender and feminist perspectives in tourism research. In A. A. Lew, C. M Hall, & A. M. Williams (Eds.), *The Wiley Blackwell Companion to Tourism* (pp. 314–324). Chichester: Wiley Blackwell.

Pritchard, A., Morgan, N., Ateljevic, I., & Harris, C. (2007). *Tourism and Gender: Embodiment, Sensuality and Experience*. Wallingford: CABI.

Rogerson, C. M., & Rogerson, J. M. (2020). Inclusive tourism and municipal assets: Evidence from Overstrand local municipality, South Africa. *Development Southern Africa, 37*(5), 850–854. doi:10.1080/0376835X.2020.1796597

Rogerson, C. M., & Saarinen, J. (2018). Tourism for poverty alleviation: Issues and debates in the Global South. In C. Cooper, S. Volo, B. Gartner, & N. Scott (Eds.), *The SAGE Handbook of Tourism Management: Applications of Theories and Concepts to Tourism* (pp. 22–37). London: SAGE.

Scheyvens, R., & Biddulph, R. (2018). Inclusive tourism development. *Tourism Geographies, 20*(4), 589–609. doi:10.1080/14616688.2017.1381985

Shen, F., Hughey, K. F., & Simmons, D. G. (2008). Connecting the sustainable livelihoods approach and tourism: A review of the literature. *Journal of Hospitality and Tourism Management, 15*(1), 19–31. doi:10.1375/jhtm.15.19

Sinclair, M. T. (1997). Gendered work in tourism: Comparative perspectives. In M. T. Sinclair (Ed.), *Gender, Work and Tourism* (pp. 1–15). London: Routledge.

UN Women (2011). *Women's Empowerment Principles – Equality Means Business*. Retrieved from https://www.unwomen.org/-/media/headquarters/attachments/sections/library/publications/2011/10/women-s-empowerment-principles_en%20pdf.pdf?la=en&vs=1504

UN Women (2015). *Zimbabwe*. Retrieved from https://africa.unwomen.org/en/where-we-are/eastern-and-southern-africa/zimbabwe

UN Women (2020). *Facts and Figures: Economic Empowerment*. Retrieved from https://www.unwomen.org/en/what-we-do/economic-empowerment/facts-and-figures

UNDP (2019). *Human Development Report 2019: Beyond Income, Beyond Averages, Beyond Today: Inequalities in Human Development in the 21st Century*. New York: United Nations Development Program. Retrieved from http://hdr.undp.org/en/content/human-development-report-2019

UNESCO (1989). *Decision: CONF 004 XV.A Inscription: Victoria Falls/Mosi-oa-Tunva (Zambia/Zimbabwe)*. Retrieved from http://whc.unesco.org/en/list/509

UNICEF Zimbabwe (2016). *Extended Analysis of Multiple Indicator Cluster Survey (MICS) 2014: A Story of Inequality and Inequity in Zimbabwe Harare*. Harare: UNICEF.

United Nations (2012). *Africa Renewal*. Retrieved from https://www.un.org/africarenewal/sites/www.un.org.africarenewal/files/Africa-Renewal-Womens-Issue-2012-en.pdf

United Nations (2018). *Leaving No One Behind - Excerpt from Committee for Development Policy, Report on the Twentieth*. Retrieved from https://sustainabledevelopment.un.org/content/documents/2754713_July_PM_2._Leaving_no_one_behind_Summary_from_UN_Committee_for_Development_Policy.pdf

UNWTO (2019). *Global Report on Women in Tourism*. Retrieved from https://www.e-unwto.org/doi/pdf/10.18111/9789284420384

UNWTO (2020). *UNWTO Regional Conference on Sustainable Development through Tourism – Proceedings, Thimphu, Bhutan, 03 June 2019*. Madrid: UNWTO. doi:10.18111/9789284421312

World Bank (2014). *Voice and Agency Empowering Women and Girls for Shared Prosperity*. Retrieved from https://www.worldbank.org/content/dam/Worldbank/document/Gender/ALTERNATE_VOICE_AGENCY_EXECUTIVE_SUMMARY_PRINTING.pdf

World Bank (2020). *Zimbabwe*. Retrieved from https://data.worldbank.org/indicator/SP.POP.TOTL?locations=ZW

Worldometers (2020). *Zimbabwe Population (LIVE)*. Retrieved from https://www.worldometers.info/world-population/zimbabwe-population/

World Travel & Tourism Council (WTTC) (2020). *Zimbabwe 2020 Annual Research: Key Highlights*. Retrieved from https://wttc.org/Research/Economic-Impact

Zimbabwe Ministry of Tourism (2015). *Discussion Paper on Aligning the Tourism Act [Chapter 14:20] and Related Legislation to the Constitution of Zimbabwe*. Retrieved from http://www.justice.gov.zw/imt/wp-content/uploads/2016/08/TOURISM-ACT-IMT-DISCUSSION-PAPER-MINISTRY-OF-TOURISM-AND-HOSPITALITY-INDUSTRY-18-DEC-2015.pdf

Zimbabwe Tourism Authority (ZTA) (2018). *2018 Tourism Trends Statistics*. Retrieved from http://www.zimbabwetourism.net/tourism-trends-statistics/

12 Economic development, tourism and conservation in developing countries

Joseph E. Mbaiwa

Introduction

The tourism industry has grown to become the world's largest economic sector. It is one of the leading job creators in the world, creating more than 3% of all global employment, which translates to 98 million people employed in the tourism industry (Blanke & Chiesa, 2013). The tourism industry is thus a significant and growing contributor to global economies through job creation, revenue generation and infrastructure development (World Bank Group, 2018). Tourism is often seen as the most promising and dynamic power for economic growth in less developed economies and in areas that have natural beauty in abundance because it allows these countries to diversify their economies (Saarinen et al., 2009). Although the tourism industry is an important sector of economic development, it is essential to involve local communities living in wildlife-based tourism destinations in tourism development. This is because a community that is involved in the planning and implementation of tourism develops a positive attitude towards tourism development (Pretty et al., 2009). Involving local communities in tourism in their localities is crucial for sustainable conservation and consumption of natural diversity (Buckley, 1994; Hassan, 2000).

The wildlife-based tourism industry is one of the types of tourism thriving in Asia and Africa (World Bank Group, 2018). Wildlife areas such as the Serengeti in East Africa, Kruger National Park in South Africa, Kafue National Park in Zambia, Etosha National Park in Namibia and the Okavango Delta in Botswana are among the busiest tourist destinations in Africa (Mulale et al., 2013). In addition, these areas figure as globally influential sanctuaries of wildlife conservation. The impacts of wildlife-based tourism on the regional and national economy are expected to uplift the livelihoods of communities by generating income and employment, thereby contributing to rural development and making community-based tourism economically viable (Mbaiwa & Stronza, 2010; Moswete & Lacey, 2015; Mulale et al., 2013; Salazar, 2012).

The wildlife-based tourism industry does not only have impacts on the livelihoods of local communities through economic development but also has long-term natural environmental impacts (Lenao, 2014; Mbaiwa & Kolawole, 2013; Mbaiwa et al., 2011). In East and Southern Africa, the Community Based Natural

Resource Management (CBNRM) programme was developed in the 1980s as a means of co-opting rural communities into the conservation process by generating revenue primarily from safari hunting (DeGeorges & Reilly, 2009). Communities involved in the CBNRM programme have established tourism projects to derive economic benefits from wildlife-based tourism. Although the CBNRM programme was introduced to ensure local community involvement in the management of natural resources and the need to generate profit from tourism, controversy has emerged about the use of wildlife, especially the inclusion of trophy hunting as wildlife-based tourism development and a land-use option. This controversy constitutes a series of North–South debates. The differences between the Global North and Global South in wildlife utilisation and conservation requires an interrogation with the assumption of recommending a sustainable approach for the wildlife-based tourism industry especially for East and Southern Africa. The objective of this chapter, therefore, is to draw on the concept of sustainable tourism to analyse economic development, tourism and conservation in wildlife-based tourism destinations in developing countries.

Sustainable tourism

The concept of sustainable tourism is used to inform this chapter. The term "sustainable tourism" was first used almost two decades ago (Buckley, 2012). Since the 1980s, the concept of sustainable tourism has attracted significant attention in tourism studies. The concept of sustainable tourism is largely influenced by the idea of sustainable development that promotes inter-and intra-generational equity, that is, development is considered sustainable only to the extent that the needs of today can be fulfilled without prejudice to those of future generations (WCED, 1987). The World Tourism Organization (WTO, 2001) defines sustainable tourism as development that meets the needs of present tourists and host regions while protecting and enhancing opportunities for the future. The WTO envisages sustainable tourism as leading to the management of all resources in such a way that those economic, social and aesthetic needs can be fulfilled while maintaining sustainability in cultural integrity, essential ecological processes, and biological diversity and life support systems. Sustainability in tourism is driven by the following: dissatisfaction with existing products; growing environmental awareness and cultural sensitivity; realisation of the precious resources that destination regions possess and their vulnerability; and the changing attitudes of developers and tour operators (Liu, 2003; Prosser, 1994). Sustainable tourism emerged in part as a reactive concept in response to tourism issues such as environmental damage and impacts on society and traditional cultures (Bramwell & Lane, 1993). Sustainable tourism has played an important role in identifying ways to secure positive benefits and established approaches to regulation and development control (Bramwell & Lane, 2012).

Sustainable tourism is a multi-disciplinary and broad concept touching upon a wide range of issues such as economic development policy, environmentally matters, social factors and the structure of the international tourism system (Buckley,

2012; Sabir, 2019; Tosun, 2001). These scholars argue that sustainable tourism refers to all kinds of tourism developments that make a notable contribution to or, at least, do not contract the maintenance of the principles of development in an indefinite time without compromising the ability of future generations to satisfy their own needs and desires from the same resources. Sustainable tourism development should contribute to the emergence of necessary conditions in tourist destinations that will lead to local participation in tourism development (Ramos & Prideaux, 2014; Saufi et al., 2013; Tosun, 2001). Sustainable tourism should help host communities to be free or emancipated from alienating material conditions of life and from social servitude to nature, ignorance, other people, misery, institution and dogmatic beliefs (Tosun, 2001). Sustainable tourism should bring acceleration not only of national economic growth but also regional and local economic growth. This growth must be shared fairly across the social spectrum; and, finally, achieving these principles indefinitely without compromising the ability of future generations to meet their own needs (Tosun, 2001; Buckley, 2012).

The concept of sustainable tourism has its critics, especially after it became increasingly popular in the late 1980s. However, there are differences in meaning and definition attached to the concept of sustainable tourism by different academic disciplines (Mika, 2015). Sustainable tourism means different things to different disciplines. Liu (2003) argues that the sustainable tourism debate is patchy, disjointed and often flawed with false assumptions and arguments (Liu, 2003). The concept of sustainable tourism is further described as problematic because of its vagueness (Cohen, 2002). As a result, critics argue that problems associated with the implementation of the principles of sustainable tourism concern reaching a common agreement about local development (Beeton, 2006; Beritelli, 2011; Buckley, 2012; Dredge, 2006; Miller et al., 2010). Critics further argue that sustainable tourism will not be able to address the global challenges as is the case with climate change (Gössling et al., 2012; Scott, 2011; Scott et al., 2012). These scholars argue that the global tourism sector is neither prepared nor fit to face the consequences of climate change. Critics also argue that sustainable development upon which the concept of sustainable tourism is built advocates a holistic perspective; however, development can only be sustainable if it is considered within a global political, socio-economic and ecological context (Mika, 2015; Sharpley, 2000). Sharpley argues that at first sight, the underlying philosophy of sustainable tourism development appears to embrace holistic planning. However, sustainable tourism presupposes a balanced triangular relationship between host areas and host communities, tourists and the tourism industry where no one stakeholder upsets the equilibrium (Lane, 1994). Lane argues that such a scenario where one stakeholder does not upset the equilibrium is likely to be ideal than practical. The concept of sustainable tourism development is noted for being weak and contradictory (Sharpley, 2000). This is because the concept of sustainable tourism calls for both intra- and intergenerational equity, that is, fair and equitable opportunities for development for all people, both in the present and in the future. Both the flows and the structure of international tourism suggest that equitable development through tourism is unachievable (Scott, 2011; Sharpley, 2000). These problems indicate that sustainable tourism policies do not fully embrace the three principles of sustainable development, namely: ecological, equity and economic efficiency (Sharpley, 2000).

Although there are numerous criticisms of the concept of sustainable tourism, it is globally accepted that "real" development should adhere to the principles of sustainable development (WCED, 1987). Proponents of sustainable tourism development argue that it holds a considerable promise as a vehicle for addressing the problems of negative tourism impacts and maintaining its long-term viability (Liu, 2003). Sustainable tourism is therefore a positive approach intended to reduce the tensions and friction created by complex interactions between the tourism industry, tourists, the environment and the host communities so that the long-term capacity and quality of both natural and human resources can be maintained (Bramwell & Lane, 1993, 2012). The rapid growth of tourism has socio-economic and environmental problems in tourism destinations that can no longer be ignored (Mika, 2015). This chapter uses the concept of sustainable tourism to analyse economic development, tourism and conservation in wildlife-based tourism destinations in developing countries. The chapter argues that economic development, tourism development and conservation in nature-based destinations should be considered to achieve wildlife conservation and improved rural development.

Economic and tourism development

The wildlife-based tourism industry in destination areas of developing countries has socio-economic and environmental impacts. This section, therefore, provides an analysis of the wildlife-based tourism industry and its associated economic development, tourism development and conservation aspects in destination areas.

Economic development

The expansion of wildlife-based tourism in developing countries has stimulated economic development in destination areas. For example, in Botswana, the growth of the tourism industry since the 1990s has stimulated the development of a variety of allied infrastructure and facilities, such as hotels, lodges and camps, an airport and airstrips, in the Okavango region which became a key international tourist destination in the country (Mbaiwa, 2003). Through its backward linkages, wholesale and retail businesses have also been established, especially in northern Botswana in areas such as Maun and Kasane, which have come to offer various goods to the tourism industry. In addition, tarred roads and other communication facilities have also been developed in northern Botswana, partly to facilitate tourism development (Mbaiwa & Hambira, 2020). Wildlife-based tourism in northern Botswana continues to provide employment opportunities to local communities and it is a significant source of foreign exchange for Botswana and many developing countries. The wildlife-based tourism industry is the second-largest economic sector in Botswana after mining, contributing over 7.2% to the country's GDP.

Tourism is a primary source of foreign exchange earnings in 46 out of 50 of the world's least developed countries (Jiang et al., 2009).

The growth of wildlife-based tourism is thus encouraged by most governments as it promotes economic development. For the least developed countries, regions, towns and cities, tourism is seen as a fast track to development (Glasson et al., 1995). The main reason why governments, particularly in developing countries, encourage tourism investment is because of the expectations that it will contribute to economic development. The expectation has resulted in most if not all developing countries enacting tourism policies that ensure that they derive significant socio-economic benefits from tourism development. For instance, countries such as Botswana and Namibia enacted their tourism policies immediately after independence in 1990. In addition, these countries have also adopted wildlife utilisation policies to ensure the use of tourism to promote wildlife-based tourism. For example, Botswana adopted the Wildlife Conservation Policy in 1986 and the motive was to ensure the sustainable utilisation of wildlife resources and promotion of wildlife-based tourism throughout the country. This policy paved way for local communities to benefit from wildlife conservation and tourism.

The contribution of the wildlife-based tourism industry in destination areas indicates that an analysis of the backwards and forwards linkages between tourism and other sectors is critical. This is because an understanding of the spatial location of tourism activities and identification of the beneficiaries of its economic and other impacts can either result in wildlife conservation and improvement of rural livelihoods if sustainable tourism ideals are adhered to. That is, for the tourism industry to have a major influence on the economy of a country or a region, it should have strong linkages with the rest of the domestic economy. Some of these linkages include agriculture, manufacturing, construction, wholesale and retail trade, hotels and restaurants, transport, banking and insurance services, water and electricity, and social and personal services (Mbaiwa, 2003).

Tourism development and community-based tourism projects

Wildlife and the habitats on which wildlife-based tourism is based are increasingly threatened by rising human populations, economic activity, illegal poaching and lack of funding (World Bank Group, 2018). In response to this threat, different communities within wildlife-based tourism destinations in developing countries have in the last 30 years participated in tourism development to derive economic benefits from the industry. Some governments in developing countries have adopted the Community-Based Natural Resource Management (CBNRM) programme to promote rural development and wildlife conservation. The CBNRM programme is known by different names in East and Southern Africa. For example, in Zimbabwe, it is called the Communal Area Management Program for Indigenous Resources (CAMPFIRE); the Luangwa Integrated Rural Development Project (LIRDP) and the Administrative Design for Game Management Areas (ADMADE) in Zambia; the Living in a Finite Environment (LIFE) programme in Namibia; the Tchuma Tchato "Our Wealth" in Mozambique; the Conservation of Biodiversity Resource Areas Program (COBRA) in Kenya; the Ujirani Mwena "Good Neighbourliness" in Tanzania; and in Botswana, it is called the CBNRM programme (Mbaiwa, 2005). The CBNRM principles of conservation and rural livelihoods in different countries are the same. However, the success of the CBNRM programme differs from one project to the other and from one country to another.

In addition to deriving economic benefits from the wildlife-based tourism industry, both government and rural communities desire to achieve wildlife conservation in destination areas. In this regard, wildlife-based tourism is assumed to provide opportunities for rural communities to improve their livelihoods through tourism-related jobs, revenue-sharing arrangements and co-management of natural resources (World Bank Group, 2018). As a result, sustainable wildlife tourism is often considered an economic alternative or preferable land-use model to other economic activities with greater environmental impact, such as agriculture or mining (World Bank Group, 2018). The assumption by the government in promoting sustainable wildlife tourism through the CBNRM programme is that when communities derive benefits from prospering wildlife populations, they have a greater stake in protecting wildlife resources in their local environment.

Communities around the world have established different tourism projects in their respective countries. As a result, there are numerous examples of community involvement in wildlife tourism, in which community-based tourism enterprises such as eco-lodges and cultural villages were set up to generate conservation incentives and enhance tourism experiences (World Bank Group, 2018). In Namibia, small tourism enterprises were founded on communal conservancies outside national parks. For example, the Grootberg Lodge is wholly owned by a communal conservancy. The lodge has created local jobs and revenue as well as facilitated the successful reintroduction of endangered species, such as black rhinos, lions and elephants (World Bank Group, 2018). Namibia's successful model of wildlife-based tourism promotes community conservation and has contributed approximately N$5.98 billion (approximately US$488 million at 2018 exchange rates) to net national income and created 5,147 jobs from the beginning of 1990 to the end of 2016.

Uganda is famous for its gorilla tourism which has brought economic and social benefits to communities located within the Bwindi Impenetrable National Park (World Bank Group, 2018). In this regard, world-based tourism based on gorilla tourism creates employment and income for the community. In India, community-based cultural tourism experiences are created around the Pench Tiger Reserve. The Gond tribal communities living on the fringes of the reserve are assisted by the Conservation Wildlands Trust, an NGO that promotes community-based wildlife conservation. This NGO created the Tiger Tribes programme to provide communities with an opportunity to share their cultural life and traditions with tourists through guided tours, while simultaneously earning a livelihood (World Bank Group, 2018). In Kenya, communities established joint ventures with private tourism companies with the African Wildlife Foundation acting as a broker (World Bank Group, 2018).

In addition, a community-led NGO known as the Northern Rangelands Trust has set up a coalition of local leaders, politicians and conservation interests to derive tourism revenue from wildlife resources, especially tigers (World Bank Group, 2018). The World Bank Group argues that 60% of the revenue earned by the trust goes into community projects, such as school endowments, water pumps and infrastructure, while the remaining 40% of funds are dedicated to operating costs of the conservancy, particularly the salaries of the rangers. These case studies indicate that wildlife-based tourism in developing countries generates income that can simultaneously promote rural development and wildlife conservation (Table 12.1).

These results indicate that wildlife-based tourism is a key economic sector in most developing countries. As such, wildlife is a tourism product that supports tourism and generates benefits that sustain the welfare of the people living adjacent to tourism destinations. Wildlife-based tourism development is envisaged as a key pillar for national development in some East, Central and Southern African countries; governments perceive wildlife-based tourism as a mechanism to alleviate poverty, generate foreign revenue and contribute to wildlife conservation. Wildlife-based tourism is a sector that countries can use to diversify their economies while protecting biodiversity and contributing to SDGs, including SDGs 12 and 15. Successful wildlife-based experiences are emerging from across the world, especially in Southern and East Africa, Southeast Asia, Latin America, and the Caribbean. Community participation in tourism development has created better outcomes that promote the well-being of people. Some of these include employment opportunities, financial benefits, household dividends, access to game meat and other social benefits such as funeral assistance, old-age pensions, the provision of housing for the elderly and poor, transportation services, the introduction of modern technology and other intangible benefits like skill development.

Performance of community wildlife-based tourism projects

Wildlife-based community projects have mixed results; that is, some projects have collapsed while others have succeeded (Mbaiwa, 2011). Therefore, this suggests that the performance of wildlife-based tourism projects cannot be generalised. Instead, the performance of these projects should be analysed on individual bases taking into consideration the location of the project as well as the abundance of wildlife resources in the area. At the iSimangaliso Wetland Park Community Concessions in South Africa, the World Bank Group (2018) notes the following success of this community project: skills development programmes in tourism, hospitality and guiding have enabled local people to work in the tourism industry. The World Bank Group further notes that the park has attracted new investment into a range of lodges and hotels in the area. The region has seen a 14% increase in tourism beds (including the buffer zone) and an 89% growth in the number of establishments providing accommodation, food and other services. The authority established a craft programme, which supports 25 groups of mostly female artisans to design, produce and market their crafts in higher-value urban markets and

Table 12.1 Case studies of community-based tourism projects and impacts on rural livelihoods

Case study	*Country*	*Size (in square kilometres)*	*No of tourists*	*Benefits from tourism*
Okavango Delta	Botswana	16,000	259,000	14 ecotourism organisations employed 610 people and generated over US$600,000 in revenue (2012)
Periyar National Park and Tiger Reserve	India	925	800,000	US$93,000 raised through eco-charges on park entrance fees (2005)
iSimangaliso Wetland Park	South Africa	3,600	510,000	US$1.5 million in revenue plus 1,600 direct jobs and 6,000 indirect jobs (FY 2015/2016)
Majete Wildlife Reserve	Malawi	691	8,000	US$400,000 in revenue (2016)
Sanctuary at Ol Lentille, Laikipia	Kenya	40,000 acres (~160 km^2)	N/A 30% occupancy rate	US$20,000–30,000 in lodge fees
Communal Conservancies	Namibia	162,030	N/A	Community conservation facilitated 5,147 jobs and generated nearly US$9 million (N$111 million) in returns for local communities (2016)
Bwindi National Park	Uganda	331	N/A	US$573,000 (1998–2009)

Source: World Bank Group (2018).

to tourists. Finally, the World Bank Group (2018) argues that almost half of all the community residents employed are women.

In Botswana, the CBNRM programme has achieved the following: increased the number of CBNRM projects indicating acceptance of the programme; increased revenues from CBNRM projects; increased employment opportunities from CBNRM projects and associated services such as transportation, household dividends and building houses for the needy and elderly; and increased wildlife population and reduced poaching in CBNRM areas (Mbaiwa, 2018). In Namibia, a total of 86 conservancy or community-based organisations hosted 164 registered enterprises based on natural resources and community conservation that supported wildlife recovery and environmental restoration: 42% of Namibian land is reserved for conservation management; Namibia's elephant population has risen from around 7,000 to 23,000 between 2000 and 2018; increased the lion and black rhino population; and Namibia's conservancies bring in about US $4.8 million annually to rural communities. The World Bank Group (2018) notes that the lodges and luxury tented camps in Namibian communal conservancies are among the top-rated eco-lodges in the world. This type of venture brings jobs and employment and has empowered people to develop skills in negotiating with private investors and managing large sums of money. From the beginning of 1990 to the end of 2016, community conservation contributed an estimated N$5.98 billion (over US$482 million) to Namibia's net national income. During 2016, community conservation generated over N$111 million (nearly US$9 million) in returns for local communities, and it facilitated 5,147 jobs (World Bank Group, 2018).

In Malawi, a collaboration between Majete Wildlife Reserve and an NGO has increased visitation, employment and revenue (World Bank Group, 2018). The World Bank Group (2018) estimates that Majete's visitor numbers grew from 0 in 2003 to 8,000 in 2016, which generated US$400,000 in gross revenue. The collaborative partnership has also resulted in the reintroduction of wildlife species in the area including lions. A private operator, Robin Pope Safaris, opened a luxury safari lodge and works closely with local communities (World Bank Group, 2018). In this regard, conservation and anti-poaching efforts have led to an increase in the elephant population. Due to an increase in the elephant population in the area, some of the elephants were relocated to another national park in 2017 (World Bank Group, 2018). The World Bank Group (2018) also notes that health care and education have improved in the area. In Kenya, the World Bank Group (2018) notes that the community receives US$20,000–30,000 a year in lodge fees, plus a further US$250,000 in education, health, and conservation benefits, including salaries for school teachers and health workers, from the Sanctuary at Ol Lentille, Laikipia. Women's roles are gradually improving due to linkages with other communities through women's groups and enhanced financial status. Vegetation is recovering in the conservation area with an increase in mammals, such as elephants, zebras and wild dogs.

In Zimbabwe, the CAMPFIRE programme has led to socio-economic and environmental benefits in rural areas (Child, 2003). Child et al. (2003) argue that these achievements include the following: the protection of 50,000 square kilometres reserved for wildlife; wildlife populations increased by about 50%, with the elephant population increasing from 4,000 to 8,000 in CAMPFIRE areas; reduced or contained veld fires in various districts particularly Chiredzi, Chipinge and Gokwe North; reduction of land-use conflicts between agricultural production and wildlife management; reduced poaching levels and illegal wildlife off-take; acceptance countrywide as shown by a total of 23 districts that have established CAMPFIRE projects; increased membership of the CAMPFIRE Association has also increased to 52 of Zimbabwe's 57 Rural District Councils in the last 10 years; and economic benefits include an increase of revenues from safari hunting. This shows that wildlife utilisation through the CBNRM programme and sustainable hunting has successfully achieved the rewilding of Southern Africa and has improved the wildlife economy. Apart from Kenya, India and South Africa, sustainable hunting is one of the key land use options supporting wildlife-based tourism.

Despite the success stories of CBNRM in Botswana and Southern Africa in achieving improved livelihoods and increased wildlife populations, especially elephants, critics are of the view that the programme has failed (DeGeorges & Reilly, 2009). DeGeorges and Reilly (2009) argue that low household incomes from CBNRM are unlikely to change the attitudes of rural dwellers towards Western approaches to conservation. Scholars argue that there is no evidence that community conservation and development programmes are achieving their objectives of rural development. As a result, these scholars argue for a return to authoritarian and centralised forms of resource management. This argument is misleading because this chapter has shown that in countries such as Namibia and Botswana, the CBNRM programme has achieved rural development and has particularly improved livelihoods. It is, therefore, erroneous to generalise and conclude that community conservation and development projects are failing to achieve conservation and development. Programmes should be evaluated and judged based on the socio-economic and political context of areas (Wilshusen et al., 2002).

Community partnership models

To achieve sustainable wildlife conservation and improved rural livelihoods, effective planning, management skills, technology use and stakeholder collaboration in wildlife destination is essential. In areas where CBNRM projects are successful, the collaboration between communities and stakeholders, such as tourism operators, and the government as well as the availability of natural resources are sufficient. The collaborative and partnership approach with communities has thus proved effective in delivering tourism's economic and development benefits while minimising potential adverse environmental and social impacts. In this regard, a partnership between communities and related stakeholders are essential. There are several approaches to community tourism management in wildlife-based tourism destination in developing countries. As shown in Table 12.2, three of the most common include passive engagement, active engagement and co-management (World Bank Group, 2018).

Table 12.2 Approaches to community tourism management

Approach	*Description*
Passive engagement	Communities own the asset and lease to a private or public-sector partner, or the government owns the asset and provides revenue sharing to communities (a legal obligation in such countries as Kenya, Tanzania and Zambia).
Active engagement	Communities are empowered economically and socially through employment in tourism activities either independently or with a public, private or NGO partner.
Co-management	Communities have an equal decision-making role in the management of the asset with the owner. Who owns the asset, services the asset and decides and sets the rules are critical factors.

Source: World Bank Group (2018).

The successful understanding of these three approaches helps to determine the kind of tourism management model different communities in wildlife areas should adopt. Table 12.3 shows the common wildlife tourism partnership models often used in tourism destinations to achieve improved community benefits and wildlife conservation.

Community projects can be successful through joint venture partnership with experienced stakeholders such as private tourism operators (Table 12.2). This approach is appropriate mainly because, in most cases, communities lack the capacity to run tourism enterprises. Joint venture partnerships are therefore a means towards capacity building and the transfer of skills to local communities.

The North–South debate

Wildlife utilisation in developing countries historically constitutes a series of North–South disputes. As a result, the history of the North–South dynamic has a dominant repetitive pattern in thinking about biodiversity, which is simply that the North proposes and the South reacts and adapts. The North–South debate has fuelled the development of biodiversity law, and policy is largely unidirectional (Beyerlin, 2006). The North–South debate is further illustrated by the adoption of protocols and agreements mostly originating from the Global North than the Global South. International conservation organisations such as the Worldwide Fund for Nature, the International Union for Conservation of Nature and the Convention on International Trade in Endangered Species of Wild Fauna and Flora (CITES) have been at the forefront of protecting endangered species and their habitats. Bilateral and multinational agreements have been made over the last century to conserve species in danger of extinction (Mitchell, 2003). Such agreements, in addition to national legislation, restricted hunting and trading of wildlife in various ways (Pires & Moreto, 2016). In 1973, CITES was formed. CITES policies at the national level subject certain wildlife to varying degrees of

Table 12.3 Wildlife tourism partnership models

Public–community initiative	*Public–private partnership*	*Community-owned and run enterprise*	*Community–private partnership*	*Public–private–community partnership*
Cooperation between public institution and the local community to operate and manage community-based tourism venture	Agreement between public institution and private party, granting private party use rights to land or infrastructure for commercial purposes	Tourism enterprise that is wholly or mostly owned and run by the community group or member	Agreement between the community and private party, granting private party use rights to land or infrastructure for commercial purposes	Agreement between public institution, private party and community to manage and operate tourism venture
Partners' roles and responsibilities	*Partners' roles and responsibilities*	*Partners' roles and responsibilities*	*Partners' roles and responsibilities*	*Partners' roles and responsibilities*
Public institution facilitates access to and use of land or infrastructure Community operates the venture	Public authority contracts or leases rights to land or infrastructure Private party funds and operates tourism venture and receives risks and benefits	Community operates and manages the venture Donors provide initial or ongoing investment and training	Community enters agreement with a private party who would fund and/or operate tourism venture, with community sharing in benefits Private party ensures community benefits through revenue, employment, social services, etc.	Public institution facilitates the partnership with a private party who would fund and/or operate tourism venture, with community sharing in benefits Community may work in or manage the venture Parties all agree that the community should benefit If the private party is an NGO, may also provide training and funding to community

Key success factors	*Key success factors*	*Key success factors*	*Key success factors*	*Key success factors*
Capacity building for community members in business, service, marketing and sales skills Tourism is part of integrated community development	Attractive environment and terms that enable private involvement Professional private party is invested for medium to long term	Ongoing donor involvement Capacity building for community Direct market access Strong business viability	Attractive environment and terms that enable private involvement Professional private party is invested for medium to long term Community is supportive and engaged Capacity building and skills training for community members	Attractive environment and terms that enable private involvement Professional private party is invested for medium to long term Community is supportive and engaged Capacity building and skills training for community members
Challenges and limitations	*Challenges and limitations*	*Challenges and limitations*	*Challenges and limitations*	*Challenges and limitations*
Conducting effective marketing Coexisting with private-sector enterprises Fostering sustainability Managing growth	Managing the level of community involvement	Financial self-sustainability	Managing community's varying levels of readiness and engagement Dealing with tensions that arise within and outside of community	Managing community's varying levels of readiness and engagement Dealing with tensions that arise within and outside of the community
Case studies	*Case studies*	*Case studies*	*Case studies*	*Case studies*
Periyar National Park, India iSimangaliso Wetland Park, South Africa	Majete Wildlife Reserve, Malawi	Bwindi Impenetrable National Park, Uganda	Sanctuary at Ol Lentille, Kenya Okavango Delta, Botswana	Communal Conservancies, Namibia

Source: World Bank Group (2018).

controls through a licensing system (Moyle, 2003; Pires & Moreto, 2016). CITES restricts wildlife trade using appendices in which species are afforded greater protection when traded between countries (Pires & Moreto, 2016). For example, Appendix I lists species that are threatened with extinction and may be affected by international trade. All international commercial trade in such species is banned, although some trade may be allowed under exceptional circumstances.

Botswana and many other African countries are signatories of CITES – elephants in most of the countries are listed in Appendix II. Appendix II species are not always at risk of extinction but can be at risk without a system of regulations. Such species may be commercially traded only if export permits are procured, and depending on the import nation, may also require import permits. While CITES can save threatened species from extinction and protect others from being at risk, the organisation is blamed by Southern African states, including Botswana, for failing to comprehend the elephant situation in the region. In addition, the United States Fish and Wildlife Service regulations currently do not allow elephant imports from Botswana. That Botswana cannot trade in elephant products illustrates the North–South differences and the result is that Botswana can only trade on terms laid down by the North.

Trade restrictions suggest that Botswana is silenced by CITES. Wildlife trade restrictions by CITES together with the perception that some citizens in the North have about how wildlife utilisation in the South should be accomplished appear to be punishing Botswana for having achieved a success story in using wildlife to achieve conservation and improved rural livelihoods. Botswana has succeeded in conserving the elephant and other wildlife species. Through the CBNRM programme, Botswana has achieved the following: an increased number of CBNRM projects, indicating acceptance of the programme; increased revenues from CBNRM projects; increased employment opportunities from CBNRM projects and associated services such as transportation, household dividends, building houses for the needy and elderly; increased wildlife population in CBNRM areas; and reduced poaching in CBNRM areas (Mbaiwa, 2018). It is, therefore, not realistic for Botswana to be subjected to CITES trade restrictions of species even though the country has succeeded in improving rural livelihoods and the conservation of species such as elephants.

Conclusion

It is well established by theory and scientific best practice that sustainable conservation of wildlife must consider a socio-ecological framework (Ostrom, 2009; Poteete et al., 2010). This means that the conservation of wildlife resources in developing countries cannot succeed in the long term if local communities are alienated from wildlife benefits through the tourism industry. The command-and-control approach in wildlife tourism is not the best approach in wildlife conservation (Mbaiwa, 2015). While wildlife-based photo-tourism is an important contributor to GDP and conservation efforts in developing countries, it is not a panacea to rural development and conservation challenges in developing countries as the Global North seems to recommend for developing countries. Photo-tourism is viable in areas with very high densities of visible wildlife and/or spectacular scenery but not in marginal areas of Namibia and the drylands of Botswana (Mbaiwa, 2018). In this regard, sustainable wildlife conservation requires an integrated land-use planning approach that considers both photographic and hunting tourism and agricultural uses.

The growth of wildlife-based international tourism in developing countries has created a window of opportunity to implement the bottom-up approach in resource use that involves communities living in wildlife areas. Through CBNRM, communities are therefore likely to conserve wildlife resources because of the economic benefits they derive from it (Thakadu, 2005). The CBNRM programme plays a critical role in the harmonisation of various stakeholders involved in the wildlife-based tourism industry and conservation in developing countries. In addition, the success of the wildlife-based tourism industry in community areas largely depends on the approach and model adopted by the community living in the wildlife area (World Bank Group, 2018). Partnership and collaborative models depend on the location and abundance of wildlife resources in an area. For example, communities in Botswana have chosen the community–private partnership model (Centre for Applied Research, 2016). This is a model that promotes an agreement between the community and a private party, granting the private party use rights to land or infrastructure for commercial purposes. The model has been reported to be successful for most of the communities in Botswana. The CBNRM programme has therefore proved that it can improve livelihoods and conservation through both sustainable wildlife hunting and photographic tourism development (Mbaiwa & Stronza, 2010). Therefore, local communities living in remote parts of the world and in rich biodiversity areas, such as having an abundance of wildlife, have the potential to mobilise, coordinate and facilitate their communities to participate in tourism development and conservation. The transfer of some management responsibility, decision-making and benefits of resource utilisation from central government to local communities is thus critical in achieving wildlife conservation, improved livelihoods and sustainable tourism development in wildlife-based tourism destinations of developing countries.

References

Beeton, S. (2006). *Community Development Through Tourism.* Collingwood: Landlinks Press.

Beritelli, P. (2011). *Tourist Destination Governance Through Local Elites: Looking Beyond the Stakeholder Level.* Doctoral dissertation. Universität St. Gallen, Switzerland.

Beyerlin, U. (2006). Bridging the North-South divide in international environmental law. *Zeitschrift für Ausländisches Öffentliches Recht und Völkerrecht, 66*(2), 259–296.

Blanke, J., & Chiesa, T. (2013). *The Travel & Tourism Competitiveness Report 2013.* Geneva: World Economic Forum.

Bramwell, B., & Lane, B. (1993). Sustainable tourism: An evolving global approach. *Journal of Sustainable Tourism, 1*(1), 1–5. doi:10.1080/09669589309450696

Bramwell, B., & Lane, B. (2012). Towards innovation in sustainable tourism research. *Journal of Sustainable Tourism, 20*(1), 1–7. doi:10.1080/09669582.2011.641559

Buckley, R. (1994). A framework for ecotourism. *Annals of Tourism Research, 21*(3), 661–665.

Buckley, R. (2012). Sustainable tourism: Research and reality. *Annals of Tourism Research, 39*(2), 528–546. doi:10.1016/j.annals.2012.02.003

Centre for Applied Research (2016). *2016 Review of Community-Based Natural Resource Management in Botswana.* Report prepared for Southern African Environmental Programme (SAREP).

Child, B. (2003). Origins and efficacy of modern community based natural resources management (CBNRM) practices in the southern African region. In *Regional Workshop on Local Communities and conservation: Issues and Challenges Towards a More Equitable and Sustainable Future* (pp. 26–28). Pretoria.

Cohen, E. (2002). Authenticity, equity and sustainability in tourism. *Journal of Sustainable Tourism, 10*(4), 267–276. doi:10.1080/09669580208667167

DeGeorges, P. A., & Reilly, B. K. (2009). The realities of community based natural resource management and biodiversity conservation in sub-Saharan Africa. *Sustainability, 1*, 734–788. doi:10.3390/su1030734

Dredge, D. (2006). Networks, conflict and collaborative communities. *Journal of Sustainable Tourism, 14*(6), 562–581. doi:10.2167/jost567.0

Glasson, J., Godfrey, K., & Goodey, B. (1995). *Towards Visitor Impact Management: Visitor Impacts, Carrying Capacity and Management Responses in Europe's Historic Towns and Cities*. Aldershot: Avebury.

Gössling, S., Scott, D., Hall, C. M., Ceron, J. P., & Dubois, G. (2012). Consumer behaviour and demand response of tourists to climate change. *Annals of Tourism Research, 39*(1), 36–58. doi:10.1016/j.annals.2011.11.002

Hassan, S. S. (2000). Determinants of market competitiveness in an environmentally sustainable tourism industry. *Journal of Travel Research, 38*(3), 239–245. doi:10.1177/004728750003800305

Jiang, M., DeLacy, T., & Noakes, S. (2009). Tourism, the millennium development goals and climate change in the South Pacific Islands. In *International Conference on Meeting the Millenium Development Goals: Old Problems, New Challenges, 30 Nov–1 Dec 2009*. Melbourne, Australia.

Lane, B. (1994). What is rural tourism? *Journal of Sustainable Tourism, 2*(1–2), 7–21. doi:10.1080/09669589409510680

Lenao, M. (2014). Rural tourism development and economic diversification for local communities in Botswana: The case of Lekhubu Island. *Nordia Geographical Publications, 43*(2), 53–53.

Liu, Z. (2003). Sustainable tourism development: A critique. *Journal of Sustainable Tourism, 11*(6), 459–75. doi:10.1080/09669580308667216

Mbaiwa, J. E. (2003). The socio-economic and environmental impacts of tourism development on the Okavango Delta, North-western Botswana. *Journal of Arid Environments, 54*(2), 447–467. doi:10.1006/jare.2002.1101

Mbaiwa, J. E. (2005). Enclave tourism and its socio-economic impacts in the Okavango Delta, Botswana. *Tourism Management, 26*(2), 157–172. doi:10.1016/j.tourman.2003.11.005

Mbaiwa, J. E. (2011). Changes on traditional livelihood activities and lifestyles caused by tourism development in the Okavango Delta, Botswana. *Tourism Management, 32*(5), 1050–1060. doi:10.1016/j.tourman.2010.09.002

Mbaiwa, J. E. (2015). Ecotourism in Botswana: 30 years later. *Journal of Ecotourism 14*(2–3), 1–19. doi:10.1080/14724049.2015.1071378

Mbaiwa, J. E. (2018). Effects of the safari hunting tourism ban on rural livelihoods and wildlife conservation in Northern Botswana. *South African Geographical Journal, 100*(1), 41–61. doi:10.1080/03736245.2017.1299639

Mbaiwa, J. E., & Hambira, W. L. (2020). Enclaves and shadow state tourism in the Okavango Delta, Botswana. *South African Geographical Journal, 102*(1), 1–21. doi:10.1080/03736245.2019.1601592

Mbaiwa, J. E., & Kolawole, O. D. (2013). Tourism and biodiversity conservation: The case of community-based natural resource management in Southern Africa. *CAB Reviews, 8*(10), 1–10. doi:10.1079/PAVSNNR20138010

Mbaiwa, J. E., & Stronza, A. L. (2010). The effects of tourism development on rural livelihoods in the Okavango Delta, Botswana. *Journal of Sustainable Tourism, 18*(5), 635–656. doi:10.1016/j.jenvman.2011.03.009

Mbaiwa, J. E., Stronza, A., & Kreuter, U. (2011). From collaboration to conservation: Insights from the Okavango Delta, Botswana. *Society and Natural Resources, 24*(4), 400–411. doi:10.1080/08941921003716745

Mika, M. (2015). Sustainable tourism: A critique of the academic feasibility of the concept. *Turyzm, 25*(1), 9–17. doi:10.2478/tour-2014-0015

Miller, G., Rathouse, K., Scarles, C., Holmes, K., & Tribe, J. (2010). Public understanding of sustainable tourism. *Annals of Tourism Research, 37*(3), 627–645. doi:10.1016/j.annals.2009.12.002

Mitchell, R. B. (2003). International environmental agreements: A survey of their features, formation, and effects. *Annual Review of Environment and Resources, 28*(1), 429–461. doi:10.1146/ANNUREV.ENERGY.28.050302.105603

Moswete, N., & Lacey, G. (2015). "Women cannot lead": Empowering women through cultural tourism in Botswana. *Journal of Sustainable Tourism, 23*(4), 600–617. doi:10.2139/ssrn.2205354

Moyle, B. (2003). Regulation, conservation, and incentives. In S. Oldfield (Ed.), *The Trade in Wildlife Regulation for Conservation* (pp. 41–51). London: Earthscan.

Mulale, K., Matema, C., Funda, X., Slater-Jones, S., Njovu, D., Kangueehi, G., Hay, D., & Crookes, V. (2013). *Community - Based Natural Resource Management in Southern Africa*. Florida: University of Florida, Centre for African Studies.

Ostrom, E. (2009). A general framework for analyzing sustainability of social-ecological systems. *Science, 325*(5939), 419–422. doi:10.1126/science.1172133

Pires, S. F., & Moreto, W. (2016). The illegal wildlife trade. *Oxford Handbooks Online*. doi:10.1093/oxfordhb/9780199935383.013.161

Poteete, A. R., Janssen, M. A., & Ostrom, E. (2010). *Working Together: Collective Action, The Commons, and Multiple Methods in Practice*. Princeton: Princeton University Press.

Pretty, J., Adams, B., Berkes, F., Ferriera de Athayde, S., Dudley, N., Hunn, E., Maffi, L., Milton, K., Rapport, D., Robbins, P., Sterling, E., Stoltin, S., Tsing, A., Vintinner, E., & Pilgrim, S. (2009). The intersections of biological diversity and cultural diversity: Toward integration. *Conservation and Society, 7*(2), 100–112. doi:10.4103/0972-4923.58642

Prosser, R. (1994). Societal change and the growth in alternative tourism. In E. Cater & G Lowman (Eds.), *Ecotourism: A Sustainable Option* (pp. 19–37). New York: John Wiley and Sons Ltd.

Ramos, A. M., & Prideaux, B. (2014). Indigenous ecotourism in the Mayan rainforest of Palenque: Empowerment issues in sustainable development. *Journal of Sustainable Tourism, 22*(3), 461–479. doi:10.1080/09669582.2013.828730

Saarinen, J., Becker, F., Manwa, H., & Wilson, D. (Eds.) (2009). *Sustainable Tourism in Southern Africa: Local Communities and Natural Resources in Transition*. Bristol: Channel View Publications.

Sabir, B. (2019). Developing a multidisciplinary tourism planning approach on cultural routes. *Journal of Multidisciplinary Academic Tourism*, *4*(1), 37–47. doi:10.31822/jomat.543891

Salazar, N. B. (2012). Community-based cultural tourism: Issues, threats and opportunities. *Journal of Sustainable Tourism*, *20*(1), 9–22. doi:10.1080/09669582.2011.596279

Saufi, A., O'Brien, D., & Wilkins, H. (2013). Inhibitors to host community participation in sustainable tourism development in developing countries. *Journal of Sustainable Tourism*, *22*(5), 801–820. doi:10.1080/09669582.2013.861468

Scott, D. (2011). Why sustainable tourism must address climate change. *Journal of Sustainable Tourism*, *19*(1), 17–34. doi:10.1080/09669582.2010.539694

Scott, D., Hall, C. M., & Gössling, S. (2012). *Tourism and Climate Change: Impacts, Adaptation and Mitigation*. Abingdon: Routledge.

Sharpley, R. (2000). Tourism and sustainable development: Exploring the theoretical divide. *Journal of Sustainable Tourism*, *8*(1), 1–19. doi:10.1080/09669580008667346

Thakadu, O. T. (2005). Success factors in community-based natural resource management projects' mobilization in northern Botswana: Lessons from practice. *Natural Resources Forum*, *29*(3), 199–212. doi:10.1111/j.1477-8947.2005. 00130.x

Tosun, C. (2001). Challenges of sustainable tourism development in the developing world: The case of Turkey. *Tourism Management*, *22*, 289–303. doi:10.1016/S0261-5177(00)00060-1

Wilshusen, P. R., Brechin, S. R., & West, P. C. (2002). Reinventing a square wheel: Critique of a resurgent "protection paradigm" in international biodiversity conservation. *Society and Natural Resources*, *15*, 17–40. doi:10.1080/089419202317174002

World Bank Group (2018). *Supporting Sustainable Livelihoods through Wildlife Tourism*. Washington DC: World Bank Group.

World Commission on Environment and Development (WCED) (1987). *Our Common Future*. Oxford: Oxford University Press.

WTO (2001). *Cultural Heritage and Tourism Development - A Report on the International Conference on Cultural Tourism*. Madrid: WTO.

13 Municipal assets for tourism and small-town local economic development

Caravan parks in South Africa

Christian M. Rogerson and Jayne M. Rogerson

Introduction

Globally, one important new direction in planning for local economic development is an emphasis on identifying and mobilising local or endogenous assets (Pike et al., 2015, 2017; Tomaney, 2010). This new direction denotes a shift away from an earlier tradition of local economic development planning that focused mainly on the attraction of external "footloose" investment to now emphasise the recognition and maximisation of the potential of local assets (Tomaney, 2010). Place-based planning stresses the use of local assets as part of an approach of endogenous planning which builds on the comparative advantages of a locality and its human and natural assets as well as institutional capabilities (Barca, 2019). The rise of place-based development planning in recent years has been given impetus by support from the European Union and OECD, which have resuscitated and re-purposed the concept of "endogenous growth" (Barca et al., 2012; Rodriguez-Pose & Wilkie, 2017). Beyond its European heartland, the principles of place-based development planning have diffused and been taken up in planning that is occurring in several other parts of the world (Nel et al., 2019; Todes & Turok, 2018). In the context of the Global South, the relevance and need for extended place-based development initiatives is highlighted by several researchers (Duranton & Venables, 2019; Pike et al., 2014; Rogerson, 2014, 2019a). Place-based planning is about strategies that are derived from contextual specificity and the assessment of local conditions, including for the leveraging of a local asset base (Lincoln, 2019; Rodriguez-Pose & Wilkie, 2017). The rise and international spread of place-based development interventions in part is a reaction against the disappointments that were associated with the implementation of the earlier tradition of local economic development planning (Pike et al., 2014, 2015). Place-based planning is increasingly acknowledged as a strategy with the potential to address local development backlogs, particularly in the Global South, where traditional "top-down" approaches generally have failed to effect significant local change (Barca et al., 2012; Rodriguez-Pose & Palavicini-Corona, 2013).

The mobilisation of local assets for energising local economic development is a particular challenge for subnational levels of government (Rogerson, 2020).

It is observed that local governments themselves hold or control significant baskets of physical assets such as land, buildings and infrastructure, which can be valuable for promoting local economic well-being (Kaganova, 2010; Kaganova & McKellar, 2006; Phelps, 2011). Arguably, the good management of these physical assets is critical for leveraging opportunities for local economic development (Kaganova & Kopanyi, 2014). Among others, Phelps (2011, p. 417) considers such assets potentially as "a catalyst" for local development. In many cities in the Global North experiencing economic decline and high levels of unemployment, property owned by local government is allocated (often at below market value) for "the sake of local economic development" (Kaganova, 2010, p. 2). Kaganova and Nayyar-Stone (2000) observe that while municipal governments generally attend – to a greater or lesser degree – to the day-to-day issues surrounding their assets, few think of their holdings as a "portfolio" that requires a coherent asset management strategy. Since the 2008 global financial crisis that compelled many local governments to re-examine their assets and seek to better improve their performance, there is evidence of expanding interest in the systematic management of municipal assets (Kaganova & Telgarsky, 2018). In many cases, however, knowledge deficits exist as "many local governments do not even realize the full extent of what assets they own" (Kaganova & Telgarsky, 2018, p. 146). This is the case especially in the context of local governments in the Global South where basic inventories of assets are often lacking (Hanis et al., 2010). The practice of municipal asset management is unevenly developed across the world with Australia, New Zealand and the UK viewed as the most advanced (Kaganova & McKellar, 2006; Phelps, 2011).

By contrast, with capacity constraints and limited knowledge of existing asset bases in the transition economies of Eastern Europe and the less developed context of sub-Saharan Africa, there is often little focus on assets and a low-level awareness of asset management in municipalities (Sedmihradská, 2010). South Africa represents one country in the Global South where issues of asset management have crept onto the policy agenda of both national and local governments. The Constitution mandates South African local government to provide equitable and efficient services, build local democracy and promote economic and social development (Venter, 2020). In addition, the national government launched initiatives for what is termed "immovable asset management", which is considered to represent the planned alignment of "immovable assets" with the developmental mandate of local governments (Buys & Mavusa, 2007). It is recognised that the essential work of local governments is an asset-intensive business (Pretorius, 2014). In terms of the 1996 Constitution, the Department of Public Works became the custodian of the government's immovable assets and undertook the completion (over several years) of the first national register of state assets. This department led the establishment of an immovable asset management framework which was formalised in 2007 with the launch of the Government Immovable Asset Management Act (Phatela & Cloete, 2018). Among its several goals are support for socio-economic objectives of job creation and poverty alleviation, and to advance the objectives for Black Economic Empowerment in South Africa (Department of Public Works, 2005, 2008).

Overall, the international debates surrounding place-based planning and municipal assets are highly relevant for tourism scholarship (Rogerson, 2019b). Zhang et al. (2019, p. 2) view asset management as a "critical issue" for the tourism industry. At the outset, it is well-known that for tourism to be a lead sector in local development there is a need for maximising the local tourism asset base, which could be around natural beauty, heritage or wildlife, in order for destinations to evolve a total tourism product or portfolio of products (Davidson & Maitland, 1997). In this regard, the role of the private sector and local entrepreneurs is vital for local development futures. Nevertheless, as is revealed in Australian research by Ruhanen (2013), the role of local government can also be crucial for ensuring sustainable trajectories of destination development. It is contended that

> local governments are often best placed to support tourism development within a destination due to their local knowledge; something that is often lacking in distant capital cities and among leaders who are less familiar with regional cultures and local conditions.
>
> (Ruhanen, 2013, p. 82)

In those countries, such as South Africa, where the local government bears the responsibility for driving local economic development, the mobilisation of all local assets is crucial for evolving successful pathways both for tourism-led development as well as for greater inclusivity (Rogerson & Rogerson, 2020a). A recent national investigation of the strategies for local development planning undertaken by 278 South African local governments recorded that 227 or 81.7% had identified tourism as a lead sector (Nel & Rogerson, 2016). Indeed, tourism promotion is the most widespread form of local economic development intervention across contemporary South Africa and particularly significant in the country's small towns (Donaldson, 2018; Kontsiwe & Visser, 2019; Rogerson, 2020; Rogerson & Rogerson, 2019a, 2020a).

Against this backcloth, the objective of this chapter is to examine the challenges of leveraging one particular form of tourism asset which is in direct municipal ownership in several countries including Australia, New Zealand and South Africa. The specific focus is on caravan parks, their role as municipal assets and the planning challenges around their leveraging for local development in contemporary South Africa. Three sets of discussion follow. The first section briefly locates the study as part of a broader international literature on drive tourism and specifically on caravanning tourism. The second section reviews the historical evolution of caravan parks as part of the asset base of many South African municipalities. Finally, attention turns to present key issues surrounding the challenges of local government planning for caravan parks.

Drive tourism and caravan parks

For many observers, caravan parks are a component of the broader categories of camping tourism or outdoor hospitality and recreation (Brooker & Joppe, 2013,

2014; Rogerson & Rogerson, 2020b). However, as pointed out by Lashley (2015, p. 115), unlike other campers, caravanners "combine mobility with accommodation that incorporates many of the tangible comforts of home". Caravanning tourism is a distinctive form with the caravan as part of both tourism transportation as well as accommodation (Doğantan & Emir, 2019; Patterson et al., 2015). Blichfeldt (2009, p. 420) avers that in Denmark the country's caravan parks not only function as accommodation sites but also can function as "experiencescapes and/or even as attractions in themselves". Lashley (2015, p. 122) opines that because of several unique dimensions, caravanning and caravanners ought to be disaggregated from the wider body of scholarship concerning outdoors hospitality. Prideaux and McClymont (2006, pp. 45–46) define the activity of caravanning "as a subset of tourism where the main form of accommodation used during the trip is a caravan". Caravanning and caravan parks are thus understood as a subset of "drive tourism" that involves the bringing of a home in the form of a diverse range of recreational vehicles, including caravans (Blichfeldt & Mikkelsen, 2014; Prideaux, 2020; Prideaux & McClymont, 2006; Van Heerden, 2010a).

From research in Australia, Caldicott (2011, p. 13) observes the caravan sector as "the Cinderella sub-sector of tourism". Indeed, as compared with other forms of tourism, caravanning has attracted only relatively small research literature (Mikkelsen & Cohen, 2015; Rogerson & Rogerson, 2020b, 2020c). Lashley (2015, p. 121) avers "the amount of research undertaken with the specific intention of exploring caravanning and caravaners is limited". Likewise, Mikulić et al. (2017, p. 226) assert that caravan parks are "a rather neglected element of the tourism and hospitality industry". The findings of the recent bibliometric analysis produced by Okumus et al. (2019) signal that caravanning is one of the least well-researched of several different forms of lodging. Demand-side research has revealed aspects of the demographics and motivations of caravanners and their satisfaction with choices of particular campsites (Brooker & Joppe, 2014; Mikkelsen & Blichfeldt, 2015; Prideaux & McClymont, 2006; Van Heerden, 2010b, 2020). In Denmark, caravan sites are visited by both extended and nuclear families (Blichfeldt & Mikkelsen, 2013; Mikkelsen & Blichfeldt, 2015, 2018), whereas in Australia the observed trend is away from families and towards retirees or "grey nomads" (Patterson et al., 2015; Prideaux & McClymont, 2006).

Supply-side investigations of caravan parks and caravan tourism are less in evidence. The most notable studies are the rich cluster of works produced by Caldicott (2011), who examines the parallel histories of the development of the country's caravan manufacturing sector and the caravan park sector. The historical record of Australia and New Zealand confirms the assertion made by Simeoni and Cassia (2019) that the makers of caravans are "co-creators" of the tourism experience of caravanning. As Brooker and Joppe (2014) record, a range of public agencies and private family businesses own and operate caravan parks. For Australia, local councils "operate approximately 20 % of the parks, although the management may be leased to private interest groups" (Brooker & Joppe, 2014, p. 338). The life cycle of caravan parks in Australia is interrogated within the framework of Butler's classic tourism area life cycle model (Caldicott & Scherrer, 2013a).

In both Australia and New Zealand, the establishment of caravan parks has been undertaken both by private entrepreneurs and local governments (Caldicott, 2011; Collins & Kearns, 2010).

Research in Australia and New Zealand tracks the downturn in the numbers of caravan parks and causes of caravan park redevelopment. Many parks in coastal areas have been converted to residential or commercial uses or redeveloped by upgrading into tourist resorts or holiday parks (Gilbert, 2013). It is evidenced that caravan parks were vulnerable to pressures from developers and the demise of caravan parks linked to their conversion to higher-yielding land uses most especially of beachfront land to high rise units and hotels (Caldicott & Scherrer, 2013b; Collins & Kearns, 2010). In Australia, Gilbert (2013) shows that the closure and/or redevelopment of parks was triggered by the decision that was made by several municipal authorities that caravan park management simply was not their core business. This finding reflects an under-represented theme in the extant literature which surrounds the tourism developmental role of caravan parks and local planning. Of interest here are the impact and planning challenges surrounding publicly owned caravan parks as occurs in South Africa

The municipal asset of caravan parks in South Africa

Historically, the first development of caravan parks began in South Africa during the 1930s (Rogerson & Rogerson, 2021). A burst of new developments occurred after the Second World War and by 1950 the major share of listed caravan parks actually functioned on public land (Caravan Club of Southern Africa, 1950). The 1950 national guide listed 63 municipal sites, mostly in coastal areas of the Western Cape but with a spread of municipal parks across many small towns. Further expansion of the caravan sector occurred in the 1950s. The provision of new caravan park facilities was taking place both by the growing numbers of private entrepreneurs entering the sector and also with the addition of new municipal caravan parks. For 1961, the national guidebook provides details for 113 municipal sites, which is a near-doubling in supply as compared with 1950 (Caravan Club of South Africa, 1961). The continued expansion in the numbers of municipal caravan parks and their facilities is striking. The growth must be understood in terms of what Grundlingh (2006, p. 118) argues was the desire by the apartheid government to make tourism in South Africa more accessible and "more affordable for the white lower-middle classes". This prompted a burst of new openings of public resorts in many parts of the country, a trend that continued into the 1970s. However, the role and overall market share of municipal caravan parks were increasingly eroded by the growth in the numbers of parks that were operated by private entrepreneurs (Rogerson & Rogerson, 2021).

At the time of democratic transition in South Africa, the country's approximate 700 caravan parks comprised a mix of majority privately controlled and a smaller group of municipal-operated parks. During the 1990s, the caravan park sector of South Africa faced a rapidly shifting market environment with a major change taking place as a result of the increased sophistication of the caravan market that was becoming part of an emerging economy of "drive tourism" (Prideaux, 2020). No longer was the activity of caravanning in South Africa a cheap holiday option. It was acknowledged that "caravanning can be a relatively economical holiday but set-up costs are not negligible and investing in a caravan or motorhome can be hefty" (Van Heerden, 2008, pp. 125–126). Evidence of this change in South Africa's caravan sector is given in a series of works produced by Van Heerden (2008, 2010a, 2010b) disclosing the shifts taking place which, in many respects, mirror those impacting the caravan sectors of the UK or Australia (Caldicott, 2011; Lashley, 2015).

From 2015, mounting criticism emerged, particularly from leaders of the opposition Democratic Alliance (DA), about the failures of local government in South Africa towards tourism as a whole and specifically about the neglect of numerous tourism sites (Hatchuel, 2016). Several press reports appeared concerning "misuse of facilities" and of the lacklustre approaches of national and local government to addressing the issues of turning around under-performing municipal tourism assets (Hatchuel, 2016). For example, the Pilgrims Rest caravan park in Mpumalanga was reported to have gone to total ruin with millions of Rands worth of damage done to a public asset that was once functional. It had a series of sales and unsatisfactory lease agreements since 1994 that resulted in it becoming derelict and vandalised after it was taken back into the "care" of the Mpumalanga Department of Public Works, Roads and Transport in 2015 (*Lowvelder*, 8 August 2018).

Many municipal parks in South Africa were acknowledged to be in a degraded state because of neglect and lack of maintenance. This situation was inseparable from the financial collapse of many South African local authorities, which, in part, was a consequence of insufficient resourcing by way of skills and organisational systems but, in most cases, was exacerbated by, if not directly, the result of the rampant culture of corruption and mismanagement that characterised the years of President Zuma (Lincoln, 2019). It was pointed out that the building of caravan parks and resorts by South African municipalities during the apartheid period often had been "at the most desirable locations" and that these facilities "were deemed strategic assets of municipalities" (Hatchuel, 2016). Leaders of the opposition DA maintained that there was a need for "new and innovative concepts to take existing assets forward" and that potentially there were as many as 700 municipal tourism sites "that are not being optimally used" (Vos, 2017).

It was suggested in a research study commissioned by the Industrial Development Corporation that in certain provinces of South Africa there was a potential "over-supply" of caravan sites (Mthente Research and Consulting Services, 2013, p. 112). Suggestions were made for the utilisation and re-purposing of certain under-utilised state assets through their conversion into a planned network of budget resorts, akin to Butlins style holiday camps, which would be targeted at the market of low income (mainly Black) domestic leisure seekers (Mthente Research and Consulting Services, 2013). Among the list of identified under-utilised resorts that might be potentially converted into low-budget resort facilities were noted several municipal campgrounds and caravan parks. Although feasibility studies were undertaken, and discussions took place about pilot projects for budget destinations for low-income travellers, no implementation of these plans occurred.

Leveraging assets: Local government planning challenges

The research on the local planning challenges of South African caravan parks was conducted in a policy environment in which the national government was searching for new drivers for a more inclusive tourism economy (Giddy et al., 2020). During 2017, South Africa's Department of Tourism signalled that policy consideration was being given to leverage the opportunities of state-owned assets for tourism development and in particular for the upgrading of opportunities linked to tourism for Black entrepreneurs by applying public procurement as a tool to preference opportunities for Black SMME entrepreneurs (Rogerson & Rogerson, 2019b). A first step was a national audit of state tourism-related assets (conducted by the Department of Tourism), which revealed an array of important assets that might be levered for more inclusive tourism development. At the municipal or local level of government, alongside small nature reserves, conference centres, museums and even lighthouses, among the most widespread of municipal state assets was the residual network of municipal-owned and operated caravan parks/resorts (Rogerson & Rogerson, 2019b).

In terms of examining the challenges surrounding local planning for municipal caravan parks in South Africa, the focus was upon the rump of operational and functioning parks, the overwhelming majority of which are concentrated in the province of Western Cape. Figure 13.1 maps the geographical spread of the municipal caravan parks across the province. In total, it shows the location of 57 municipal parks. It is evident from Figure 13.1 that in terms of numbers of municipal parks, the most important local municipalities are the City of Cape Town, Saldanha Bay and Mossel Bay followed by Hessequa, Cederberg, Cape Agulhas, Berg River and Overstrand. These eight local governments are responsible for 70% of the total municipal caravan parks in the province, a finding that points to the geographically concentrated character of municipal caravan parks. In terms of their location, municipal caravan parks are a component of the coastal tourism economy of South Africa and more particularly in the Western Cape, which is one of the country's major areas for coastal tourism (Rogerson & Rogerson, 2020d).

The summary analysis that follows on local planning and caravan parks draws from two sets of material.[1] The first is a set of documentary sources on these parks including planning reports, newspaper reports and internet sources. Second, and most importantly, are the results of a set of structured interviews that were conducted during 2019 with 14 municipal officials and stakeholders who together account for the running of over 40% of the operational municipal parks in the province. More specifically, in terms of the small towns of the Western Cape, localities where a successful caravan park can make a difference to the local tourism economy, the sample of interviews represents 52% of the total.

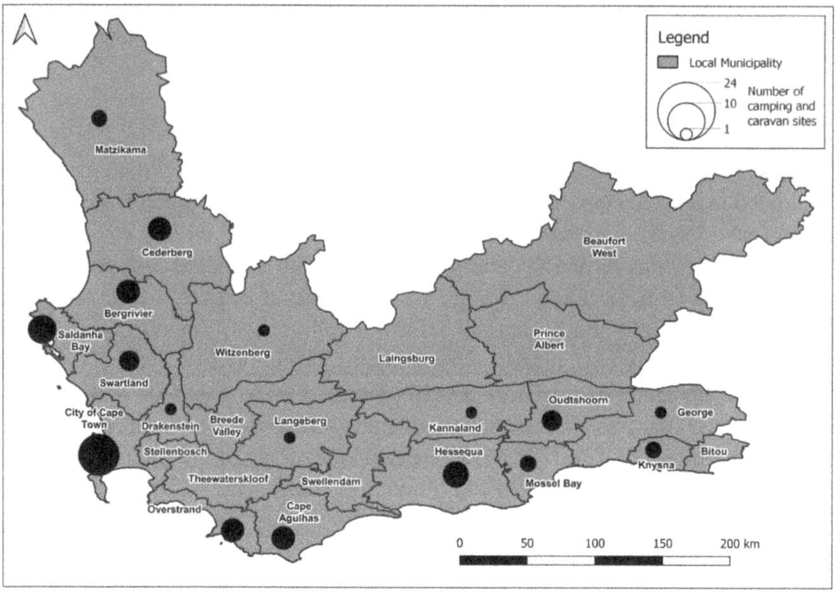

Figure 13.1 The geography of municipal caravan parks in the Western Cape on a local municipality basis. Source: Authors.

Administration

A key finding from the interviews was of the diverse position of caravan parks within the administrative structure of local municipalities. The management of caravan parks falls under no common department within municipalities in the Western Cape. In the case of Cape Agulhas Local Municipality, the responsibility falls to the Department of Public Services for day-to-day management. At Hessequa, management is within the domain of the department of technical services. At Overstrand Local Municipality, the administration of the four caravan parks is within the portfolio of Community Services. The Directorate for Community Services of the Cedarberg municipality has six sections (such as libraries, traffic) of which one is allocated responsibility for the municipality's four resorts. This small town municipality offers one of the few cases where there was a particular subdepartment that was responsible for caravan parks.

The general pattern observed, however, was that the management of the municipal assets of caravan parks was part of a bundle of diverse responsibilities that would fall within the portfolio of specific sections of municipal administration. Indeed, the study disclosed a disconnect between institutional planning for potential tourism assets and local economic development in the case of caravan parks. The lack of a dedicated focus on caravan parks within local municipalities can be illustrated by the findings from two particular cases. At Cape Agulhas, the director's responsibility is for general public services as well as the camping and caravan sites, and during the interview, he was called by a community member to facilitate the mowing of the grass on a residential pavement. In the case of the Overstrand Local Municipality where caravan parks fall under community services, the responsibility of the director also includes roads, water and sewage as well as cemeteries. The consequences of this bundling of caravan parks within a diverse portfolio of responsibilities are that in many cases it is marginalised because it is seen as of minor significance as compared with issues of water supply or road infrastructure. The marginalised position of caravan parks is therefore reflected in budgetary allocations within municipalities. In one interview it was pointed out that caravanning was not a primary issue as compared with roads and water and this low prioritisation was evident when it was competing for funding for these other issues.

Role in the local economy

Notwithstanding its somewhat marginal position within the administrative structure of local municipalities, the interviews disclosed that several of those municipalities with caravan parks were seen as significant for the local economy and most especially for tourism development. Typically, one interviewee stated that the caravan and camping parks in Saldanha "are very important for the local tourism economy as they provide affordable options for accommodation and provide access to the sea, not all campsites do" (Leentjiesklip Resort Manager). At other local municipalities where financial costs of caravan parks were closely monitored and a revenue surplus was generated for the municipality from park operations, the positive income generation flow from the asset was stressed (Siyanda Business Solutions, 2019). At Hessequa, the interviewee also emphasised the income generated from the parks, as it was stated that the "caravan and camping parks are a big asset in terms of income, just over R12 million". Further, at Hessequa, there was wider recognition of the local multiplier effects on the local economy of the caravan sector. The municipal director for Hessequa views the "parks as important as they have a direct and indirect investment. Tourism spend within the town is more important than the direct spend with the resorts itself for accommodation".

Overall, the most significant demonstration of the local recognition of the importance of caravan parks and camping occurred in the case of the municipal resorts of Cape Agulhas Local Municipality. In a detailed status quo assessment, which was prepared for the municipality, it was stated as follows: "The inclusion of resorts forms an integral part of Council's strategy to promote local and economic development. Resorts in general is (*sic*) utilised as gateway to the municipal area especially among the middle to higher income classes" (Siyanda Business Solutions, 2019, p. 1). The additional spin-offs from the caravan sector for the local economy were further elaborated. It was reported: "This influx of people, especially during December/January and Easter, provides much needed financial stimulus to the business sector, while it could potentially also lead to future investments within the municipal area" (Siyanda Business Solutions, 2019, p. 1). In turn, the impacts for the municipality as well as the municipal area as a whole, albeit difficult to quantify, were seen as highly beneficial in this particular municipality. The local importance of the caravan parks was evidenced for this municipality by its marketing initiatives that include a brochure highlighting the area's attractions for sea angling and proclaiming, "Come and Enjoy your Holiday at the Southern Most Caravan Park in Africa!!" (Cape Agulhas Municipality, n.d.).

Inclusion

In the international context of research on tourism small enterprises, the most distinctive aspect of the South African experience relates to the national government's focus on "transformation" and of ensuring the greater participation in the tourism industry of communities that were formerly disadvantaged under apartheid (Giddy et al., 2020; Rogerson, 2004). As mentioned earlier, the introduction of the government's immovable asset management policy was aligned to the goal of strengthening Black economic empowerment (BEE) in South Africa (Department of Public Works, 2005, 2008). A theme of particular interest that was explored in the interviews was the extent to which outsourcing opportunities might exist linked to the caravan parks and which could provide opportunities for the entry and development of Black-owned SMMEs.

The overall situation was that in virtually all municipalities most of the construction and maintenance work around caravan parks is done in-house. Nevertheless, certain opportunities do arise, and the process is to use the municipal procurement system with the hiring of businesses run by Black people that would be targeted for economic empowerment. The process is explained for Saldanha Bay as follows. The resort manager takes responsibility for initiating maintenance, identifies the needs and sends a procurement request for contractors to the council. If work is not done in-house, a municipal tender process is initiated, and work outsourced and "generally black and coloured small businesses from Saldanha that are registered on the municipal database will carry out the work" (Leentjiesklip Resort Manager). Security is outsourced and runs through a tender process from the municipality. At the Tietiesbaai resort, there are three full-time staff, "but when it fills up we get more people from the municipality. These are mostly maintenance and cleaning jobs. Any business that is outsourced is done utilising the municipal system and municipal finance and procurement processes".

At Cape Agulhas, a similar work arrangement exists. The Cape Agulhas municipal director explained:

> We have an in-house maintenance team that will work on the resorts. For big capital development jobs such as building work we have a tender process where we advertise to contractors. Local workers are employed but if there is something bigger to complete than what we are capable of the work is outsourced to Bredasdorp for a team that does a wider range of maintenance. Outsourcing is through government tenders.

At Saldanha Bay, the seasonal peak provides temporary opportunities for SMMEs. It was observed as follows:

> Over December a tender is put out for cleaning services and gate guards. It adds up to an additional 140 temporary job opportunities in the season. 95% of the contractors used within the parks are local. It upskills a lot of local workers that are then able to complete private work. A lot of contractors don't perform well and this is an opportunity for Local Economic Development Department to develop them.

Overall, therefore, in terms of Black SMME development, the network of caravan parks in the Western Cape offers a number of (mainly short-term) business opportunities with the most significant long-term opportunities relating to the outsourcing of security.

Planning directions

The future development of and planning for caravan parks elicited several responses. The interviews revealed that different choices are under consideration and have been in many cases under discussion for many years. Several possible options exist going forward for the management of caravan parks (Seaton Thomson & Associates, 2012).

First, a no-change situation with municipalities continuing to manage the parks as they have been doing for many years. Second, the option of privatisation and the selling off of assets, which usually involves a change of land use, often for commercial or residential development. Third, to consider developing a joint venture partnership with the provincial government and/or private developers/operators to facilitate the development, management and operation of the resorts. Four, there is the option of institutional restructuring and establishing a unit within the local municipality that would be dedicated to caravan parks (Seaton Thomson & Associates, 2012).

Finally, there is the option of leasing out parks to the private sector for a defined time period and on a tender basis. For many local government representatives, this last option is attractive and preferred as caravan parks are seen as "not our basic service" in municipalities. Nevertheless, as municipalities do not want to lose ownership of the property but instead want to get an optimal return on investment, the decision is often to outsource as the private sector lessee is mandated to upgrade and improve the parks in a way that municipalities do not have the funding to do so.

Conclusion

Place-based planning and development interventions formed the conceptual base for this investigation of one facet of tourism for change in modern-day South Africa. For successful local development led by tourism, there is a recognition of the need to maximise the potential of local assets (Rogerson & Rogerson, 2020a). One component of local assets is municipal assets that can be levered for tourism development. In the South African case, the developmental mandate of local governments under the Constitution provides the context for municipalities to seek to maximise the impact of their "immovable asset base". The trajectory of the caravan sector in South Africa has followed trends in other countries as it is shifting in its make-up with different segments, some linked to a low-cost offering but others much more upmarket in terms of facilities and offerings.

The history of caravanning in South Africa reveals that it emerged as a low cost or budget option for leisure (Rogerson & Rogerson, 2021). The expansion of municipal caravan parks was given impetus from the 1960s by government concern and support to extend the country's tourism domestic market by offering low-cost options for budget family holidays (mainly oriented at the white market). Over the past 30 years, however, South Africa has experienced a shift that parallels that in Europe, Australia and New Zealand with a change towards upmarket offerings. This upward shift in the caravan segment has mainly been capitalised upon by the further growth of private sector caravan grounds that in much of the country have taken over from municipal-owned caravan parks. It is evident that across most of South Africa, the municipal assets of caravan parks have been under-performing, and in many localities, caravan parks are in a state of decay and disrepair (Rogerson & Rogerson, 2020c).

The coastal province of the Western Cape is the largest geographical focus for caravan parks in South Africa, both for private and municipal-owned parks/resorts. A critical finding from our investigation was that the administrative responsibility for these municipal tourism assets was not always linked to tourism planning or local economic development planning but instead to parks and recreation or community development. This location of the responsibility for caravan parks certainly limits the prospects for these potentially valuable assets to be applied and linked to broad-based local tourism growth and place-based planning. In several of the small towns in the Western Cape, municipal caravan parks are viewed as contributing positively to local development prospects. Only minor leveraging opportunities exist, however, to align them to maximise empowerment opportunities for small business development by groups previously disadvantaged under apartheid. Finally, in the fluid market environment that challenges the changing caravan sector across the small towns of the Western Cape, an array of options are under consideration for the future planning of municipal caravan parks. In 2020, COVID-19 is an additional element impacting the caravan economy of South Africa and one that potentially may act – at least in the short term - to encourage the growth of this form of drive tourism.

Note

1 Space limitations only permit a summary presentation in this chapter. Full details of the survey and findings are available in University of Johannesburg 2020.

References

Barca, F. (2019). Place-based policy and politics. *Renewal: A Journal of Labor Politics, 27*(1), 84–95.

Barca, F., McCann, P., & Rodriguez-Pose, A. (2012). The case for regional development intervention: Place-based versus place-neutral approaches. *Journal of Regional Science, 52*, 134–152. doi:10.1111/j.1467-9787.2011.00756.x

Blichfeldt, B. S. (2009). Innovation and entrepreneurship in tourism: The case of a Danish caravan site. *Pasos, 7*(3), 415–431.

Blichfeldt, B. S., & Mikkelsen, M. V. (2013). Vacability and sociability as touristic attraction. *Tourist Studies, 13*(3), 235–250. doi:10.1177/1468797613498160

Blichfeldt, B. S., & Mikkelsen, M. V. (2014). Camping, tourism. In J. Jafari & H. Xiao (Eds.), *Encyclopedia of Tourism (Living Edition)*, (pp. 1–2). Dordrecht: Springer.

Brooker, E., & Joppe, M. (2013). Trends in camping and outdoor hospitality – An international review. *Journal of Outdoor Recreation and Tourism, 3–4*, 1–6. doi:10.1016/j.jort.2013.04.005

Brooker, E., & Joppe, M. (2014). A critical review of camping research and direction for future studies. *Journal of Vacation Marketing, 20*(4), 335–351. doi:10.1177/1356766714532464

Buys, F., & Mavasa, T. (2007). The management of government immovable assets. *Acta Structilia, 14*(1), 81–92.

Caldicott, R. W. (2011). *Supply-side Evolution of Caravanning in Australia: An Historical Analysis of Caravan Manufacturing and Caravan Parks*. Lismore, NSW: School of Tourism and Hospitality, Southern Cross University.

Caldicott, R. W., & Scherrer, P. (2013a). The life cycle of caravan parks in Australia: The case of northern New South Wales. *Australian Geographer, 44*(1), 63–80. doi:10.1080/00049182.2013.765348

Caldicott, R. W., & Scherrer, P. (2013b). Facing divergent supply and demand trajectories in Australian caravanning: Learnings from the evolution of caravan park site-mix options in Tweed Shire. *Journal of Vacation Marketing, 19*(2), 117–131. doi:10.1177/1356766712457672

Cape Agulhas Municipality (n.d.). *L'Agulhas Caravan Park*. Bredasdorp: Cape Agulhas Municipality.

Caravan Club of Southern Africa (1950). *Guide to the Principal Caravan Parking Sites in Southern Africa*. Johannesburg: Caravan Club of Southern Africa.

Caravan Club of Southern Africa (1961). *Guide to the Principal Caravan Parking Sites in Southern Africa*. Johannesburg: Caravan Club of Southern Africa.

Collins, D., & Kearns, R. (2010). Pulling up the tent pegs?: The significance and changing status of coastal campgrounds in New Zealand. *Tourism Geographies, 12*(1), 53–76. doi:10.1080/14616680903493647

Davidson, R., & Maitland, R. (1997). *Tourism Destinations*. London: Hodder & Stoughton.

Department of Public Works (2005). *Government-wide Immovable Asset Management Policy*. Pretoria: Department of Public Works.

Department of Public Works (2008). *Immovable Asset Management in National and Provincial Government: Guideline for Custodians*. Pretoria: Department of Public Works.

Doğantan, E., & Emir, O. (2019). Determination of trailer park criteria in rural areas. *Journal of Tourism and Gastronomy Studies, 7*(4), 2383–2398. doi:10.21325/jotags.2019.477

Donaldson, R. (2018). *Small Town Tourism in South Africa*. Cham: Springer.

Duranton, G., & Venables, A. (2019). *Place-based Policies: Principles and Developing Country Applications*. Department of Economics Discussion Papers No. 893. University of Oxford, Oxford.

Giddy, J. K., Idahosa, L. O., & Rogerson, C. M. (2020). Leveraging state-owned nature-based assets for transformation and SMME development: Pilanesberg National Park, South Africa. In J. M. Rogerson & G. Visser (Eds.), *New Directions in South African Tourism Deographies* (pp. 299–316). Cham: Springer International.

Gilbert, H. (2013). The loss of low cost caravan parks – causes, cases and social consequences. In K. Ruming, B. Randolph, & N. Gurran (Eds.), *Proceedings: State of Australian Cities Conference, 2013*. Sydney: State of Australian Cities Research Network.

Grundlingh, A. (2006). Revisiting the "old" South Africa: Excursions into South Africa's tourist history under apartheid, 1948–1990. *South African Historical Journal, 56*, 103–122. doi:10.1080/02582470609464967

Hanis, M. H., Trigunarsyah, B., & Susilawati, C. (2010). Elements of public asset management framework for local governments in developing countries. In *Proceedings, 8Th International Conference on Construction and Real Estate Management (ICCREM)*, 1–3 December, Brisbane.

Hatchuel, M. (2016). Newsletter, commentary and online discussion about tourism in South Africa. *This Tourism Week*, 4 April.

Kaganova, O. (2010). *Government Management of Land and Property Assets: Justification for Engagement by the Global Development Community*. Washington DC: Urban Institute Center on International Development and Governance.

Kaganova, O., & Kopanyi, M. (2014). Managing local assets. In C. Farvacque-Vitkovic & M. Kopanyi (Eds.), *Municipal Finances: Guidebook for Local Governments* (pp. 275–324). Washington DC: World Bank.

Kaganova, O., & McKellar, J. (Eds.) (2006). *Managing Government Property Assets: International Experiences*. Washington DC: Urban Institute.

Kaganova, O., & Nayyar-Stone, R. (2000). Municipal real property asset management: An overview of world experience, trends and financial implications. *Journal of Real Estate Portfolio Management, 6*(4), 307–326.

Kaganova, O., & Telgarsky, J. (2018). Management of capital assets by local governments: An assessment and benchmarking survey. *International Journal of Strategic Property Management, 22*(2), 143–156. doi:10.3846/ijspm.2018.445

Kontsiwe, N., & Visser, G. (2019). Tourism as a vehicle for local economic development in small towns? When things go wrong: The case of Aliwal North, South Africa. *GeoJournal of Tourism and Geosites, 27*(4), 1334–1346. doi:10.30892/gtg.27418-437

Lashley, C. (2015). Researching snails on holiday: An agenda for caravanning and caravanners? *Research in Hospitality Management, 5*(2), 115–122. doi:10.1080/22243534.2015.11828336

Lincoln, G. M. (2019). *Competitive Regions in South Africa: A Study of Regional Planning in the iLembe District in the Province of KwaZulu-Natal*. DPhil dissertation (Management Sciences), Durban University of Technology, Durban.

Lowvelder (Newspaper, Nelspruit), 8 August 2018.

Mikkelsen, M. V., & Cohen, S. A. (2015). Freedom in mundane mobilities: Caravanning in Denmark. *Tourism Geographies, 17*(5), 663–681. doi:10.1080/14616688.2015.1084528

Mikkelsen, M. V., & Blichfeldt, B. S. (2015). "We have not seen the kids for hours": The case of family holidays and free-range children. *Annals of Leisure Research, 18*(2), 252–271. doi:10.1080/11745398.2014.999342

Mikkelsen, M. V., & Blichfeldt, B. S. (2018). Grand parenting by the pool. *Young Consumers, 19*(2), 127–140. doi:10.1108/YC-04-2017-00675

Mikulić, J., Prebežac, D., Šerić, M., & Krešić, D. (2017). Campsite choice and the camping tourism experience: Investigating decisive campsite attributes using relevance-determinance analysis. *Tourism Management, 59*, 226–233. doi:10.1016/j.tourman.2016.07.020

Mthente Research and Consulting Services (2013). *Towards Developing a Budget Travel Destination Chain in South Africa.* Report prepared for the Industrial Development Corporation, Johannesburg.

Nel, E., & Rogerson, C. M. (2016). The contested trajectory of applied local economic development in South Africa. *Local Economy, 31*(1–2), 109–123. doi:10.1177/0269094215623729

Nel, E., Connelly, S., & Stevenson, T. (2019). New Zealand's small town transition: The experience of demographic and economic change and place based responses. *New Zealand Geographer, 75*(3), 163–176. doi:10.1111/nzg.12240

Okumus, F., Köseoglu, M. A., Putra, E. D., Yildiz, M., & Dogan, I. C. (2019). Conceptual structure of lodging-context studies: 1990–2016. *Journal of Hospitality & Tourism Research, 43*(2), 210–225. doi:10.1177/1096348018823912

Patterson, I., Pegg, S., & Mahedevan, R. (2015). The benefits of short stay caravan travel based on the lived experiences of grey caravanners in Australia. *Tourism Analysis, 20*(5), 539–549. doi:10.3727/108354215X14411980111497

Phatela, A. V., & Cloete, C. E. (2018). The impact of the Government Immovable Asset Management Act (GIAMA) on the Department of Public Works, South Africa. *RELAND - International Journal of Real Estate and Land Planning, 1*, 285–291.

Phelps, A. (2011). Municipal property asset management – A comparative study of UK and Russia. *International Journal of Strategic Property Management, 15*(4), 416–437. doi: 10.3846/1648715X.2011.642537

Pike, A., Rodriguez-Pose, A., & Tomaney, J. (2014). Local and regional development in the global north and south. *Progress in Development Studies, 14*(1), 21–30. doi:10.1177/1464993413504346

Pike, A., Rodriguez-Pose, A., & Tomaney, J. (Eds.) (2015). *Local and Regional Development.* 4 Vols, London: Routledge.

Pike, A., Rodriguez-Pose, A., & Tomaney, J. (2017). Shifting horizons in local and regional development. *Regional Studies, 51*(1), 46–57. doi:10.1080/00343404.2016.1158802

Pretorius, W. (2014). The municipal engineer's responsibility to practice life cycle infrastructure asset management and integrated communication to the financial asset register. In *Proceedings 2014 IMESA Conference* (pp. 130–133). Durban: Institute of Municipal Engineering of Southern Africa.

Prideaux, B. (2020). Drive and car tourism: A perspective article. *Tourism Review, 75*(1), 109–112. doi:10.1108/TR-05-2019-0162

Prideaux, B., & McClymont, H. (2006). The changing profile of caravanners in Australia. *International Journal of Tourism Research, 8*, 45–58. doi:10.1002/jtr.546

Rodriguez-Pose, A., & Palavicini-Corona, E. I. (2013). Does local economic development really work?: Assessing LED across Mexican municipalities. *Geoforum, 44*, 303–315. doi:10.1016/j.geoforum.2012.07.010

Rodriguez-Pose, A., & Wilkie, C. (2017). Revamping local and regional development through place-based strategies. *Cityscape: A Journal of Policy Development and Research, 19*(1), 151–170.

Rogerson, C. M. (2004). Tourism, small firm development and empowerment in post-apartheid South Africa. In R. Thomas (Ed.), *Small Firms in Tourism: International Perspectives* (pp. 13–33). Amsterdam: Elsevier.

Rogerson, C. M. (2014). Reframing place-based economic development in South Africa: The example of local economic development. *Bulletin of Geography. Socio-Economic Series, 24*, 203–218. doi:10.2478/bog-2014-0023

Rogerson, C. M. (2019a). Revamping local economic development policy in South Africa. In J. Knight & C. M. Rogerson (Eds.), *Geography of South Africa: Contemporary Changes and New Directions* (pp. 253–259). Cham: Springer International.

Rogerson, C. M. (2019b). Globalisation, place-based development and tourism. In D. Timothy (Ed.), *Handbook of Globalisation and Tourism* (pp. 44–53). Cheltenham: Edward Elgar.

Rogerson, C. M. (2020). Using municipal tourism assets for leveraging local economic development in South Africa. *Bulletin of Geography. Socio-Economic Series, 48*, 47–63. doi:10.2478/bog-2020-0013

Rogerson, C. M., & Rogerson, J. M. (2019a). Tourism, local economic development and inclusion: Evidence from Overstrand local municipality, South Africa. *GeoJournal of Tourism and Geosites, 25*(2), 293–308. doi:10.30892/gtg.25202-360

Rogerson, C. M., & Rogerson, J. M. (2019b). Public procurement, state assets and inclusive tourism: South African debates. *GeoJournal of Tourism and Geosites, 26*(3), 686–700. doi:10.30892/gtg.26301-389

Rogerson, C. M., & Rogerson, J. M. (2020a). Inclusive tourism and municipal assets: Evidence from Overstrand Local Municipality, South Africa. *Development Southern Africa, 37*(5), 840–854. doi:10.1080/0376835X.2020.1796597

Rogerson, C. M., & Rogerson, J. M. (2020b). Camping tourism: A review of recent international scholarship. *GeoJournal of Tourism and Geosites, 28*(1), 349–359. doi:10.30892/gtg.28127-474

Rogerson, C. M., & Rogerson, J. M. (2020c). Municipal caravan parks in South Africa: Geography, management and planning. *Bulletin of Geography. Socio-Economic Series, 49*, 109–124. doi:10.2478/bog-2020-0027

Rogerson, C. M., & Rogerson, J. M. (2020d). Coastal tourism in South Africa: A geographical perspective. In J. M. Rogerson & G. Visser (Eds.), *New Directions in South African Tourism Geographies* (pp. 227–247). Cham, Switzerland: Springer International.

Rogerson, C. M., & Rogerson, J. M. (2021). Mundane urban tourism; The historical evolution of caravan parks in South Africa 1930–1994. In C. M. Rogerson & J. M. Rogerson (Eds.), *Urban Tourism in the Global South: South African Perspectives.* Cham: Springer International.

Ruhanen, L. (2013). Local government: Facilitator or inhibitor of sustainable tourism development? *Journal of Sustainable Tourism, 21*(1), 80–98. doi:10.1080/09669582.2012.680463

Seaton Thomson & Associates (2012). *Tourism Development Plan for the Saldanha Bay .unicipality Phase 3.* Knysna: Seaton Thomson & Associates.

Sedmihradska, L. (Ed.) (2010). *Municipal Asset Management in Transition Countries: Selected Case Studies.* Bratislava: NISPAcee Press.

Simeoni, F., & Cassia, F. (2019). From vehicle suppliers to co-creators in tourism: The evolving role of Italian motorhome manufacturers. *Current Issues in Tourism, 22*(2), 218–236. doi:10.1080/13683500.2017.1293619

Siyanda Business Solutions (2019). *Status quo Assessment – Cape Agulhas Municipal Reports.* Report prepared for the Cape Agulhas Local Municipality.

Todes, A., & Turok, I. (2018). Spatial inequalities and policies in South Africa: Place-based or people-centred? *Progress in Planning, 123*, 1–31. doi:10.1016/j.progress.2017.03.001

Tomaney, J. (2010). *Place-based Approaches to Regional Development: Global Trends and Australian Implications*. Sydney: Australian Business Foundation.

University of Johannesburg (2020). *Leveraging State-Owned Assets for Black Small, Medium and Micro-Enterprises Development: A Case of State-Owned Parks, Lodges and Attractions Phase 2*. Report for the Department of Tourism, Pretoria.

Van Heerden, C. H. (2008). Leisure motorhoming: The case of the motorhome club of South Africa. *South African Journal for Research in Sport, Physical Education and Recreation, 30*(1), 125–136.

Van Heerden, C. H. (2010a). Self drive tourism in South Africa with specific emphasis on caravanning. In B. Prideaux & D. Carson (Eds.), *Drive Tourism: Trends and Emerging Markets* (pp. 85–97). London: Routledge.

Van Heerden, C. H. (2010b). An exploratory analysis of leisure caravanning in the Kruger National Park in South Africa. *Innovative Marketing, 6*(1), 66–72.

Van Heerden, C. H. (2020). Motivation and involvement in camping - a gender perspective. *African Journal of Hospitality, Tourism and Leisure, 9*(1), 1–14.

Venter, M. (2020). *Exploring Local Economic Development in South Africa*. Cape Town: Juta.

Vos, J. (2017). Millions allocated to "failed" tourism projects - Shadow Tourism Minister. *Tourism Update*, 23 February.

Zhang, J., Yin, M., Han, J., & Aroskar, R. (2019). Why is asset-light strategy necessary?: An empirical analysis of cost stickiness. *Tourism Management Perspectives, 32*, 100571. doi:10.1016/j.tmp.2019.100571

14 The gig economy and employment in tourism in Southern Africa

A global finger in the informal sector pie?

Tom Baum and Julia K. Giddy

Introduction

Our narrative in this chapter is to cast a critical eye on the gig economy in tourism in the context of Southern Africa. We contend that, in many respects, gig platforms are in direct competition with the established informal economy and this competition is particularly effective because of the influence and global importance of gig brands such as Uber, Taxify/Bolt, Airbnb and Uber Eats. In fact, highly publicised violent encounters have erupted in Southern Africa between traditional informal operators that play a significant role in the travel and tourism sectors (Carmody & Fortuin, 2019; Giddy, 2019). Freelance operators in these sectors exhibit many of the characteristics of their counterparts in the existing informal economy but they have to pay a significant levy to their brand owners, which are tied into global companies and global capital, for the privilege of doing so. In addition, as is the case with many cities in the Global North, workers on some gig economy platforms in Southern Africa have attempted to confront these global players, seeking acknowledgement of their status as employers, a role the companies consistently deny, in order to benefit from national labour laws (Adebayo, 2019; Geitung, 2017). As of yet, these efforts have been largely unsuccessful (van der Merwe & Dias, 2018). The platform economy, therefore, is the cause of significant controversy in South Africa, both in general and specific travel and tourism contexts.

Workers in both the informal and gig sectors of Southern African tourism are in precarious situations and are particularly vulnerable to employment abuse and frequently do not enjoy the rights that employees within the formal sector have (Adebayo, 2019). People who work in these sectors, according to the OECD (2001, p. 5),

> typically operate at a low level of organization, with little or no division between labour and capital as factors of production and on a small scale. Labour relations – where they exist – are based mostly on casual employment, kinship or personal and social relations rather than contractual arrangements with formal guarantees.

Although the gig economy differs slightly in the way "employees" are sourced and "contracted", the issue of casual employment and lack of formal guarantees still applies (Geitung, 2017). In contrast to descriptors that are used by gig economy corporations, such as Airbnb, the vast majority of those offering service subcontract workers to generate income through these platforms, though with little to no protection and less profit for the workers. Therefore, uneven power dynamics emerge between asset owners and informally contracted workers within the gig economy. This is yet another indicator of the precarious employment situation in travel and tourism in Southern Africa, as well as the immense inequality between those with capital (i.e. assets) and those without (Booyens, 2020).

This chapter, therefore, seeks to examine labour dynamics in the emergent gig economy in Southern Africa by analysing both core sources, such as media coverage, national legislation and the policies of gig economy actors in the region, as well as other secondary sources including previous academic work both on the gig economy in the region and the informal sector. We adopt the following definition of the gig economy from the UK Department of Business, Energy and Industrial Strategy (Lepanjuuri, Wishart, & Cornick, 2018, p. 12), "The gig economy involves exchange of labour for money between individuals or companies via digital platforms that actively facilitate matching between providers and customers, on a short-term and payment by task basis" in so far as these operate within travel and tourism. In doing so, we draw comparisons between the extant informal sector in tourism and the emergent gig economy, which interfaces widely with tourism. They are both significant components of Southern African economies and we seek to provide an initial understanding of the dynamics between them.

The informal economy in tourism

The informal sector can also be described as the underground economy, black economy, shadow economy or grey economy (Hart, 2006) and is characterised by diversity but, generally, workers find themselves in the informal economy by necessity, not choice (ILO, 2018). The informal economy is not subject to the bureaucratic rules and regulations that are generally associated with the formal economy in most countries (Hart, 2006). The informal economy was long perceived to be a shrinking legacy of pre-industrialised societies (Geertz, 1963; Lewis, 1959), but such views have been open to increasing criticism as a result of the persistence of informal economic activity. The informal economy is also a sector that is not generally subject to regulation and control with respect to areas, *inter alia*, that include health and safety, quality control, consumer rights, employment rights and environmental protection. The informal sector includes both legal activities, such as the preparation and sale of street food and construction, and activities such as drug dealing and people trafficking, activities that are generally deemed illegal. Employment in the informal economy frequently does not meet ILO criteria with regard to "decent work" but, nonetheless, more than six workers among ten and four enterprises among five in the world operate in the informal economy (ILO, 2020). Thus, more than 60% of the world's employed population earn their livelihoods in the informal economy (ILO, 2018). Williams (2017, p. 146) refers to narratives (from, among others, Davis, 2006) that represent informality as "an integral component of contemporary capitalism and a central feature of the new outsourcing, downsizing and subcontracting practices that have appeared under deregulated global capitalism". Williams (2017) challenges the neo-liberal argument that employment in the informal economy is a consequence of a high tax regime combined with public sector corruption and state interference in the free market. He also suggests that the evidence of his research is that employment in the informal economy appears to reduce when higher levels of regulation and state intervention are in place.

The informal economy exists in all countries regardless of their level of socio-economic development, although it is more prevalent in the Global South. The ILO (2018) note that the 2 billion women and men who make their living in the informal economy are deprived of decent working conditions. Evidence shows that most people do not enter the informal economy by choice. Rather, it is a consequence of the lack of opportunities in the formal economy, poor educational chances and the absence of other means of livelihood. Contrary to some assumptions, the presence of an informal economy is by no means confined to the Global South. Leonard (2000) rightly highlights the common cause of those on low income in the developed world and the poor in many developing countries and sees the response and route to the survival of both as lying within the informal economy. She compares the resilience of informal economic activity among low-income populations in both contexts and finds that their coping strategies show many of the same characteristics.

There has long been an interest in the informal economy in the context of tourism. For example, Wahnschafft (1982) investigated the living and working conditions of those in the informal tourism economy of Pattaya, Thailand, defined as those working in petty enterprises in tourism. He asks key questions about their business prospects and the socio-economic potential for its further development. Wahnschafft (p. 430) describes the informal sector as including the smallest of businesses "where the conditions of poverty are revealed in their sharpest forms". Crick (1992, p. 135), in discussing fieldwork conducted in Sri Lanka some ten years earlier, addresses the challenges but also the mitigation strategies employed by those whose work is located in "that arena beyond the effective control of the tourism authorities: street corners; unlicensed guesthouses; cheap cafes etc." because of the absence of bureaucratic controls. Damayanti et al. (2017) note the prevalence of informal activities in tourist spaces, especially in the Global South, pointing to businesses in food, souvenir sales, accommodation and tourist guiding as examples. Their work is part of a growing interest in informal economy tourism activity from multiple perspectives and much of this work is conducted as case studies located in the Global South (see, for example, Çakmak et al., 2018; Henderson & Smith, 2009; Slocum et al., 2011).

The Southern Africa context

In the African context of this chapter, understanding of the informal tourism sector is informed by the work of, among others, Bah and Goodwin (2003), Brown et al. (2010), Slocum, Backman and Robinson (2011) and Rogerson (2014). In

Southern Africa, the informal economy takes on a number of different facades and contributes to the tourism economy in several different ways. There are differences in the way that the informal tourism sector manifests and also its relationship with the formal tourism sector (Kirsten & Rogerson, 2002). For example, street vendors are a very significant component of the informal tourism economy in Southern Africa (and throughout the continent). In the context of leisure tourists, one of the most obvious goods provided by the informal sector (as is the case in many countries of the Global South) is the provision of craft and souvenirs (Peberdy, 2000). In addition to street vendors who supply the leisure tourism sector, many informal workers also supply goods to formal tourism businesses (Giddy et al., 2020). Another important way in which the informal feeds into the informal sector is through the provision of services which tourists, on organised tours, engage with, for example, street vendors. As discussed in more detail in the following sections, this is commonly in the supply of domestic workers (many of whom remain informal) such as cleaners, gardens, maintenance people, etc., which are often used by small-scale tourism businesses (Kirsten & Rogerson, 2002).

In terms of the body of literature on tourism and the informal sector in South Africa, two primary avenues of research are apparent. The first is the role of the informal sector in pro-poor tourism initiatives (Ashley & Roe, 2002; Mtapuri & Giampiccoli, 2016; Rogerson, 2006). Some of these studies have called for increased formalisation of many informal tourism workers, particularly those supplying the formal tourism sector (Ashley & Roe, 2002). The majority of this work, however, has focused on leisure tourism. The second avenue of research has been an examination of a particular tourism subsector significant particularly in Southern African cities, that of cross-border business tourism, which is dominated by the informal economy (Peberdy, 2000; Rogerson, 2014, 2018). Basically, these studies have demonstrated the prevalence and significance of microentrepreneurs who cross borders, often to procure and sell goods that will be sold in neighbouring countries. Some of the most significant sites of these transactions in Southern Africa are those close to borders such as Gaborone. Rogerson (2014) demonstrated not only the positive impact of these migrant entrepreneurs on, in this case, the South African tourism economy but also the fact that this sector is overlooked, in policy, planning and literature.

The gig economy and employment in tourism

The emergent gig or sharing economy is a source of growing importance but also controversy and debate within tourism (Martin, 2016; Wachsmuth & Weisler, 2018). There is a wide range of employment found within the gig economy which is often divided into what is considered place-based work and remote work (Graham & Anwar, 2018). The latter encompasses components such as call centres, online teaching, translation, transcription and other tasks that can be handled remotely (Anwar & Graham, 2019). Place-based work, however, is more relevant to tourism and involves work that is facilitated by a platform but executed in a specific location (Graham & Anwar, 2018). Gig operations in tourism typically encompass accommodation (for example, Guttentag, 2015; Guttentag et al., 2018), local transport (for example, Cramer & Krueger, 2016; Pollio, 2019) and food delivery services (for example, Drahokoupil & Piasna, 2019; Todoli-Signes, 2018). These online services and labour platforms are associated with significant potential for changing the way work is organised in society. In particular, they represent a shift from accessing labour through employment towards relying on controversial forms of self-employment (Drahokoupil & Fabo, 2016). In the context of tourism, the gig economy is closely linked with the so-called sharing economy, defined as "consumers granting each other temporary access to their under-utilized physical assets ('idle capacity'), possibly for money" (Meelen & Frankel, 2015). Although the marketing of the gig/sharing economy discusses the potential for self-employment and earning money from underutilised assets, the inherent working community involved within the sharing economy, at different levels, is often neglected (although Kessler [2018] and others have sought to redress this imbalance).

There has been extensive critical debate in discussions of the gig economy in a broad sense, addressing a range of themes and issues (Calo & Rosenblat, 2017; Martin, 2016). Within these discussions, growing critical appraisals of the gig environment have focused on the precarious nature of work that is offered as well as on how businesses operate on the margins of the formal economy and, in many countries, replicate many of the features of the informal economy (Graham & Woodcock, 2018; Martin, 2016). In fact, most gig economy platforms have successfully skirted employment regulations in several countries throughout the world through the deliberate misclassification of employment status (Moon, 2015; Wiessner, 2019). The majority of gig economy platforms are run by large corporations (primarily located in the Global North), but due to the nature of these platforms (i.e. that they are digital), they have expanded and operate worldwide (Giddy, 2019; Graham & Anwar, 2018). This presents challenges in terms of regulation, as the corporations have operations throughout the world and differing labour laws between countries of operation and the country (countries) in which the headquarters are based (Cetin & Deakin, 2017; Rayle et al., 2016). Therefore, in the countries in which these platforms are present, there are difficulties in upholding labour laws since the corporations are not headquartered in that country and so national regulations cannot be enforced (Anwar & Graham, 2019). In addition, one of the primary arguments among gig economy platforms is that they are simply a mechanism to facilitate these transactions, and vehemently deny that they should be considered a direct employer (Moon, 2015; Redlich, 2019; Wiessner, 2019). This, coupled with the aforementioned issues around regulation enforcement, has, for the most part, left gig economy workers with little to no legal protection (Anwar & Graham, 2019). This is another way in which gig economy employment mirrors characteristics seen in informal economies.

Several of these platforms are particularly relevant to the tourism industry. The most widely discussed type of platform is short-term housing rentals, primarily Airbnb. Airbnb, which offers peer-to-peer accommodation, has been the focus of a wide body of literature among tourism and hospitality scholars and forms a significant component of the debate surrounding the sharing economy (Guttentag, 2015; O'Regan & Choe, 2017; Wachsmuth & Weilser, 2018; Zervas et al., 2014). The impact of these platforms, in particular, on the tourism and hospitality sectors, has been well-documented. One of the primary benefits noted by travellers is the affordability of the accommodation options in contrast to traditional options such as hotels and lodges (Guttentag, 2015; Visser et al., 2017). However, Airbnb promotes the supposed "authenticity" of the experience offered by staying in someone's house, typically in residential neighbourhoods, in contrast to tourist precincts in which hotels and other formal accommodation are located (Airbnb, 2020; Stors & Kagermeier, 2015). Criticisms of Airbnb are far-ranging, most significantly the impacts on rental housing markets in many cities throughout the world (Garcia-Ayllon, 2018; O'Regan & Choe, 2017; Wachsmuth & Weisler, 2018). In recent years, some research has indicated that rental housing markets are becoming eliminated as large developers buy up properties traditionally serving (primary lower and middle class) rental housing markets and redeveloping these properties for the specific purpose of short-term rentals (Wachsmuth & Weisler, 2018).

This chapter, however, focuses on employment in the context of the gig economy. Therefore, when considering tourism employment in the context of Airbnb, two important components emerge. The first is, clearly, the impact that the emergence of Airbnb has had on the formal accommodation sector as a whole, and thus on employees within this sector who are already incredibly vulnerable as demonstrated in myriad studies (e.g. Fang et al., 2018; Oskam & Boswijk, 2016). The second, however, which has received less attention in the literature and media is the work that goes into operating an Airbnb. This has resulted in two forms of labour depending on the nature of the Airbnb establishment itself. The first, which is associated with Airbnbs operated as they are marketed (out of homes in which people actually reside regularly), is the emotional labour of Airbnb "hosts" who consistently have strangers in their home (Roelofsen, 2018). This emotional labour is closely associated with that discussed in other tourism subsectors, as the need to ensure the satisfaction of clients by consistently appearing happy/in good spirits and available to assist at all times. In the case of Airbnb hosts, this can become a constant state, as the client is based within one's home space (Roelofsen, 2018; Roelofsen & Minca, 2018). The other form of work associated with Airbnb has not readily appeared in the literature and that is those "behind the scenes" such as domestic staff (i.e. cleaners, gardeners). Although these may be present in the case of what we are calling the Airbnb-marketed "ideal", where the offering is the "hosts" home, and often involves cohabitation. However, it is even more prevalent in the type of Airbnb that has become more common and the source of much scrutiny, properties that have been specifically purchased for the sole purpose of developing Airbnb rentals (Cocola-Gant & Gago, 2019). This is most common among large developers (Cocola-Gant & Gago, 2019; Martin, 2016; Wachsmuth & Weisler, 2018). In this case, there is a need for a unique range of services to ensure the success of the Airbnb property, since there is no onsite "host". Thus, a new labour market emerges, closely related to some of the responsibilities typically held by employees of traditional accommodation (i.e. cleaners, property managers). However, there is a much greater potential in the context of Airbnb (and other short-term rentals) for these employees to operate informally, and thus without worker protections.

The second type of platform which is of significance in the tourism sector and of significance in discussions of employment is e-hailing, also sometimes referred to as ridesharing (Azmi et al., 2016) and widely used by tourists when visiting destinations (Giddy, 2020a). It is typically defined along the lines of "transportation services that connect community drivers – people who drive private cars instead of commercial vehicles – with passengers via mobile devices and applications" (Jin et al., 2018, p. 96). E-hailing has been another highly controversial platform, but, in this case, the major controversies are, in fact, directly related to issues of employment (Calo & Rosenblat, 2017; Martin, 2016; Moon, 2015; Rosenblat & Stark, 2015). There has been much debate surrounding e-hailing platforms due to the differences in regulation and standards required by e-hailing platforms and drivers in contrast to traditional metered taxis. In many cities, globally, private driver services, typically offered by metered taxis, are highly regulated and have strict training requirements and vehicle standards (Calo & Rosenblat, 2017; Graham & Woodcock, 2018; Rosenblat & Stark, 2015). However, e-hailing has largely operated in many cities without adhering to these standards and are therefore able to offer lower rates. As a result, in many places throughout the world, legal action has been taken against e-hailing platforms for operating below these standards. This also has allowed e-hailing platforms to bypass many labour regulations and, importantly, protections offered among other private driver services (Carmody & Fortuin, 2019; Graham & Woodcock, 2018; Moon, 2015). The best known of these platforms is Uber, which has continuously argued that it facilitates connecting drivers to riders and cannot be considered an employer (Calo & Rosenblat, 2017; Moon, 2015). Thus, drivers for Uber have no employee rights and they are considered "independent contractors" in most places throughout the world. Numerous lawsuits have been lodged against the company, though they have often been able to win these cases as the law, in many countries, struggles to keep up with this rapidly changing nature of work (van der Merwe & Dias, 2018). Many of the debates around e-hailing are also found in the third type of platform, mentioned above, that of food delivery, though much less research currently exists on this type of platform.

These two sectors of the gig economy are significant because they offer services that have direct parallels with those on offer in the extant informal economy in Southern Africa and constitute direct competition to them.

The Southern Africa context

The concerns over the working conditions in the context of the gig economy are particularly significant in the Global South including Southern Africa (Anwar & Graham, 2019; Carmody & Fortuin, 2019; Henama & Sifolo, 2017). Southern

Africa is also a region with high rates of un- and underemployment. The precarity of work in the region is closely related to the vast informal sector and tourism more specifically, as discussed previously. The literature on employment and tourism in Southern Africa is also minimal (Rogerson, 2020). However, those studies that have emerged have provided evidence for some important trends. Basically, it shows that tourism employment is incredibly precarious and therefore little protection is given to employees, often since workers are employed seasonally or are freelance contractors (Ndiuini & Baum, 2020; Rogerson, 2012, 2020; Visser & Rogerson, 2004). In addition, major concerns emerge over training, particularly in certain contexts, where some governing bodies exist for specific subsectors or activities, but there is little legal regulation of tourism employee qualifications and even when it does exist, there is often a lack of enforcement among regulators (Giddy, 2020b; McKay, 2013). Furthermore, due to high unemployment rates throughout the country, operators are typically able to find employees who will work under poor conditions with high staff turnover. As is the case with many service industries, tourism employees are often highly dependent on per-trip commission and/or tips (De Beer et al., 2014). Only when companies are either relatively large or have initiated stringent corporate social responsibility policies themselves does decent work appear to be present in the Southern African tourism industry (Giddy, 2020b). Due to the precarious nature of the work, much of which is ad hoc and based on the number of clients/excursions, it does tend to mimic the informal economy in many ways.

Limited research currently exists focusing on gig economy work in Southern Africa and even less on those platforms that are relevant to the tourism sector (Anwar & Graham, 2019; Giddy, in press; Henama & Sifolo, 2017). However, a few studies have emerged with a focus on assessing the nature of work associated with the gig economy in the region. Anwar and Graham (2019) discuss concerns over agency and resistance among (albeit among remote) gig economy workers, which demonstrate the many barriers put in place to prevent workers from having any kind of agency over their working conditions. This issue has also been raised in Giddy (in press) and Carmody and Fortuin (2019) both of which focused on e-hailing operators and found that there was little to no response or feedback from platforms, even in cases where employees (in this case drivers) had been physically harmed while working. Interestingly, however, some gig workers have come together to try and demand rights. This has resulted in the development of organisations advocating for gig workers (Anwar & Graham, 2019; Carmody & Foruin, 2019; Giddy, in press). This work continues in assessing the nature of employment, specifically in the context of e-hailing, and demonstrates that the work emerging among e-hailing drivers is incredibly concerning and cannot be considered decent work (Giddy, in press). Some major concerns are issues around driver safety, as major violent altercations have occurred between e-hailing drivers and traditional metered taxi operators. The driver-partner issue is also significant, as the majority of drivers were found to be subcontracted by vehicle owners, often without contracts (Carmody & Fortuin, 2019; Giddy, in press; Henama & Sifolo, 2017). This meant that in several cases, the actual take-home earnings of drivers were incredibly low and their working hours unsustainable (some were working more than 12-hour days, seven days a week) in order to meet the necessary targets required by vehicle owners (Giddy, in press). The Fairwork Foundation conducted a broad study on place-based digital work, much of which is relevant to the tourism and hospitality sectors and included South Africa as a case study (Fairwork Foundation, 2010; Graham & Woodcock, 2018). They sought to rate different platforms based on certain aspects of decent work: pay, working conditions, contracts, management and representation (Graham & Woodcock, 2018). Among the 11 platforms which were rated in 2020, those most directly relevant to the tourism industry (e-hailing, including Uber and Bolt; and food delivery, including MrDelivery) were the lowest-rated (Fairwork Foundation, 2010). In the case of e-hailing, Uber rated four out a possible ten and Bolt at one; while food delivery services rated, overall, lowest with Uber Eats and OrderIn at three and the local Mr D at zero, meaning it did not meet any of the above-mentioned standards of decent work. Those with the highest ratings were freelance platforms that were only very adjacently be related to tourism. However, interestingly, two of the platforms which offer domestic worker services rated relatively high (both at eight), though the third also came in at four (Fairwork Foundation, 2010). However, these services are only relevant if, in fact, operators that require domestic services actually utilise the apps rather than take the traditional route of informal workers.

There has been some coverage in the local media on the employment dynamics of some gig economy operators in Southern Africa (Burke, 2017; Mabuza, 2017; Malingo, 2018; van der Merwe & Dias, 2018). The primary type of platform, in the tourism-related context, in which this has emerged is e-hailing operations. As mentioned previously, concerns over the regulation of e-hailing operations (most notably Uber) have been widespread, particularly their status as an employer (Martin, 2016; Moon, 2015). In South Africa, there have been publicised cases that have focused on e-hailing. In the first case, the Commission Consolation Mediation and Arbitration (CCMA) ruled in July 2017 that Uber should, in fact, be considered an employer and thus Uber drivers should be awarded the protections of employees guaranteed by South African labour laws. However, this ruling was overturned soon after by the South African Labour Court in January 2018 (van der Merwe & Dias, 2018). This case (which at the moment is the definitive ruling on the matter) found that because Uber operates under two different corporations, it is not considered an employer (van der Merwe & Dias, 2018). The first is Uber SA, which provides training and oversees regulation of drivers and cars that are admitted to the platform but does not facilitate monetary transactions. This is done by the second corporation Uber BV, which is based in the Netherlands. Thus, as the company is not registered in South Africa, drivers were not awarded protections (van der Merwe & Dias, 2018). In addition, customer violence against e-hailing drivers has been widely publicised, and though some changes appear apparent, these altercations appear like they will continue (Burke, 2017; Henama & Sifolo, 2017). Thus, gig economy work in the Southern African context is highly problematic and faces many challenges.

Is the gig economy the new informal?

The big question we seek to pose in this chapter is whether the "traditional" informal economy in tourism in Southern Africa is being replaced by rampant growth in the gig economy in three key areas: accommodation, food delivery and e-hailing rides. Alternatively, has the gig economy found and developed an entirely new market in Southern Africa (and elsewhere) which exists alongside, even complementary to, the existing informal (and formal) economies in these sectors? These are important questions for which little or no evidential data exist and, in many respects, are the key to addressing trends with respect to the existence of parallel or competing sectors of the tourism economy in Southern Africa.

As we have shown in this chapter, in many respects, the two sectors, the traditional and the rapidly emergent, have key characteristics in common with regard to operations but also, in our area of interest, in employment terms. Both sectors are characterised by forms of employment that are precarious and offer little by way of job security. People who work in both sectors do not benefit from employment contracts of any discernible nature and those who work in subcontracting roles in both (e.g. cleaners, gardeners, groundskeepers who work for accommodation operators) are generally anonymous and invisible as employees to the guests. Neither areas of work are subject to close scrutiny or accountability through national or local employment protection laws or collective agreements. They do not offer defined progression or career paths. At the same time, both offer the potential for flexible engagement in a way that can allow those working in the sectors to build alternative lives around commitments to the job although, in practice, this notional ideal may be difficult to attain.

The big difference between the "traditional" informal economy in tourism and the emergent gig economy relates to the ultimate ownership of the brand and financial accountability. The informal economy in Southern Africa is a loose amalgam of independent operators where there is little or no evidence of anything other than localised coopetition – working with neighbouring vendors in a souvenir market to cover stalls during slack periods, for example. By contrast, operators in the gig economy depend on platforms that are owned by global corporations with little or no local ownership, identity or accountability which, in turn, collect a significant percentage of the money paid by consumers by way of a brand and platform service charge. In a market access sense, they also benefit from access to the global platform, its market and marketing presence and, in tourism terms, familiarity with the brand by international visitors to Southern Africa who may be reassured by their familiarity when in the destination.

Conclusions and a future research agenda

Our core question, therefore, remains largely unanswered both with respect to Southern Africa and the wider world of global tourism in the absence of reliable statistics on both areas of economic activity. Addressing the core questions and attendant issues relating to their employment impact provides a challenging research agenda going forward and one that needs to be considered at a macro (economic and policy) level but also through interrogation of the lived experiences of those working in both the "traditional" informal economy and the growing gig sector in tourism. Attempts to address such future research is currently fraught with additional complications by the global COVID-19 pandemic which, as we write this chapter, has altered the face of international tourism to such an extent that it is impossible to predict what the industry, post-pandemic, will look like. Intuitively, a pre-pandemic perspective would have led us to conclude that, in all probability, the gig economy is eating into aspects of what the informal economy (along with operations in the formal economy) have traditionally offered to tourists. Whether we will be able to support a similar argument in the post-pandemic era remains to be seen.

References

Adebayo, J. (2019). "Who stole my passengers?": Uber cabs, metered taxis and the search for common ground. *Conflict Studies Quarterly*, *27*, 3–20. doi:10.24193/CSQ.27.1

Airbnb (2020). United States vacation rentals & homes. *Airbnb*. Retrieved from http://www.airbnb.com/united-states/stays

Anwar, M. A., & Graham, M. (2019). Hidden transcripts of the gig economy: Labour agency and the new art of resistance among African gig workers. *Environment and Planning A: Economy and Space*. doi:10.1177/0308518X19894584

Ashley, C., & Roe, D. (2002). Making tourism work for the poor: Strategies and challenges in southern Africa. *Development Southern Africa*, *19*(1), 61–82. doi:10.1080/03768350220123855

Azmi, A., Kamal, S. B. M., & Abdullah, D. (2016). A framework of customer's intention to use Uber service in tourism destination. *International Academic Research Journal of Businnes and Technology*, *2*(2), 102–106.

Bah, A., & Goodwin, H. (2003). *Improving Access for the Informal Sector to Tourism in Gambia*. London: ODI.

Booyens, I. (2020). Education and skills in tourism: Implications for youth employment in South Africa. *Development Southern Africa*, *37*(5), 825–839. doi:10.1080/0376835X.2020.1725447

Brown, A., Lyons, M., & Dankoco, I. (2010). Street traders and the emerging spaces for urban voice and citizenship in African cities. *Urban Studies*, *47*(3), 666–683. doi:10.1177/0042098009351187

Burke, J. (2017). Violence erupts between taxi and Uber drivers in Johannesburg. *The Guardian*. Retrieved from https://www.theguardian.com/world/2017/sep/08/violence-erupts-taxi-uber-drivers-johannesburg

Çakmak, E., Lie, R., & McCabe, S. (2018). Reframing informal tourism entrepreneurial practices: Capital and field relations structuring the informal tourism economy of Chiang Mai. *Annals of Tourism Research*, *72*, 37–47. doi:10.1016/j.annals.2018.06.003

Calo, R., & Rosenblat, A. (2017). The taking economy: Uber, information and power. *Columbia Law Review*, *117*(6), 1623–1690.

Carmody, P., & Fortuin, A. (2019). "Ride-Sharing": Virtual capital and impacts on labor in Cape Town, South Africa. *African Geographical Review* *38*(3), 196–208. doi:10.1080/19376812.2019.1607149

Cetin, T., & Deakin, E. (2017). Regulation of taxis and the rise of ridesharing. *Transport Policy*, 1–10. doi:10.1016/j.tranpol.2017.09.002

Cocola-Gant, A., & Gago, A. (2019). Airbnb, buy-to-let investment and tourism-driven displacement: A case study in Lisbon. *Environment and Planning A: Economy and Space*, 1–18. doi:10.1177/0308518X19869012

Cramer, J., & Krueger, A. (2016). *Disruptive Change in the Taxi Business: The Case of Uber*. NBER Working Paper No. 22083. National Bureau of Economic Research, Cambridge, MA.

Crick, M. (1992). Life in the informal sector: Street guides in Kandy, Sri Lanka. In D. Harrison (Ed.), *Tourism and the Less Developed Countries* (pp. 135–147). London: Belhaven Press.

Damayanti, M., Scott, N., & Ruhanen, L. (2017). Cooperative behaviours in an informal tourism economy. *Annals of Tourism Research*, *65*, 25–35. doi:10.1016/j.annals.2017.04.007

Davis, M. (2006). *Planet of Slums*. London: Verso.

de Beer, A., Rogerson, C. M., & Rogerson, J. M. (2014). Decent work in the South African tourism industry: Evidence from tourist guides. *Urban Forum*, *24*, 89–103. doi:10.1007/s12132-013-9199-8

Drahokoupil, J., & Fabo, B. (2016). *The Platform Economy and The Disruption of the Employment Relationship*. Brussels: European Trade Union Institute (ETUI). Policy Brief 5.

Drahokoupil, J., & Piasna, A. (2019). *Work in the Platform Economy: Deliveroo Riders in Belgium and the Smart Arrangement*. Brussels: European Trade Union Institute (ETUI).

Fairwork Foundation (2010). *Fairwork Ratings: Uber in South Africa*. Retrieved from https://fair.work/ratings/

Fang, Z., Su, R., & Huang, L. (2018). Understanding the effect of an e-hailing app subsidy war on Taxicab operation zones. *Journal of Advanced Transportation*, 1–14. doi:10.1155/2018/7687852

Garcia-Ayllon, S. (2018). Urban transformations as an indicator of unsustainability in the P2P mass tourism phenomenon: The Airbnb case in Spain through three case studies. *Sustainability*, *10*(8), 2933. doi:10.3390/su10082933

Geertz, C. (1963). *Old Societies and New States: The Quest for Modernity in Asia and Africa*. Glencoe, IL: Free Press.

Geitung, I. (2017). *Uber Drivers in Cape Town: Working Conditions and Worker Agency in the Sharing Economy*. MA thesis, Department of Sociology and Human Geography, University of Oslo.

Giddy, J. K. (2019). The influence of e-hailing apps on urban mobilities in South Africa. *African Geographical Review*, *38*(3), 227–239. doi:10.1080/19376812.2019.1589732

Giddy, J. K. (2020a). Travel patterns of ride sharing users: Uber in South Africa. In J. M. Rogerson & G. Visser (Eds), *New Directions in South African Tourism Geographies* (pp. 267–282). Cham: Springer.

Giddy, J. K. (2020b). Insight into adventure tourism employment in South Africa. In A. Ndiuini & T. Baum (Eds), *Sustainable Human Resource Management in Tourism: African Perspectives* (pp. 189–204). Cham: Springer.

Giddy, J. K. (in press). Uber, employment and decent work: The South African experience. *Tourism Geographies*, 1–39.

Giddy, J. K., Idahosa, L., & Rogerson, C. M. (2020). Leveraging state-owned nature-based assets for transformation and SMME development: Pilanesberg National Park, South

Africa. In J. M. Rogerson & G. Visser (Eds), *New Directions in South African Tourism Geographies* (pp. 299–316). Cham: Springer.
Graham, M., & Anwar, M. A. (2018). Digital labour. In J. Ash, R. Kitchin, & A. Leszczynski (Eds), *Digital Geographies* (pp. 177–187). London: Sage.
Graham, M., & Woodcock, J. (2018). Towards a fairer platform economy: Introducing the Fairwork Foundation. *Alternate Routes, 29*, 242–253.
Guttentag, D. (2015). Airbnb: Disruptive innovation and the rise of an informal tourism accommodation sector. *Current Issues in Tourism, 18*(12), 1192–1217. doi:10.1080/13683500.2013.827159
Guttentag, D., Smith, S., Potwarka, L., & Havitz, M. (2018). Why tourists choose Airbnb: A motivation-based segmentation study. *Journal of Travel Research, 57*(3), 342–359.
Hart, K. (2006). Bureaucratic form and the informal economy. In B. Guha-Khasnobis, R. Kanbur, & E. Ostrom (Eds), *Linking the Formal and Informal Economy: Concepts and Policies* (pp. 21–35). Oxford: Oxford University Press.
Henama, U. S., & Sifolo, P. P. S. (2017). Uber: The South Africa experience. *African Journal of Hospitality, Tourism and Leisure, 6*(2), 1–10.
Henderson, J., & Smith, R. (2009). The informal tourism economy at beach resorts: A comparison of Cha-Am and Laguna Phuket in Thailand. *Tourism Recreation Research, 34*(1), 13–22. doi:10.1080/02508281.2009.11081571
ILO (2018). *Women and Men in the Informal Economy: A Statistical Picture*. Retrieved from https://ilo.userservices.exlibrisgroup.com/discovery/delivery/41ILO_INST:41ILO_V2/1252879760002676
ILO (2020). *Informal Economy (Employment Promotion)*. Retrieved from https://www.ilo.org/global/topics/employment-promotion/informal-economy/lang--en/index.htm
Jin, S. T., Kong, H., Wu, R., & Sui, D. Z. (2018). Ride sourcing, the sharing economy, and the future of cities. *Cities, 76*, 96–104. doi:10.1016/j.cities.2018.01.012
Kessler, S. (2018). *Gigged: The Gig Economy, the End of the Job and the Future of Work*. London: Penguin.
Kirsten, M., & Rogerson, C. M. (2002). Tourism, business linkages and small enterprise development in South Africa. *Development Southern Africa, 19*(1), 29–59. doi:10.1080/03768350220123882
Leonard, M. (2000). Coping strategies in developed and developing societies: The workings of the informal economy. *Journal of International Development, 12*, 1069–1085.
Lepanjuuri, K., Wishart, R., & Cornick, P. (2018). *The Characteristics of Those in the Gig Economy*. London: UK Department for Business, Energy and Industrial Strategy.
Lewis, A. (1959). *The Theory of Economic Growth*. London: Allen and Unwin.
Mabuza, E. (2017). *CCMA Says SA's Uber Drivers Are Employees, but Uber Still Disagrees*. Retrieved from https://www.businesslive.co.za/bd/companies/transport-and-tourism/2017-07-25-ccma-says-sas-uber-drivers-are-employees-but-uber-still-disagrees/
Malingo, B. (2018). Uber increases fares ahead of "big days". *The Citizen*. Retrieved from https://citizen.co.za/news/south-africa/general/2047028/uber-increases-fares-ahead-of-big-days/
Martin, C. J. (2016). The sharing economy: A pathway to sustainability or a nightmarish form of neoliberal capitalism? *Ecological Economics, 121*, 149–159. doi:10.1002/jid.696
McKay, T. (2013). Adventure tourism: Opportunities and management challenges for SADC destinations. *Acta Academica, 45*(3), 30–62.

Meelen, T., & Frenken, K. (2015). *Stop Saying Uber Is Part of the Sharing Economy.* Retrieved from http://www.fastcoexist.com/3040863/stop-saying-uber-is-part-of-the-sharing-economy
Moon, Y. (2015). Uber: Changing the way the world moves. *Harvard Business Case, 44,* 1–19.
Mtapuri, O., & Giampiccoli, A. (2016). Towards a comprehensive model of community-based tourism development. *South African Geographical Journal, 98*(1), 154–168. doi:10.1080/03736245.2014.977813
Ndiuini, A., & Baum, T. (2020). *Sustainable Human Resource Management in Tourism in Africa.* Cham: Springer.
OECD (2001). *OECD Glossary of Statistical Terms - Informal Sector – ILO Definition.* Retrieved from https://stats.oecd.org/glossary/detail.asp?ID=1350
O'Regan, M., & Choe, J. (2017). Airbnb and cultural capitalism: Enclosure and control within the sharing economy. *Anatolia, 28*(2), 163–172. doi:10.1080/13032917.2017.1283634
Oskam, J., & Boswijk, A. (2016). Airbnb: The future of networked hospitality businesses. *Journal of Tourism Futures, 2*(1), 22–42. doi:10.1108/JTF-11-2015-0048
Peberdy, S. (2000). Mobile entrepreneurship: Informal sector cross-border trade and street trade in South Africa. *Development Southern Africa, 17*(2), 201–219. doi:10.1080/713661400
Pollio, A. (2019). Forefronts of the sharing economy: Uber in Cape Town. *International Journal of Urban and Regional Research, 43*(4), 760–775. doi:10.1111/1468-2427.12788
Rayle, L., Dai, D., Chan, N., Cervero, R., & Shaheen, S. (2016). Just a better taxi?: A survey-based comparison of taxis, transit, and ridesourcing services in San Francisco. *Transport Policy, 45,* 168–178. doi:10.1016/j.tranpol.2015.10.004
Redlich, H. (2019). *Recent Uber Decision Takes the Employer/Contractor Distinction for a Ride - Lexology.* Retrieved from https://www.lexology.com/library/detail.aspx?g=0ea37f6c-8116-4d04-bece-eab00237f14f
Roelofsen, M. (2018). Performing "home" in the sharing economies of tourism: The Airbnb experience in Sofia, Bulgaria. *Fennia, 196*(1). doi:10.11143/fennia.66259
Roelofsen, M., & Minca, C. (2018). The superhost: Biopolitics, home and community in the Airbnb dream-world of global hospitality. *Geoforum, 91,* 170–181. doi:10.1016/j.geoforum.2018.02.021
Rogerson, C. M. (2006). Pro-poor local economic development in South Africa: The role of pro-poor tourism, *Local Environment, 11*(1), 37–60. doi:10.1080/13549830500396149
Rogerson, C. M. (2012). The tourism-development nexus in sub-Saharan Africa: Progress and prospects. *Africa Insight, 42*(2), 28–45.
Rogerson, C. M. (2014). Informal sector business tourism and pro-poor tourism: Africa's migrant entrepreneurs. *Mediterranean Journal of Social Sciences, 5*(16), 153–161.
Rogerson, C. M. (2018). Informal sector city tourism: Cross-border shoppers in Johannesburg. *Geo Journal of Tourism and Geosites, 22*(2), 381–391. doi:10.30892/gtg.22210-297
Rogerson, C. M. (2020). Sustainable tourism research in South Africa: In search of a place for work and the workplace. In A. Ndiuini & T. Baum (Eds), *Sustainable Human Resource Management in Tourism in Africa* (pp. 45–59). Cham: Springer.
Rosenblat, A., & Stark, L. (2015). Algorithmic labor and information asymmetries: A case study of Uber's drivers. *International Journal of Communications, 10,* 3758–3784.
Slocum, S., Backman, K., & Robinson, K. (2011). Tourism pathways to prosperity: Perspectives on the informal economy in Tanzania. *Tourism Analysis, 16*(1), 43–56. doi:10.3727/108354211X12988225900045

Stors, N., & Kagermeier, A. (2015). Motives for using Airbnb in metropolitan tourism– Why do people sleep in the bed of a stranger? *Regions*, *1*, 17–19. doi:10.1080/13673882.2015.11500081

Todoli-Signes, A. (2018). Judgment designating Deliveroo “rider” an employee and analysis of its impact on the “gig economy”. *Transfer: European Review of Labour and Research*, *24*(4), 487–490. doi:10.1177/1024258918801516

van der Merwe, P., & Dias, J. (2018). *An Analysis of the Recent Ruling of Uber South Africa Technology Services (Pty) Ltd v National Union of Public Service and Allied Workers (NUPSAW) and Others (C449/17)*. Retrieved from http://www.schindlers.co.za/news/are-uber-drivers-employees/

Visser, G., & Rogerson, C. M. (2004). Researching the South African tourism and development nexus. *GeoJournal*, *60*(3), 201–215. doi:10.1023/B:GEJO.0000034728.45071.b7

Visser, G., Erasmus, I., & Miller, M. (2017). Airbnb: The emergence of a new accommodation type in Cape Town, South Africa. *Tourism Review International*, *18*, 151–168. doi:10.3727/154427217X14912408849458

Wachsmuth, D., & Weisler, A. (2018). Airbnb and the rent gap: Gentrification through the sharing economy. *Environment and Planning A: Economy and Space*, *50*(6), 1147–1170. doi:10.1177/0308518X18778038

Wanschafft, R. (1982). Formal and informal tourism sectors: A case study in Pattaya, Thailand. *Annals of Tourism Research*, *9*(3), 429–451. doi:10.1016/0160-7383(82)90022-6

Wiessner, D. (2019). U.S. judge says Uber drivers are not company’s employees. *Reuters*. Retrieved from https://www.reuters.com/article/us-uber-lawsuit/u-s-judge-says-uber-drivers-are-not-companys-employees-idUSKBN1HJ31I

Williams, C. (2017). Tackling employment in the informal economy: A critical evaluation of the neoliberal policy approach. *Economic and Industrial Democracy*, *38*(1), 145–169. doi:10.1177/0143831X14557961

Zervas, G., Proserpio, D., & Byers, J. W. (2014). *The Rise of the Sharing Economy: Estimating the Impact of Airbnb on the Hotel Industry*. Boston: Boston University.

15 "African" representations

Culturalisation of Poverty in Township tourism

Malte Steinbrink and Ina Voshage

Introduction: A questionable kind of tourism

The slum has always symbolised the dark, low and unknown side of the city. The western imaginary geography of "the slum" is that of another world – chaotic, uncivilised and horrifying. At the same time, this threatening, strange *urban terra incognita* "where the other half lives", also promises adventures and tempts bourgeois curiosity. This curiosity is about as old as "the slum" itself: shortly after the term *slum* was introduced into Standard English in the Victorian era, the pejorative term *slumming* also appeared in London's "West-Side Lingo". The term described the burgeoning – and often criticised – leisure activities of the upper classes who set off for the "undiscovered land of the poor" (Steinbrink, 2012, p. 8) of mid-19th century London's East End. Thus, Victorian slumming can be seen as the precursor of today's slum tourism. Slumming is therefore an early part of city tourism in the old industrial nations. In the Global South, however, this phenomenon – starting in South Africa and Brazil – only developed at the end of the 20th century, but then rapidly gained momentum. Organised visits to poor urban districts are becoming a popular tourist practice in an increasing number of cities. Slum tourism is now a global phenomenon. It can be assumed that well over one million tourists – mainly from the Global North – take part in slum, favela or township tours annually. Approximately 80% of these tours take place in South Africa and Rio de Janeiro. Initially, a tour was a niche offer for politically interested travellers and commercial offer structures for tourist visits to urban poverty areas developed very quickly – and the market is booming: new destinations have emerged, and more and more tourism companies are offering an increasingly diversified portfolio (Frenzel et al., 2015). Although slum tourism is still generally advertised as "off the beaten track" (Dyson, 2012; Meschkank, 2011), in some destinations it has long since become part of mainstream tourism.

Nevertheless, slum tourism remains an extremely controversial type of tourism that gives rise to strong ethical scepticism. The moral concerns relate primarily to two aspects: first, to the motivations of tourists and, second, to the form of staging and representation. Journalists, in particular, were quick to express their reservations regarding the motives of tourists and soon made their conjectures the basis of their moral judgements: in media coverage, slum tourists are often criticised as gawkers and the slum tourist's gaze is depicted as disrespectful social voyeurism. Comparisons with visits to zoos or terms like "human safaris", "poorism" or "poverty porn" dominated media coverage for a long time. Only recently has slum tourism occasionally received media recognition for its potential to contribute to intercultural understanding and to create awareness of global inequality by bringing about direct confrontations with poverty. However, despite the media's tendency towards more differentiated coverage of slum tourism, the reporting is still strongly characterised by a moralising perspective.

The scientific community discovered the topic of global slum tourism much later than the media,[1] and most studies are also framed by a moralising perspective. Thus, it is not surprising that slum tourism is sometimes unquestioningly conceptualised as a form of "dark tourism". This in turn implicitly draws conclusions about the (immoral) motives of the tourists and the content of what is shown ("slums as places of suffering, crisis and catastrophe"). In other studies, the topic of slum tourism is also often closely linked to the general – highly normative – topic of "development". Against the background of possible positive socio-economic effects, it is discussed whether slum tourism is an ethically justifiable tourism practice (Frenzel et al., 2015). Implicitly, the question is whether the moral end (economic benefit of "the poor") justifies the potentially immoral means (unethical tourist gaze and the exploitation of the slum for leisure purposes). As the starting point for this chapter, we want to state that the moralising view not only frames the observation of global slum tourism in public circles and academia but also serves as a constitutive aspect of the form and practice of tourism itself.

Given the striking dominance of the moralising view of the phenomenon as well as the assumptions about the motives of tourists on the one hand and the content and form of the touristic representation on the other, it seems appropriate to concentrate on these aspects. Based on a study project by the University of Osnabrück, this article focuses on township tourism in Windhoek (Steinbrink et al., 2016). Guided sightseeing tours of the former township of Katutura have been offered since the mid-1990s (for baseline information on the history of Katutura, see Textbox 1). City tourism in the Namibian capital thus became part of the trend of "Slumming in the Global South" at an early stage (Textbox 2 briefly outlines the development of Katutura tourism).

Using Katutura tourism as a case study, this chapter discusses the interplay of (1) township images and expectations of tourists, (2) township representation in tour practice and (3) the possible image change through this kind of tourism. We will focus on how certain geographical imaginations (township images), which predominated prior to the tour, are taken up, strengthened or revised in a communicative-performative manner in the course of the tours and on how this township representation affects the perception and evaluation of the *destination "township"*. The methodological design of the study combined quantitative and qualitative methods such as standardised surveys, problem-centred interviews, expert discussions, participant observations and mapping (for details, see Steinbrink et al., 2016, pp. 18–21).

Textbox 1: Katutura's history

Due to its history, the former township of Katutura is not only a central narrative point of reference for the construction of Namibia's national identity; but also reflects the social changes and challenges since independence in 1990. After the end of the German colonial period, South-West Africa was placed under South African administration in 1915. Soon, demands from many "white" residents of Windhoek for an "orderly racial segregation" (Pendleton, 1996, p. 29) followed. Consequently, the central elements of the South African apartheid system were adopted from the 1950s onwards. The spatial separation of the urban populations according to skin colour was then laid down by law and provided the legal basis for the resettlement of non-white population groups (Simon, 1988, p. 53). The *Odendaal Plan,* developed as a result, envisaged the establishment of two so-called *townships*: *Katutura* for inhabitants classified as "Black" and *Khomasdal* for those classified as "Coloured". The Odendaal Plan led to uprisings in 1959 in which a dozen people were shot dead by the police and many more injured (Simon, 1988, p. 55). Nevertheless, the non-white population groups were resettled forcefully.

The first housing units in Katutura were built five kilometres northwest of Windhoek City. In the local language, *Otjiherero*, the name Katutura means "the place where people do not want to live", thus expressing people's aversion to resettlement (Pendleton, 1996, p. 29). A motorway and an industrial area were built between the city centre on one side, which was marked out as "White Group Area", and the two townships of Khomasdal and Katutura on the other. According to apartheid planning ideology, this was to serve as a so-called "buffer zone" (Simon, 1988, p. 54) to reduce contact between the population groups.

Even after apartheid, residential segregation remained, since hardly any Katutura residents could afford to live outside the township. Since then, the population of Katutura has been growing steadily. Settlement growth and informal densification were the results.

After 1990, substantial investments were made in Katutura's infrastructure, which improved basic services and connection to the city centre. However, the legacy of South African apartheid still affects the daily life of Katutura's inhabitants and is reflected in many unresolved urban-planning and social problems. The residents of the informal settlement areas on the outskirts of Katutura are particularly hard hit by the extremely poor living conditions. Against this backdrop of decades of oppression and socio-spatial marginalisation, racial segregation continues, in part under the guise and in the mode of the "free market". The social order of a dehumanising racist regime is still articulated in the extremely uneven distribution of wealth and poverty and the massively unequal allocation of the urban infrastructure in Windhoek. The socio-economic disparities between different population groups – which received their group status through racist assignments – continue to manifest themselves in the deeply fragmented urban fabric. Despite the government's efforts, there is still a strong link between skin colour and life chances.

Textbox 2: Development and structure of township tourism in Windhoek

Namibia's tourism sector has experienced tremendous growth since the declaration of independence. At the same time, Namibia's tourism sector is developing in the complex context of the socio-political and socio-economic transformations following the apartheid era. Against this background, township tourism has developed into an integral part of city tourism in Windhoek.

Namibia is one of the four fastest-growing tourism markets worldwide. In 2018, around 1.24 million international guests visited Namibia (WTTC, 2018, p. 5) – a remarkable number considering the country only has 2.4 million inhabitants. Tourism is a key pillar of the economy: its share of GDP is around 15% and almost 25% of the working population is directly or indirectly dependent on tourism (WTTC, 2018).

Namibia offers a variety of tourist attractions. While the focus has long been on nature and wildlife tourism, cultural tourism activities such as ethnic tourism and city tourism have recently gained importance. However, the potential of international tourism for Windhoek has not yet been fully exploited. City tourism has long marketed the capital's architectural relics of the German colonial era (e.g. *Alte Feste, Christuskirche, Tintenpalast, Windhoek Castles*). In recent years, the city has tended to focus on the period after independence. The National Museum and the Heroes' Acre have been added; the Independence Memorial Museum, which focuses on the history of apartheid and the city, opened in 2014 (Steinbrink & Voshage, 2020).

Namibia's capital is still considered a boring, "un-African, German" city that is not particularly attractive to international tourists. According to an analysis published in 2006, Windhoek's tourism sector is identified as displaying a "lack of cultural aspects and African image" (Basilio et al., 2006, p. 6). In response, the city is now promoting the *Cultural and Heritage Sites in Windhoek*. Windhoek's tourism campaign aims to keep tourists longer in the city, and at the same time to let more neighbourhoods and other population groups benefit from the economic advantages (City of Windhoek, n.d., p. 45) in order to balance the Namibian tourism industry in favour of previously disadvantaged groups. In the mid-2000s, city marketing decided to use the slogan *City of Many Faces*. This image strategy aims to emphasise urban heterogeneity in terms of ethnic-cultural diversity. This can be understood as an attempt to present the city's enormous socio-economic disparities not as a serious social problem, but as attractive urban diversity.

In the years after independence, not only the annual number of tourists in Namibia increased significantly; international governmental and non-governmental organisations (NGOs) involved in development cooperation also arrived in Namibia in increasing numbers. Most of them opened bases in Windhoek and have since been working in Katutura as well. At the same time, the government of the young nation was focusing its efforts on the development of the tourism sector as an important mainstay of its national economy. In doing so, it was also pursuing the goal of ensuring greater participation of disadvantaged population groups in tourism. At a very early point in time, the concept of *Community Based Tourism* was adopted as an integral part of Namibia's tourism policy (Novelli & Gebhardt, 2007). These factors, in conjunction with the establishment of township tourism in neighbouring South Africa, contributed significantly to the development of township tourism in Windhoek.

The beginnings of tourism in Katutura can be dated back to 1992, when the NGO *Penduka*, which was the first institution in Katutura to explicitly gear its activities towards the needs of tourists, began to produce craftwork for Western visitors. Founded by a Dutch woman, the organisation pursued the goal of helping socially disadvantaged women. The products are offered in a shop on the project site, which is located on the outskirts of Katutura at the picturesque Goreangab reservoir. In addition to workshops and showrooms, the women's project now has a restaurant, a backpacker hostel and several thatched chalets for tourists. Almost simultaneously, *Mama Melba* opened a small restaurant in her private home in Katutura. She was the first to offer the traditional Herero cuisine to international guests; since 2002, township tours stop here for a culinary experience, as she is particularly known for the preparation of so-called *smileys* (boiled goats' heads).

The first informally guided township tours were organised in 1998 by Abiud Karongee, who at the time worked for the NGO NACOBTA (Namibia Community Based Tourist Assistance Trust). The inspiration came from a township tour he had experienced during a training trip in

Johannesburg. As a developer and consultant for rural tourism projects, he recognised township tourism as a promising niche for urban tourism in Namibia. His first customers were employees of development organisations and embassies. In 2001, he founded the company *Face to Face*. The company subsequently worked closely with the City of Windhoek: Karongee successfully campaigned for the city to include township tourism in the city's tourism marketing. In this early phase of Katutura's destination development, the initial impulses came from NGOs in cooperation with ambitious individuals from Katutura, whose activities were supported by the City of Windhoek. The first customers were not so much "typical Namibia holidaymakers" but rather employees of international organisations, their relatives and guests.

The first purely commercial actor to operate in the Katutura tourism market was *Bwana Tucke Tucke*, a company founded by German Carsten Möhle in 1996. Today, it is one of the largest tour operators in Namibia. From the beginning, the main target group has been German vacationers, who book Katutura tours in addition to their safari trip. The customer profile thus differs from that of the early phase.

Ever since this first step of commercialisation, Katutura tourism has been booming: between 2000 and 2015, at least one new provider of Katutura tours has appeared on the scene every year. There are currently more than 20 companies offering tours, most of them owned by (white) people from outside Katutura. The estimated annual number of participants in Katutura tours is slightly over 20,000 (Steinbrink et al., 2016, p. 36).

Almost all providers report a steady increase in demand. At the same time, they are noticing a change in the customer profile. The tours are often booked by individual travellers with cultural interests. Another increasingly important segment is the steadily growing group of volunteer tourists working in social or ecological projects in Namibia. Overall, Windhoek is witnessing a growing, diversifying market for township tourism, with guided tours still dominating. Recent developments also show that tourism in Katutura has potential beyond classic tours due to a diversifying demand structure (for details see Steinbrink et al., 2016).

African ambivalence: Geographical imaginations and representations in township tourism

Tourism is always looking for new places, inventing sights and places of interest, which are then marked and marketed as tourist attractions. What is meant here, however, is not "places" in a physically material sense, but places as the product of social construction: "Places of tourism" (Pott, 2007) are brought about through social practices (especially communication). In the development and marketing of destinations, tourism draws on available place constructs (*images*) that were initially produced outside tourism, especially in the media (Urry, 2002). Often, tourism organisations (re)produce certain place-related virtual contents by drawing upon generally known place constructs external to tourism. These "geographical imaginations" are consumed by tourists and communicatively confirmed and/or modified during their trip. Tourist destinations are communicatively produced and created through practice. This applies to the destination type *township* as well as to Katutura as a specific destination within city tourism in Windhoek.

In the context of this study, we are not analysing the emergence and spreading of tourists' preconceptions regarding townships and/or Katutura. Media coverage, school experiences, travelogues and travel guides, as well as personal travel experiences and their medialisation in the context of social media, for example, Instagram, are of crucial importance in this context (Gräf & Steinbrink, 2021). Rather, we look into the interplay of (1) existing township images, (2) township representation and (3) the possible image change. Thus, we apply an analytical three-step process (*before – during – after*): In the first step, we examine tourists' imaginations and expectations *before the tour*. In a second step, we investigate how tour operators serve these stereotypes performatively *during their tour* and, finally, we discuss the influence of township representations on the tourists' perceptions and how they affect the township image *after the tour*.

Before the tour: Township tourists' expectations

When approaching the question of why people visit a township during their holidays, it seems plausible to assume that they want to see and experience what they expect to see and experience there (*Would they go there otherwise?*). Motives and geographical imaginations are closely connected. In the tourist context, "the space" functions both as a medium for generating expectation and as a promise of difference.

For this reason, as part of the study (Steinbrink et al., 2016), 70 township tourists were asked about their expectations of a township tour and their ideas about the township. This was done with the help of a semi-standardised questionnaire immediately before the tour started.

Tourists' expectations of a township tour

In order to gain insight into the motivations and interests of the township tourists, we asked them prior to the tour how important certain aspects of the tour were to them when booking. We addressed the following aspects: a) *entertainment*; b) *adventure*; c) *the wish to see something different*; d) *the wish to do something outside (the scope of) tourism*; e) *contact to local people*; f) *living conditions* in Katutura; g) *history*; h) *local culture*; and (i) *authenticity/reality* (Figure 15.1).

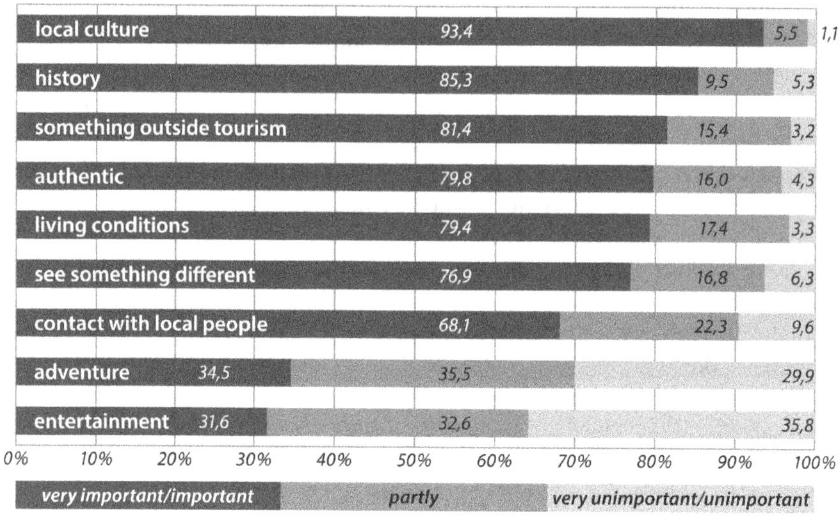

Figure 15.1 Expectations of the tour. Source: Adapted from Steinbrink et al., 2016, p. 46.

Comparatively few tourists said that *adventure* and *entertainment* were important to them. The figures for these two aspects appear rather low, especially compared with the figures for the other aspects. They also contradict the assumptions repeatedly made in the media about the motives of township tourists. Most tourists seem to have other expectations than "to be thrilled" or "to have fun" – or at least have the feeling that stating these aspects would not be socially acceptable.

More than half of the tourists interviewed (68%) explicitly expressed the desire for *contact with the local population*; and less than 10% stated that this aspect was not important for them. It is noteworthy that *local culture* is clearly the focus of interest for virtually all tourists. *History, something outside tourism, reality, living conditions* and the *desire to see something different* – these, too, are only very rarely considered unimportant. It can be assumed that these six aspects, which seem to be the main reasons for tourists' participation in the tours, are essentially closely related. This connection can be formulated as follows.

The striving to *see something different* is an almost universally recognised motivation in tourism; the tourist basically aims to create distance from everyday life by experiencing the difference. Clients of township tours also want to see *something different* or in a *different way* during the tour – but not as tourists (*something outside tourism*). Thus, township tourism is usually about what has not explicitly been staged for tourism. This aspect is presumably also expressed in the desire for a *realistic, authentic* experience. However, the desire for reality can also be understood as the desire to experience the *reality of others*. From this, it can be concluded that the tour participants are not interested in experiencing the *reality of others* (as displayed) on the stage of tourism, but that they hope to for once *see it in a different, authentic way*. In the context of tourism, the township is therefore regarded as the *place of the other* (different to one's own place) and, at the same time, as *authentically different*. The Other, meant to be experienced as *realistically* as possible, obviously relates to *local culture* and *living conditions*. What many township tourists have in mind is, therefore, the unstaged experience of the *reality of the living conditions of others*, which are, above all, considered to be *culturally different*. Therefore, *township tourism* is *cultural tourism*.

But what exactly do tourists think about the living conditions and local culture of "Others"? What do tourists expect to see and experience in Katutura? First, it must be demonstrated which geographical township clichés circulate among the tourists before the tour.

Tourists' township imaginations

Immediately before their tour started, the 70 interviewees were asked to spontaneously name the associations that the term *township* evoked in them. The result can be seen in the word cloud, which represents the semantic field of the term (Figure 15.2).

Overall, it is striking that negative associations clearly dominate the semantic field: 51% of the township tourists mention the term *poverty*; and less than 15% of the terms mentioned have a predominantly positive connotation.

In interpreting these findings, we must assume that (1) the attractiveness of Katutura as a tourist destination is linked to the associations and images of the place to be visited and (2) that the expectations of what can be seen in the townships are largely identical with what the tourists hope to see there. This suggests that Katutura is a destination where poverty is expected to be experienced above all else. From this, certain conclusions can be drawn regarding the motivations of the tourists. Since the concept of *poverty* is at the centre of the semantic field, township tourism could be interpreted as a kind of poverty tourism. For township tourists, the township is evidently first and foremost a place of poverty with all

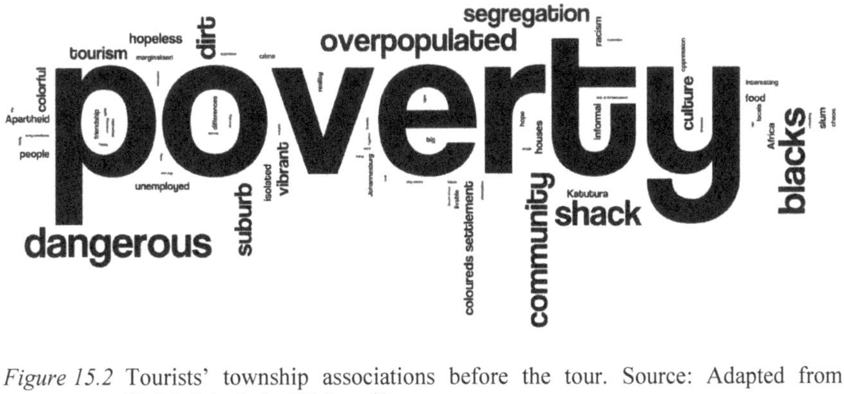

Figure 15.2 Tourists' township associations before the tour. Source: Adapted from Steinbrink et al., 2016, p. 49.

the negative qualities commonly associated with poverty. From this perspective, township tourism does indeed seem like a specific form of "negative sightseeing" (Welz, 1993, p. 43), a kind of "social bungee jumping" (Rolfes & Steinbrink, 2009), and the tourists appear as bourgeois thrill-seekers who, driven by a lust for angst, disgust, pity or similar emotions, seek to experience social depth. The township tours seem to permit tourists to explore the global social abyss with their own senses without running the risk of toppling into it. Thus, some scholars see urban poverty tourism as a form of *dark tourism*: they point out that tourists today are much less afraid of negative experiences and that interest in "the perfect scene of glamour or in kitschy Disneyland scenery" (Münder, 2013, p. 1; own translation) has increasingly been giving way to an interest in events or places that were once considered taboo. According to this view, the tour participants are striving for new, *different experiences* and are in search of *the true*, *the real* and *the unstaged.*

If the quest for the authentic is a main motive for the tour participants, we can conclude that township tourism is also a form of "Reality Tourism" (Dyson, 2012; Meschkank, 2011). It is evident that certain notions of "poverty" and "reality" are inextricably linked to one another. Gazing at poverty seems to go hand in hand with a promise of authenticity, i.e. poverty has become the guarantor for an authentic tourist experience. Seen in this light, poverty is, therefore, not the actual attraction, but, primarily, the medium through which the difference of the *reality (of the others)* is experienced.

To gain further insight into the township images of the tourists, the method of the semantic differential was applied before the tour started. In the questionnaire, 23 dichotomous word pairs were presented, which served as a five-level scale to assess the tourists' preconceptions of townships. Figure 15.3 shows the mean values for the different items.

By far the greater part of the individual mean values is located on the right side, i.e. on the negative side of the middle line. The greatest amplitude towards the negative side is represented by the term *poor*; followed by the terms *underdeveloped, noisy, unhealthy, dirty* and *ugly*. However, the results also show that the image is not limited to negative aspects, because a *community-centred, vibrant, friendly, active* and *hard-working* township also exists in the tourists' imaginations. The township image is obviously ambivalent.

Thus, the semantic differential complements the findings of the association survey (Figure 15.2) because it indicates that township tourism is not only about authentically experiencing fear-instilling, repulsive or lamentable elements of the township. Township tourism also strives for an authentic experience of positively connoted aspects of townships, for example, being *vibrant*, *friendly* and *dynamic*. Seen in this light, "poverty" serves as a guarantor of authenticity for the whole spectrum of the tourists' ideas of "township reality" – be they positive or negative.

There is another factor that merits consideration in an explanation of tourists' ambivalent images of the township. In numerous qualitative interviews, township tourists have said that they want to experience *the real Africa* during their

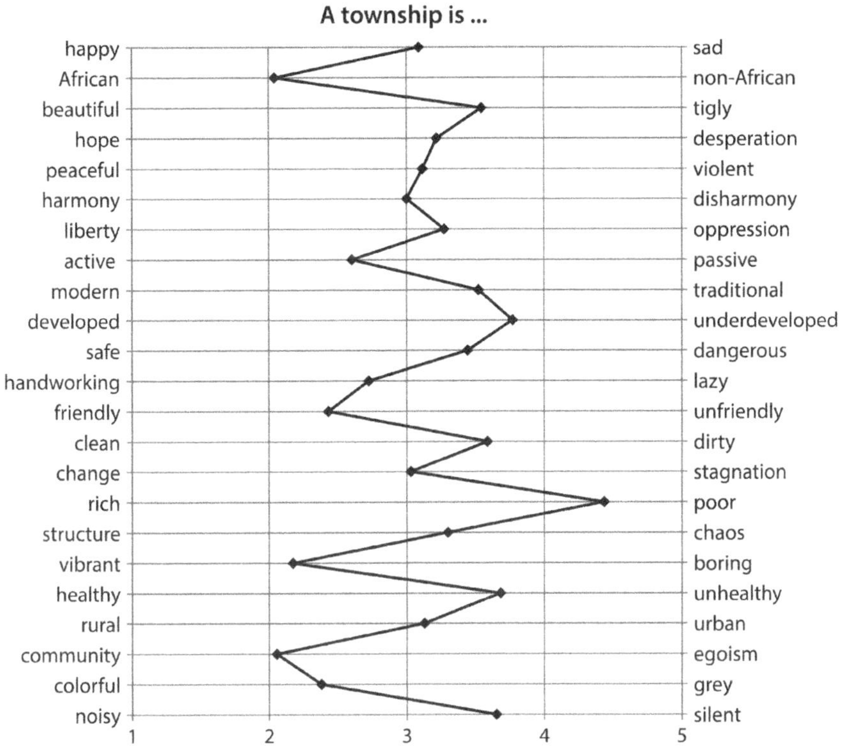

Figure 15.3 Semantic differential: Tourists' notions before the tour, n=70. Source: Adapted from Steinbrink et al., 2016, p. 51.

tours, an Africa that they obviously do not believe they can find in the centre of Windhoek. Together with the fact that the amplitude of the semantic profile swings very clearly towards *African (*Figure 15.3*)*, we can assume that township tourism is also and above all about experiencing *the real Africa*. It follows that the expectations of tourists seem to be tied to certain ideas of a somehow *real Africa*.

Numerous studies have shown that the Africa image that prevails in Europe continues to be characterised by ethnocentric, post-colonial notions, which themselves are marked by an ambivalence between horror on the one side and folkloristically romanticising exoticism on the other: on the one hand, the Africa image is dominated by wars, disasters, crime, hunger and diseases; on the other hand, the image projected by marketing media is particularly characterised by exotic fauna, myths about discoverers and adventurers as well as wild and genuinely original nature (including the pristine "noble, savage" nature of Africa's inhabitants and their "traditional" cultural contexts). It can be concluded that what the tourists expect to experience as "real" during their tours – both in its positive and negative aspects – is a representation of certain fragments of old, but still widespread, colonial stereotypes.

In summary, in the context of Windhoek's city tourism, Katutura Township is not just the "place of poverty" that guarantees authenticity; rather, Katutura represents a tourist destination that promises an authentic experience of Africa. In the tourists' imagination, however, that "real Africa", does, indeed, not look like the *Christuskirche*, Independence Avenue or *Tintenpalast* in the centre of Windhoek; in their imagination, Africa is characterised by *poverty, dirt* and *violence*, and also by *colours, community spirit* and *vibrant, noisy happiness* instead.

During the tour: Township representations

In the following, we will discuss the actual practice of township tours in order to give some insights into how Katutura is stage-managed during the tours and which representational patterns and presentation forms are employed in the process.

Basically, the conception of a guided city tour always presupposes a selection decision by the operators of what is worth seeing and, thus, worth showing. This selection significantly influences the representation of Katutura and therefore has a pivotal impact on the tourist's perception and interpretation of the township.

Based on participatory observation of 16 tours, we identified the main *places of interest* (POI) to which the tourists' attention is directed. We used this as a starting point for the analysis of the practice of township tourism (for details see Steinbrink et al., 2016). The POI fulfils the function of bearers of spatial significance, in which the contents conveyed during the tours are made perceptible to the senses (Pott, 2007, p. 165). We identified six stations (*Big Six*, I–VI) that form part of almost all tours. Since these POIs are of particular importance for the staging of Katutura, they are briefly described here; the modes of representation are interpreted against the background of the tour dramaturgy.

I Vantage point: Heterogenisation, demarcation!

The first stop is usually a lookout point in the upmarket Klein-Windhoek district near the city centre. The panoramic view (with Katutura in the distance) is mainly used to provide preliminary information about the urban structure. This "introduction to place" draws the spatial and linguistic demarcation line between Katutura and the rest of the city. The tourists are introduced to the heterogenising and contrasting perspective, which then forms the basic framework for the staging of the tour. In the guides' description, the tourist looks down on the "other side" of the city, which in turn is explicitly equated with the side of "the Blacks". In semantics, Katutura is primarily assigned the attributes "poor" and "Black". This spatial distinction – up here the "rich, white upper class", down there the "poor, Black lower class" – provides the basic structuring of the *tourist gaze*.

The introduction of the main differences links directly into the tourists' township associations, which simultaneously secures connectivity with their authenticity expectations (see above). "Cultural Otherness" is pinpointed primarily via the association of the terms "Black" and "poor", the aspects of skin colour and poverty are, thus, semantically linked and marked as characteristics of alterity and authenticity.

II Old Location: Historisation!

The next stop on the way to Katurura is the *Old Location* (west of CBD), a former residential area, which was built by the German colonial administration in 1912 for the Black population and from which forced resettlement to Katutura took place. Explanations are given at the *1959 Heroes and Heroines Memorial Grave* near the cemetery. During the visit, the main focus lies on apartheid history, the eviction and, thus, the history of Katutura's foundation. In addition to the *regionalising* and *heterogenising* perspective (see I), a *historicising* perspective is established and Katutura is represented as a place charged with history. Before the actual tour of the township, the tour participants are once again attuned to a "place of the poor and Black". Charged with "history", the meaning assignments "Black" and "poor" now appear as historical facts, thus also reinforcing their semantic connection.

III Oshetu Market: Interaction and (everyday) culture!

In tourism, markets are generally regarded as places that enable travellers to experience everyday life in a pristine way. *Oshetu Market* fulfils this function for the Katutura tours. Not least because of the more intense background, noise and smells, it offers a particularly "authentic experience" to tourists. The market visit also opens up the frequently requested interaction options via sales situations. The consumer role not only helps the tourists establish direct contact with *locals* but provides a familiar frame that structures the social interaction and gives the feeling of security of action; the first contact with *local culture* is facilitated and gives the feeling of being immersed in everyday township life.

Tasting the culinary specialities often praised by the guides as exceptionally "typical" and "authentically African" (e.g. dried Mopane worms; Oshikundu drink), offers the opportunity to explore otherness through the sense of taste. This has its own appeal: the behaviour of the tourists, who literally "incorporate the strange", revealed an atmosphere of tension between curiosity and repulsion. The storage of beef has a similar effect: the mountains of meat swarmed by flies and the severed cattle skulls are popular photo motifs. These motifs seem to be particularly connectable with certain images of an authentic Africa since they do not correspond to Western ideas of hygiene. Newspapers as a substitute for plates and eating with bare hands also deviate from Western habits and can thus be experienced as primitive, but at the same time very sensual, cultural practices.

By focusing attention not only on *filth* and *disgust* but also on features like *liveliness*, *sensuality* and *exoticism*, the tour guides enable tour participants to more strongly associate the positive meanings of the ambivalent Africa image.

IV Eveline Street: Danger and vitality!

Eveline Street is on every township tour's programme. It is known as Katutura's "party street" and "entertainment district". The lively streetscape is mainly characterised by small restaurants, street stalls, *shebeens* and informal carwash enterprises; loud music dominates its soundscape.

Here, too, the guides paint an ambiguous picture. On the one hand, social problems such as unemployment, alcohol and drug abuse, prostitution, violence and crime are addressed. The guides explicitly warn against visiting the place alone and point out that Eveline Street is dangerous, wild and actually a "no-go area" for tourists. On the other hand, the young, dynamic and cheerful Katutura is shown here. The loud music, the creative names of the shebeens on brightly hand-painted boards, the cheerful celebrations – these, too, represent messages transmitted here. Thus, the representation of Eveline Street has two sides: loud, wild, dangerous, but at the same time, lively, colourful, rhythmic, young, dynamic and creative. This also suggests semantic links to the existing ambivalence of stereotypes about Africa.

V Informal settlements: "True poverty" and authentification!

The dusty informal settlements located on the outskirts of town and consisting largely of corrugated iron huts constitute another important attraction of Katutura tourism. The visit here serves as an illustration of the living conditions of the poorest segments of Windhoek's population.

While at the beginning of the tour, all of Katutura was referred to as the "other side of the city" (see I), now only the informal settlements are delimited from the rest of the city as "the other side": "So, this side is the other side of Windhoek, the poor side of Windhoek" (informal guide). This new regionalisation proposal appears to be a reaction to the fact that tourists are often surprised by the unexpectedly good structural and infrastructural condition of the formal areas of Katutura. To put it bluntly, Katutura seems less different because it is less poor. However, the visible signs of poverty in the informal settlements came closer to the tourists' expectations and their demands for "otherness".

This part of the tour seems to be a special experience of contrast, and for many tourists, this confrontation is perceived as emotionally stressful. The atmosphere in the vehicle and the behaviour of the tourists are changing. There is less talking and the photographic behaviour changes as well: while some participants stop taking photos and express ethical scruples, others press the shutter release more often – as if they had finally found the desired motifs. These behavioural changes can each be interpreted as signs of an extraordinary experience of difference and otherness.

While positive aspects are also discussed at the other stations, the informal settlements merely represent the "truly dark side of poverty". What is shown here corresponds most closely to the expectations of the tourists before starting the tour. The intense experience of difference is based on the experience of an expected difference. In this respect, the important function of the informal settlements in the dramaturgy of the tours is to establish a correspondence between the pre-existing images of townships and what was shown during the tours – with the intensity of the experience of poverty ensuring that the tour as a whole fulfils the tourists' quest for authenticity. This, in turn, lends credibility to the programm2 as a whole.

VI Penduka: Help, tradition and relaxation!

The last stop of almost all tours is the Penduka women's project, which is located on the premises of the former Windhoek Sailing Club at the Goreangab Reservoir outside the built-up area of Katutura. Here, handmade textiles, jewellery and pottery are produced and sold by disabled and disadvantaged women. The tourist offer is complemented by folkloric dance and singing performances, accommodation in thatched huts and a restaurant. Two aspects are central to the township representation: (1) the presentation as an *aid project* and (2) the staging of *local African traditions*.

The semantic link between *poverty* and the *need for help* ties in with people's expectations. The visit of the workshop strengthens the feeling of authenticity: the tourists see disabled women at work, often in traditional clothing, which permits them to see for themselves on site the authenticity of the products as "original African" handicrafts produced by "real disabled African" women. The tourists are also given the impression that they "really help" by buying souvenirs. This good feeling has a positive effect on the tour experience because especially after the disturbing impressions in the informal settlements (see V), this *moral consumption* has the potential to reduce emotional stress.

Penduka promotes a "cultural shopping experience". The tourist offer explicitly takes up Africa stereotypes with positive connotations. In particular, the folkloric dances, singing, drumming and dancing in colourful costumes serve ideas of traditional culture. This also applies to the architectural and landscape design with grass-covered round huts against the picturesque backdrop of the lake with exotic birds. This rural idyll forms a stark contrast to the townships. The tourists are very aware that Penduka is staged for tourism. But due to its status as an aid project for "authentically poor African women" who can "really be helped", this does not disrupt the tourists' perception of authenticity. The "African need for help" continues to form the framework; poverty continues to function as a guarantor of authenticity – despite the obvious stage-managed nature of its representation.

Penduka also fulfils the function of a "recreation zone" in the dramaturgy: in the end, the tourists are led from the African township to an African idyll, from the sometimes threatening pace of urban life into the peaceful tranquillity of the countryside, from the confinement of the urban environment to the openness of nature, and from the hopeless misery of the slum to the hope-giving aid project. The fact that the tourists can help here contributes to a relaxing soothing of conscience. Penduka's clear and explicit focus on tourism means relaxation, too: after the intensive and partly confusing experiences in the township, the tourists are carefully returned to the familiar context of tourism, where the scope for action and role expectations are known to them and therefore provide security.

After the tour: Township tourists' impressions

In order to investigate the impressions of the tourists after the tour through Katutura and to draw conclusions about what township tours can contribute to changing the image of the township, we interviewed 67 tourists immediately after the tours. First, the tour participants were asked to mention up to three points that particularly impressed them during the tour (Figure 15.4).

Many of the tourists (26%) were particularly impressed by the inhabitants' *friendliness*. Other positive human qualities were frequently mentioned, such as *happiness* (13%) and *openness* (9%). Furthermore, *living conditions*, *way of life* and *community aspects* as well as the residents' *initiatives* (e.g. *Penduka*) made a particularly strong impression on the tourists.

These findings indicate that the positive aspects definitely outweigh the negative ones. Only 5% of the impressions mentioned point to negative features. In particular, the term *poverty* was surprisingly rarely mentioned after the tour (only 8% of the post-tour interviewees mentioned it). The perceptual contents expressed after the tour differ substantially from the extremely negative preconceptions before the tour. This can be interpreted as a change in the interviewees' township image during the tour.

In order to specify the township tours' potential for improving the image of Katutura, after the tour, the tourists questioned were asked to fill in the semantic differential again. The two profiles (*pre-tour* and *post-tour*) are shown in Figure 15.5.

Figure 15.4 Tourists' township impressions after the tour. Source: Adapted from Steinbrink et al., 2016, p. 89.

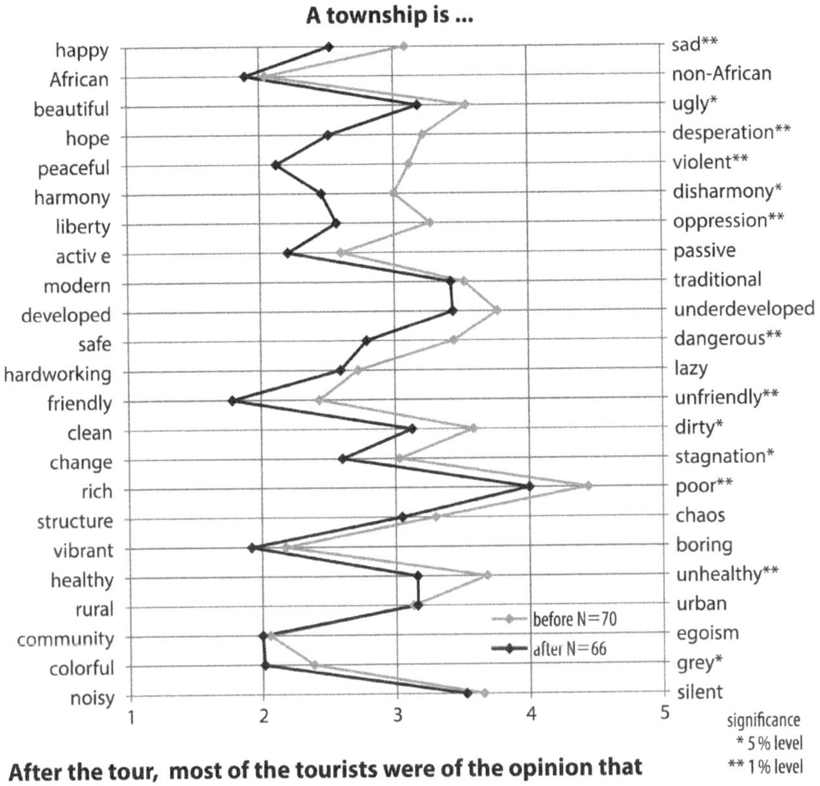

After the tour, most of the tourists were of the opinion that

... Katutura is as *underdeveloped* und *noisy* as expected und the population is so *traditional* and *community-centred* (no significant deviations).

... the residents of Katutura are about as *hardworking* as expected (no significant difference).

... Katutura is – as expected – very *African* (no significant difference).

... Katutura is – as expected – very *vibrant* (no significant difference).

... Katutura is *poor*, but *not quite as poor* as expected (significant difference).

... Katutura is *less dirty* and *less unhealthy* than assumed (significant differences).

... Katutura is *less ugly* than preconceived (significant difference).

... the inhabitants are – contrary to expectations – rather *happy, hopeful* and *free* than sad, desperate and oppressed (significant differences).

... Katutura is more *safe* than dangerous and the population is more *peaceful* than violent (significant difference).

... Katutura is *more harmonious* than expected (significant difference).

... Katutura is even *more colorful* than expected (significant difference).

... the residents of Katutura are in fact *much friendlier* than assumed (significant difference).

Figure 15.5 Semantic differential: Tourists' preconceptions before versus their impressions after the tour. Source: Adapted from Steinbrink et al., 2016, p. 90.

Here, too, the partly considerable differences are immediately obvious: the vast majority of item pairs show a clearly more positive evaluation after the tour; for none of the items a negative deviation is recognisable.

The findings show that the place that the tourists had booked for their visit was apparently not perceived as they had previously imagined, i.e. the destination did not meet their expectations. In tourism, such an experience usually results in disappointment or even anger (*That isn't what I booked, is it?*). Hence, if township tourism were indeed a form of "negative sightseeing", tours offered with a special emphasis on the positive aspects of the destination would miss the point. This, in turn, would have to be reflected in customers' dissatisfaction. But, when asked about their satisfaction with the offer immediately after the tour, most tourists responded very positively: 95% stated that they would recommend the tour to others (Steinbrink et al., 2016, p. 100). Our results by no way indicate that the tourists were disappointed because their Katutura experiences did not meet their pre-tour expectations. A possible explanation for this apparent paradox can be found in the aforementioned ambivalence of the Africa image among the tourists in combination with the tour practice.

The representation of Katutura during the tours did not contradict expectations. In fact, some of the central aspects of the tour dramaturgy do not fit into one side of the Africa image, but they fit quite well into the other (i.e. the positive side) of that ambivalent image. For instance, before the tour, connotations such as *cheerfulness*, *friendliness* and *liveliness*, which, in the end, had pushed the negative elements of the image – such as *filth*, *crime* and *misery* – into the background, already belonged in the semantic field of the township image shaped by stereotypes about Africa (even though to a much lesser extent than after the tour). That is why the tourists obviously did not have the feeling that they were presented with something that had nothing to do with what they expected to see – it was only the weighting that was different.

During the course of the tour, the negative township image is modified through an emphasis on positively connoted aspects of the ambivalent Africa stereotype by the presentation and the weighting of the sights. This enables tour participants to build on positive notions that complement and relativise negative stereotypes of townships. This display style is an essential feature of all tours.

Overall, the tours are designed to meet customers' expectations of alterity and authenticity linked to the concept of poverty, but the tours can simultaneously be encountered as a positive holiday experience. Poverty, however, continues to function as a guarantor of authenticity. The poverty perceived by the tourists was indeed not quite as visible as expected; yet, in their perception, Katutura still appeared to be "poor enough" to serve their image of Africa. In addition, this still permitted them to interpret the visible signs of poverty as an authentic experience outside the stage of tourism.

Accordingly, the tour participants' satisfaction with having seen a part of "the real Africa" in Katutura was often accompanied by the reassuring awareness that the township, or Africa, was in reality not as wretched, dirty and dangerous (as many had thought), but above all just as lively, colourful, friendly and cheerful (as many had thought). For they who had *really* been there now knew that, and they were *really* satisfied.

Conclusion: Ambivalent representations and image transformations

The aim of this chapter was to use the example of township tourism in Katutura to illustrate the interplay of geographical imaginations of the township that are shared by tourists and (re)produced by tourist representations. Using an analytical three-step – before, during and after the tour – the township tours' influence on the township image was examined.

Visiting Katutura evidently led to remarkable changes: the tourists' previously extremely negative images are now much more positive. What is surprising on the surface is that the obvious discrepancy between the expectations of the tourists and what is shown, i.e. between the existing township images and the representations, is resolved. The dark grey image is not simply displayed and reproduced but enriched with lighter and friendlier colours. However, this break does not lead to disappointment and dissatisfaction among tourists – on the contrary. Thus, township tourism is – contrary to what is often claimed – not a form of "dark tourism"!

The effect of township tourism as described here is consistent with observations documented in various studies on poverty tourism in the Global South. The studies by Rolfes et al. (2009) in Cape Town and by Meschkank (2011, 2012) and Dyson (2012) in Mumbai, as well as the studies by Freire-Medeiros (2012) in Rio de Janeiro demonstrate that tourists' preconceptions undergo noticeably positive changes as a result of participation in comparable tours (Rolfes & Burgold, 2013). It is therefore obvious that "poverty tourism" (which – as we argued – is actually a type of *cultural tourism*) in general functions as a means to change the way of looking at urban poverty or poor urban areas, and thus contributes to image improvement.

The fact that touristic reframing of "the township" functions without disappointing customers can be explained by the ambivalence of prevailing African stereotypes: to some degree, the representations tie in with central touristic imaginations. Through the style of depiction and the weighting of the sights, positively connoted aspects are emphasised and brought to the fore; the negative aspects of the picture are partly faded out. The township now symbolises not only a squalid, dirty and dangerous but also a lively, colourful, friendly and happy Africa. Customers' expectations of otherness and authenticity are still met, and poverty acts as a guarantor for this. The tours still present Katutura as "sufficiently poor" to serve the image of Africa and at the same time convey an experience of *real Africa*, beyond touristic staging.

The effects described should be entirely in the interests of the City of Windhoek, as it promotes township tourism, not least to improve the image of Katutura and, thereby, also of Windhoek. But how can the image-transforming potential of this tourism form be assessed beyond the economic logic of destination marketing?

On the one hand, one could argue that the tours create a counter-image to the township horror propagated by the media. By presenting Katutura as an urban area whose residents pursue everyday activities, homogenising and essentialising perspectives are challenged and conventional township clichés are relativised. On the other hand, however, touristic reframing can also be critically assessed. There is indeed a tendency in township tourism to trivialise. The selection of what is presented in Katutura and the way it is shown and commented on can easily lead to living conditions in Katutura being perceived as "typically local" and as an expression of an interesting "African lifestyle" and not as the result of social inequality and injustice in Namibian society as well as in the (world) economy. This carries the danger of a social or cultural-romanticising depoliticisation – which is often linked to the reproduction of certain (post) colonial stereotypes. In the context of township tourism, poverty is all too often implicitly communicated as a quasi-natural characteristic of being Black or as an expression of "Black African culture" – township tourism, thus, all too easily contributes to the ethnicisation/racialisation and depoliticisation of poverty.

It is assumed that the *culturalisation of poverty* observed in the context of "poverty tourism" is not least an outcome of the tour operators' efforts to morally legitimise this suspect and often criticised form of tourism (e.g. "This tour is *not* about gazing at people's misery, it's about their authentic culture!"). From this, it can be concluded that the fundamental ethical concerns about slum tourism shape tour practice in such a way that the actual representation is incorporated into an ethically highly dubious narrative about poverty in the Global South.

Note

1 The following text collections and state-of-the-art articles provide an overview of the research field: Frenzel, Koens & Steinbrink (2012): "Slum Tourism: Poverty, Power and Ethics"; Special Issue Tourism Geographies 14 [2] (2012); Themed Issues "Slum Tourism – Part I & 2" in Tourism Review International (2015, 2016); Frenzel et al. (2015); Tzanelli (2018).

References

Basilio, A., Byrne, M., Kelly, L., & Lavado, R. (2006). *Tourist Information Kiosks in the City of Windhoek.* Retrieved from http://www.wpi.edu/Pubs/E-project/Available/Eproject-051606-212023/unrestricted/Tourist_Information_Kiosks_in_Windhoek.pdf

City of Windhoek (n.d.). *The City of Windhoek Local Economic Development: Strategy 2010–2015.* Windhoek: City of Windhoek.

Dyson, P. (2012). Slum tourism: Representing and interpreting "reality" in Dharavi, Mumbai. *Tourism Geographies, 14*, 254–274.

Freire-Medeiros, B. (2012). Favela tourism: Listening to local voices. In F. Frenzel, K. Koens, & M. Steinbrink (Ed), *Slum Tourism: Poverty, Power and Ethics* (pp. 175–192). Abingdon: Taylor and Francis.

Frenzel, F., Koens, K., & Steinbrink, M. (Eds) (2012). *Slum Tourism: Poverty, Power and Ethics.* Abingdon: Taylor and Francis.

Frenzel, F., Koens, K., Steinbrink, M., & Rogerson, C. (2015). Slum tourism: State of the art. *Tourism Review International, 18*, 237–252.

Gräf, D., & Steinbrink, M. (2021). #SLUMIOTICS: Instagram und touristische Raum-Bilder aus mediensemiotischer Perspektive. In R. Kogler & J. Wintzer (Eds), *Raum und Bild. Strategien visueller raumbezogener Forschung* (pp. 183–196). Cham: Springer.

Meschkank, J. (2011). Investigations into slum tourism in Mumbai: Poverty tourism and the tensions between different constructions of reality. *Geojournal, 76*, 47–62.

Meschkank, J. (2012). Negotiating poverty: The interplay between Dharavi's production and consumption as a tourist destination. In F. Frenzel, K. Koens, & M. Steinbrink (Ed), *Slum Tourism: Poverty, Power and Ethics* (pp. 144–158). Abingdon: Taylor and Francis.

Münder, P. (2013). *Dark Tourism: Kaffeefahrt zur Hölle oder historische Edutainment-Exkursionen?* Retrieved from http://culturmag.de/litmag/dark-tourism-kaffeefahrten-zur-holle-oder-historische-edutainment-exkursionen/69337

Novelli, M., & Gebhardt, K. (2007). Community based tourism in Namibia: "Reality show" or "window dressing"? *Current Issues in Tourism, 10*(5), 443–479. doi:10.2167/cit332.0

Pendleton, W. C. (1996). *A Place Where We Stay: Life in a Post-apartheid Township in Namibia*. Ohio: Ohio University Press.

Pott, A. (2007). *Orte des Tourismus: Eine raum- und gesellschaftstheoretische Untersuchung*. Bielefeld: Transcript Verlag.

Rolfes, M., & Burgold, J. (2013). Of voyeuristic safari tours and responsible tourism with educational value: Observing moral communication in slum and township tourism in Cape Town and Mumbai. *Die Erde, 144*, 161–174.

Rolfes, M., & Steinbrink, M. (2009). Raumbilder und Raumkonstruktionen im Townshiptourismus. In M. Dickel & G. Glasze (Eds), *Vielperspektivität und Teilnehmerzentrierung: Richtungsweiser der Exkursionsdidaktik (Praxis Neue Kulturgeographie, 6)* (pp. 123–140). Münster: LIT.

Rolfes, M., Steinbrink, M., & Uhl, C. (2009). *Townships as Attraction: An Empirical Study of Township Tourism in Cape Town*. Potsdam: Universitätsverlag Potsdam.

Simon, D. (1988). Symbol städtischer Apartheid. In H. Melber (Ed), *Katutura, Alltag im Ghetto* (pp. 49–70). Bonn: Issa-Verlag.

Steinbrink, M. (2012). "We did the slum!": Urban poverty tourism in historical perspective. *Tourism Geographies, 2*, 213–234.

Steinbrink, M., & Voshage, I. (2020). Ambivalente Repräsentationen – Betrachtungsweisen des Townshiptourismus in Windhoek, Namibia. In R. Hartmann (Ed), *Tourismus in Afrika. Chancen und Herausforderungen Einer Nachhaltigen Entwicklung*. Berlin: De Gruyter.

Steinbrink, M., Buning, M., Legant, M., Schauwinhold, B., & Süssenguth, T. (2016). *Touring Katutura!: Poverty, Tourism, and Poverty Tourism in Windhoek, Namibia*. Potsdam: Potsdam Universitätsverlag.

Tzanelli, R. (2018). Slum tourism: A review of state-of-the-art scholarship. *Tourism, Culture & Communication, 18*(2), 149–155.

Urry, J. (2002). *The Tourist Gaze*. London: SAGE.

Welz, G. (1993). Slum als Sehenswürdigkeit: "Negative Sightseeing" im Städtetourismus. In D. Kramer & R. Lutz (Ed), *Tourismus-Kultur, Kultur-Tourismus* (pp. 39–53). Münster: LIT.

WTTC (World Travel & Tourism Council) (2018). *Travel & Tourism Economic Impact 2018 Namibia*. Retrieved from https://www.wttc.org/-/media/files/reports/economic-impactresearch/countries-2018/namibia2018.pdf

Index

adaptation 21, 66, 91, 93, 95, 98
Africa 24, 48, 57, 105, 111, 123–132, 172, 187, 193, 214, 224, 237, 246–251, 254–255
African: ambivalence 242; diaspora 123–124, 126–132; representations 237, 242, 255
Airbnb 160–161, 222–223, 226–228
alienation 51–54, 56, 117
alliance of small island states (AOSIS) 91, 96
ANC (African National Congress) 154
apartheid 117, 151–153, 161–162, 209–210, 216, 239–240, 249; struggle 157
Australia 34, 35, 41, 97, 101, 102, 105–108, 110–112, 114–116
authenticity 33–34, 37–39, 51–52, 54–55, 116, 143, 227, 243, 246, 248–249, 251, 254–255
authentification 250

Barbados Programme of Action 93–94
Bill Clinton 155
biodiversity conservation 95, 201
biological diversity 188
Black economic empowerment (BEE) 206, 214
Bolt 222, 230
Botswana 172, 187, 190–192, 194–196, 199–201
Butler's Tourism Area Lifecycle 151, 156

Cape Agulhas 211–214
Cape Town 17, 154–158, 160, 162, 211, 255
caravan park(s) 208–209, 211, 214
caravanning 207–208, 210, 213, 216
Caribbean 86, 93, 128, 130, 193
Cederberg 211
change management 15–19, 25
CITES 197, 200
climate change 5, 10, 20, 25, 86–87, 90–91, 93–98, 169, 182, 189
commercialisation 152, 156–157, 159, 162, 242
community based tourism 135, 147, 191–192, 194, 198, 241
Community Partnership Models 196
conceptual model 48–51, 53–54, 57
conscientisation 173–174, 181
conservation 111, 113–114, 188, 190, 195, 200; biodiversity 95; community 192, 194–196; heritage 33–34, 37, 40, 55, 71, 137; mindset 41; nature 5, 9, 33; ocean 97; sustainable 187; wildlife 9, 190–193, 197, 201
control 35, 38, 51, 65, 78, 136, 170, 172–174, 181–182, 206, 223–224; "local control" 151, 156
COVID–19 pandemic 10, 15, 25, 135, 138, 148, 163, 182, 232
cross border business tourism 225
cross-cultural learning 102, 104, 118
cultural: experiences 7, 105; tourism 9, 156, 192, 240, 245, 255
culturalisation of poverty 237, 256

decent work 22, 23, 95, 223, 229–230
destination marketing 33, 255
developing countries 4, 38, 43, 86, 90, 94, 173, 179, 187–188, 190–191, 193, 196–197, 200–201, 224
diaspora community 138, 141
discontinuity 51–54
domestic tourism 39
Dominican Republic 88, 172
drive tourism 54, 207–208, 210, 216, 221

Ebeye Atoll 91
economic development 9, 43, 91, 96, 98, 105, 116, 123, 129, 131, 135, 187–188, 190–191; community 20; department of 113; local 65–67, 69–71, 73–82, 159, 205–207, 212, 215–216; national 71; socio 65, 71, 224; tourism-related 123
Elmina 2015 strategy 65, 69–73, 75–81
Elmina Castle 71–72, 76–80
employment 4, 6, 9, 16, 22–24, 70, 79, 170, 172, 175–176, 181, 187, 190, 192, 195, 197, 200, 222–232
Etosha National Park 187
evolutionary approach 20–21

FIFA World Cup 159
Fiji 88, 96–97
Freedom Square 158–159

gender empowerment 172–173
Ghana 8, 66, 70–72, 76–77, 79–80, 123–132
gig economy 9, 222–223, 225–230
global economy 25
governance 8, 15, 17–19, 25, 65–73, 75–76, 78–81; capacity 8, 65–69, 71–73, 78–81
government planning 207, 211

heritage tourism 33–34, 36, 48, 55–56, 71, 73, 125–126, 131, 137
Hessequa 211–213
Hillary Clinton 156
history of slavery 8, 123–124, 127, 131–132
human patrimony 8

ILO 223–224
inclusiveness 23
Indian Ocean 86
indigenous: tour operators 8, 102–106, 114, 118; voices 101, 117, 173
informal: sector 182, 222–225, 229; settlements 250–251
islands of sustainability 92

Johannesburg City Council 153
justice tourism 157

Kafue National Park 187
Katutura 9, 238–239, 241–243, 245, 248–252, 254–256
Kgalagadi 102, 104, 106, 113
Kliptown Precinct 158–159
Kruger National Park 187

labour dynamics 223
large ocean states 86–87, 95, 97–98
least developed countries (LCDs) 86, 90
leisure tourism 152, 155, 162, 225
livelihood strategy 182
local assets 205, 207, 216

Marshall Islands 89, 91
mass tourism 34, 43, 143, 145
Mauritius Strategy of 2005 95
microentrepreneurs 225
migration 36, 124, 126, 129
mitigation 21, 224
mobilisation 65, 69, 76, 78, 126, 173–174, 181, 205, 207
Mossel Bay 211
municipal asset(s) 9, 206, 209

natural attractions 38
negative moods 8, 50–52, 54, 56
Nelson Mandela 154–155
New Caledonia 89, 96
New Zealand 35, 97, 115, 206–209, 216
niche tourism 36, 40
nostalgia 8, 37, 48–57; research 8, 49, 53, 56–57; trigger 51; vicarious 48–57

OECD 205, 222
Okavango Delta 187, 194, 199
Overstrand 211–213

PAC (Pan-African Congress) 154
Pacific 86–87, 96–97
Pacific Islands Forum Secretariat (PIFS) 96
Pacific Small Island Development States (PSIDS) 96
path dependency 151
patriarchal society 169
photographic tourism 201
political tourism 151–152, 155, 161–162
poverty 22–23, 25, 36, 40, 42, 65, 67–69, 71, 80–82, 157, 161–162, 171, 173, 175, 193, 206, 224, 237–238, 240, 245–246, 248–252, 254–256
propaganda visits 152–153, 162

Queen Elizabeth 155–156

reality tourism 246
regionalism 8, 96–97
resilience 15, 17, 19–21, 24–25, 90, 128, 135, 148, 182, 224
Rio de Janeiro 87, 237, 255
Robben Island 57, 155

roots tourism 8, 123–129, 131
rural livelihoods 9, 170, 191–192, 194, 196, 200

Saldanha Bay 211, 214–215
sea-level rise 91, 93, 95
Serengeti 57, 187
sharing: economy 225–227; exchange 102, 106
SIDS Accelerated modalities of action (SAMOA) 95
slum(s) 237–238, 251; tourism 160, 237–238, 256
small island developing states (SIDS) 86–87, 90, 93–97
small micro medium sized enterprise (SMME) 211, 214–215, 218
social capital 68, 76, 135–137, 140–148
social: constructivist 138; identity 52, 54, 56, 136
socio-economic development 65, 71
souq 135–149; Souq Muttrah 136–142, 144, 146–148
South Africa 17, 57, 101–102, 106, 112, 114–116, 151–155, 157, 162, 171, 187, 193–194, 196, 199, 205–207, 209 211, 214–217, 222, 225, 230, 237, 241
Soweto 153–158, 160, 162–163
street vendors 225
Sultanate of Oman 139
sustainability 5, 7, 15, 21–23, 25, 78, 86, 91–92, 97, 172, 188, 199
sustainable: development 5, 9, 15, 21–25, 43, 70, 73, 86–87, 91–97, 157, 171, 188–190; livelihoods 170–171; tourism 15, 21–22, 24, 66, 81, 95–97, 147, 174, 188–191, 201; wildlife hunting 201
Sustainable Development Goals (SDGs) 16, 22–25, 91, 95, 97, 193

Taxify 222
Tony Blair 156
tourism contexts 48, 51, 55–57, 101, 105, 151, 222
tourism development 5–6, 8–9, 15–17, 19–20, 24–25, 65–66, 69, 81–82, 95, 105, 116, 126–127, 135–136, 138, 141, 147–148, 151–152, 156, 159, 161–163, 171, 188–191, 193, 201, 207, 211, 213, 216
tourism employment 24, 227, 229
tourism investment 191
tourist: bubble 8; gaze 37, 238, 248
touristic consumption 123
township(s) 151–163, 239, 243, 246, 251–254; images 238, 243, 246, 255; representation(s) 238, 243, 251; tour(s) 243, 250; tourism 157, 160, 164, 241–243
traveller characteristics 56

Uber 222, 228, 230
UN framework convention for climate change 94
unemployment 53, 169, 173, 206, 229, 250
UNESCO 35, 38, 71, 113, 169, 173
United Nations 22, 90, 96, 176
United Nations Conference on Environment and Development (UNCED) 87
United States 7, 123, 127–128, 130–131, 136–141, 200
unsustainable 7, 21, 92–92, 135, 230

Victoria Falls 169–170, 173–176, 178–182
Victorian era 237
Vilakazi street 154, 158, 160

Western Cape 9, 17, 209, 211–212, 215–216
wildlife: conservation 9, 190–193, 196–197, 201; utilisation 188, 197, 200
Windhoek 9, 238–243, 247–248, 250–251, 255
World Bank Group 23, 187, 191–197, 199, 201
World Tourism Organization (WTO) 21, 188

Zimbabwe 8, 169, 173–175, 179–180, 191, 195

Taylor & Francis Group
an **informa** business

Taylor & Francis eBooks

www.taylorfrancis.com

A single destination for eBooks from Taylor & Francis with increased functionality and an improved user experience to meet the needs of our customers.

90,000+ eBooks of award-winning academic content in Humanities, Social Science, Science, Technology, Engineering, and Medical written by a global network of editors and authors.

TAYLOR & FRANCIS EBOOKS OFFERS:

- A streamlined experience for our library customers
- A single point of discovery for all of our eBook content
- Improved search and discovery of content at both book and chapter level

REQUEST A FREE TRIAL

support@taylorfrancis.com

Routledge
Taylor & Francis Group

CRC Press
Taylor & Francis Group